Design and Build Contracts

Design and Build Contracts

Guy Higginbottom

WILEY Blackwell

Registered Offices
John Wiley & Sons, Inc., 111 River Street, Hoboken, NJ 07030, USA
John Wiley & Sons Ltd, The Atrium, Southern Gate, Chichester, West Sussex, PO19 8SQ, UK

For details of our global editorial offices, customer services, and more information about Wiley products visit us at www.wiley.com.

Wiley also publishes its books in a variety of electronic formats and by print-on-demand. Some content that appears in standard print versions of this book may not be available in other formats.

Library of Congress Cataloging-in-Publication Data

Name: Higginbottom, Guy, author.
Title: Design and build contracts / Guy Higginbottom.
Description: Hoboken : Wiley-Blackwell, 2024. | Includes index.
Identifiers: LCCN 2023023395 (print) | LCCN 2023023396 (ebook) | ISBN
    9781119814825 (cloth) | ISBN 9781119814832 (adobe pdf) | ISBN
    9781119814849 (epub)
Subjects: LCSH: Construction contracts. | Construction industry–Law and
    legislation.
Classification: LCC K891.B8 H54 2023 (print) | LCC K891.B8 (ebook) | DDC
    343.07/8624–dc23/eng/20230802
LC record available at https://lccn.loc.gov/2023023395
LC ebook record available at https://lccn.loc.gov/2023023396

Cover Design: Wiley
Cover Images: © ivanastar/Getty Images

Set in 9.5/12.5pt STIXTwoText by Straive, Chennai, India
Printed and bound by CPI Group (UK) Ltd, Croydon, CR0 4YY

C9781119814825_200923

Contents

List of Figures *xiii*
List of Tables *xv*
List of Cases *xvii*
Preface *xix*

1 **Introduction** *1*
1.1 Some Types of 'Design and Build' *1*
1.2 The Nature of Design and Build *2*
1.3 A Brief History of Design and Build Contracts *4*
1.4 Recent Developments in Design and Build *5*
1.5 How to Use This Book *6*

2 **Construction Contracts** *7*
2.1 General *7*
2.2 The Structure of a Construction Contract *7*
2.2.1 The General Nature of Standard Forms of Design and Build Contract *8*
2.2.2 The Specific Information *8*
2.2.3 The Conditions of Contract *11*
2.2.4 The Documents Comprising the Contract *11*
2.2.5 The Relationship Between the Conditions of Contract and the Contract Documents *12*
2.3 Fairness as a Concept *14*
2.3.1 Assistance from the Contract Terms? *16*
2.3.2 Understanding the Contract and its Implications *16*

Part I *19*

3 **Design and Build as Originally Intended** *21*
3.1 The Parties' Primary Obligations Under a Design and Build Contract *21*
3.1.1 The Employer's Primary Obligations *21*
3.1.2 The Contractor's Primary Obligations *22*

3.2   The Employer's Representative   *24*
3.3   The Relationship Between the Employer's Requirements and the Contractor's
         Proposals   *26*
3.3.1   FIDIC Yellow Book 2017   *26*
3.3.2   JCT DB 2016   *26*
3.3.3   NEC4   *27*
3.4   The Employer's Requirements   *29*
3.4.1   The Employer's Requirements – Essential Content   *30*
3.4.2   The Employer's Requirements – Common Format and Contents   *30*
3.4.3   Responsibilities Retained by the Employer   *33*
3.5   The Contractor's Proposals   *34*
3.5.1   Contractor's Proposals – Common Format and Contents   *35*
3.5.2   Contractor's Proposals – Review of Contractor's Design   *35*
3.5.3   Contractor's Proposals – Programme (Documents for Managing the Works)   *39*
3.5.4   Contractor's Proposals – Documents for Managing the Works (Cost)   *40*
3.6   The Pricing Document   *41*
3.6.1   Pricing Document – Common Format and General Contents   *41*
3.6.2   Pricing Document – Contractor's Preliminaries   *46*
3.6.3   Pricing Document – Pricing Alternatives   *47*
3.6.4   Provisional Sums   *48*
3.6.5   Employer's General Analysis of the Pricing Document   *52*

4   **Consultants and the Employer's Representative**   *55*
4.1   Means of Appointing Consultants   *56*
4.1.1   Conditions of Engagement – Typical Format and Contents   *56*
4.2   Obligations and Services   *58*
4.2.1   Applicable Standard for Performing the Services   *58*
4.2.2   Typical Duties Owed by the Designer to the Employer   *58*
4.2.3   Scope of Services   *60*
4.2.4   Incorporation of Services   *62*
4.2.5   Design Services   *62*
4.2.6   Lead Designer   *64*
4.2.7   Project Lead   *66*
4.2.8   Contract Administration   *67*

5   **Procurement and Tendering**   *73*
5.1   Procurement   *73*
5.1.1   Procurement Criteria – Public Sector to 31 December 2020   *73*
5.1.2   Procurement Criteria – Public Sector from 1 January 2021   *74*
5.2   Tendering   *75*
5.2.1   Tendering Criteria – General Requirements   *77*
5.2.2   Tendering Criteria – Instructions to Tenderers   *77*
5.2.3   Information Necessary for the Tenderer to Submit a Bid   *78*
5.3   Types of Tender Processes   *79*
5.3.1   Pre-qualification   *79*

5.3.2   Single-Stage Tendering   *80*
5.3.3   Two-Stage Tendering   *80*
5.3.4   Negotiation   *81*
5.3.5   Mid-tender Events   *82*
5.4   Early Contractor Involvement   *83*
5.4.1   Early Contractor Involvement in Engineering Projects   *83*
5.4.2   Early Contractor Involvement in Construction Projects   *84*
5.4.3   Early Contractor Involvement (NEC4 Option X22)   *85*
5.4.4   Advantages of Early Contractor Involvement   *86*
5.4.5   Disadvantages of Early Contractor Involvement   *88*
5.4.6   Pre-construction Services Agreements   *89*
5.5   Tender Returns and Evaluation   *91*
5.5.1   Tender Receipt and Initial Checks   *91*
5.5.2   Tender Evaluation   *91*
5.5.3   Evaluation of the Tenderer's Design   *92*
5.5.4   Evaluation of Other Aspects of the Tender   *93*
5.5.5   Tender Interviews   *93*
5.6   Contractor Selection and Appointment   *94*
5.6.1   Inform the Successful Tenderer   *95*
5.6.2   Inform the Unsuccessful Tenderers   *95*
5.6.3   Conclude Negotiations and Enter into the Building Contract   *95*

6   Subcontracting   *97*
6.1   General   *97*
6.1.1   The General Nature of Subcontracts   *98*
6.1.2   Types of Subcontract for Each Main Contract   *100*
6.2   Domestic Subcontractors   *103*
6.2.1   Domestic Subcontractor Selection   *103*
6.2.2   Employer's Approval of Subcontractors   *104*
6.2.3   Administration of a Domestic Subcontract   *105*
6.3   'Nominated' or 'Named' Subcontractors   *106*
6.3.1   Nominated or Named Subcontractor Selection   *106*
6.3.2   Contractor's Objection to Nominated or Named Subcontractors   *108*
6.3.3   Risks Associated with Nominated or Named Subcontractors   *110*
6.3.4   Administration of a Nominated or Named Subcontract   *112*
6.3.5   Decline of Nominated or Named Subcontractors   *113*
6.4   Early Subcontractor Involvement   *114*

**7   Collateral Warranties, Third-Party Rights, Bonds, and
             Guarantees**   *117*
7.1   General   *117*
7.1.1   Collateral Warranties – General Points   *117*
7.1.2   Third-Party Rights – General Points   *118*
7.1.3   Bonds and Guarantees – General Points   *119*
7.2   Collateral Warranties   *119*

7.2.1   Collateral Warranties Under the Main and Subcontracts   *120*
7.2.2   Collateral Warranties for Designers   *122*
7.3   Third-Party Rights   *123*
7.4   Bonds and Guarantees   *125*
7.4.1   Security for Performance   *125*
7.4.2   Security for Payment   *128*
7.4.3   Retention Bonds   *130*
7.4.4   Off-Site Materials/Goods Bond   *131*
7.4.5   Other Bonds   *131*

8   **Construction**   *133*
8.1   General   *133*
8.1.1   Pre-start Obligations   *133*
8.1.2   Contract Agreement and Performance Security Obligations   *133*
8.1.3   Financial Obligations   *134*
8.1.4   Design Obligations   *135*
8.2   Commencement   *135*
8.2.1   JCT DB 2016   *135*
8.2.2   FIDIC Yellow Book 2017   *136*
8.2.3   NEC4   *137*
8.3   Completion in Sections (Subdividing the Whole Works)   *138*
8.3.1   JCT DB 2016   *139*
8.3.2   FIDIC Yellow Book 2017   *140*
8.3.3   NEC4   *141*
8.3.4   Summary of Sectional Completion Provisions   *143*
8.3.5   Summary of Milestone or Key Date Provisions   *143*
8.4   Progress   *144*
8.4.1   Rate of Progress   *144*
8.4.2   Manner of Progress   *145*
8.4.3   JCT DB 2016   *146*
8.4.4   FIDIC Yellow Book 2017   *147*
8.4.5   NEC4   *149*
8.4.6   Reporting Progress   *149*
8.4.7   Advance or Early Warning   *150*
8.5   Acceleration   *152*
8.5.1   JCT DB 2016   *153*
8.5.2   FIDIC Yellow Book 2017   *153*
8.5.3   NEC4   *154*
8.5.4   Summary of Acceleration Provisions   *155*
8.6   Programme   *156*
8.6.1   Generating the Programme   *156*
8.6.2   Programme Requirements for Each Design and Build Contract   *157*
8.6.3   Revising the Programme and Employer's Acceptance   *158*
8.6.4   Monitoring Progress on the Programme   *161*

8.7	Interim Payments	*162*
8.7.1	JCT DB 2016	*165*
8.7.2	FIDIC Yellow Book 2017	*168*
8.7.3	NEC4	*175*
8.7.4	Consultants' Conditions of Engagement	*177*
8.8	Variations	*178*
8.8.1	General Points	*178*
8.8.2	Is the 'Variation' Recognised Under the Terms of the Contract?	*179*
8.8.3	JCT DB 2016	*181*
8.8.4	FIDIC Yellow Book 2017	*183*
8.8.5	NEC4	*187*
8.9	Valuation (or Assessment) of Variations	*196*
8.9.1	JCT DB 2016	*197*
8.9.2	FIDIC Yellow Book 2017	*201*
8.9.3	NEC4	*203*
8.10	Extensions of Time	*212*
8.10.1	Ascertaining Delaying Effects	*214*
8.10.2	Procedures for the Supervising Officer	*217*
8.10.3	Extensions of Time under JCT DB 2016	*218*
8.10.4	FIDIC Yellow Book 2017	*221*
8.10.5	Extensions of Time under NEC4	*225*
8.11	Additional Costs (Loss and Expense)	*225*
8.11.1	Categories of Additional Costs (Loss and Expense)	*227*
8.11.2	JCT DB 2016	*228*
8.11.3	FIDIC Yellow Book 2017	*229*
8.11.4	NEC4	*229*
8.12	Testing and Defects	*229*
8.12.1	JCT DB 2016	*232*
8.12.2	FIDIC Yellow Book 2017	*234*
8.12.3	NEC4	*237*
8.12.4	Summary of Testing and Defects Procedures	*240*
9	**Concluding the Contract**	*243*
9.1	Termination	*243*
9.1.1	JCT DB 2016	*245*
9.1.2	FIDIC Yellow Book 2017	*253*
9.1.3	NEC4	*259*
9.2	Obligations Prior to Completing the Physical Works	*265*
9.2.1	JCT DB 2016	*265*
9.2.2	FIDIC Yellow Book 2017	*266*
9.2.3	NEC4	*268*
9.3	Meaning of 'Completion'	*269*
9.3.1	JCT DB 2016 – 'Practical Completion'	*269*
9.3.2	FIDIC Yellow Book 2017 – 'Taking Over'	*272*

9.3.3 NEC4 'Completion' and 'Taking Over' *275*

9.3.4 Summary of Procedures and Obligations on 'Completion' *276*

9.4 Damages for Late Completion *278*

9.4.1 JCT DB 2016 *281*

9.4.2 FIDIC Yellow Book 2017 *283*

9.4.3 NEC4 *284*

9.4.4 Damages for Late Completion Summary *286*

9.5 Rectification or Completing Outstanding Works After Completion *286*

9.5.1 JCT DB 2016 *287*

9.5.2 FIDIC Yellow Book 2017 *289*

9.5.3 NEC4 *296*

9.5.4 Summary of Rectification Procedures After 'Completion' *298*

9.6 Final Account *298*

9.6.1 JCT DB 2016 *299*

9.6.2 FIDIC Yellow Book 2017 *302*

9.6.3 NEC4 *306*

9.7 Resolving Disputes *313*

9.7.1 Informal Dispute Resolution Methods *314*

9.7.2 Contractual Dispute Resolution Methods *316*

9.7.3 Formal Dispute Resolution Methods *321*

9.7.4 Statutory Methods of Dispute Resolution *326*

9.7.5 Summary of Dispute Resolution Methods *329*

9.8 Concluding the Contract *330*

9.8.1 JCT DB 2016 *330*

9.8.2 FIDIC Yellow Book 2017 *332*

9.8.3 NEC4 *333*

9.8.4 Summary *333*

9.8.5 Latent Defects *334*

Part II *335*

10 Design and Build in Its Current Form *337*

10.1 Reduction of the Contractor's Design *337*

10.2 Novation of Design Consultants *339*

10.2.1 How to Achieve Novation *340*

10.2.2 Problems with Novation *342*

11 Common Amendments to Design and Build Contracts *347*

11.1 Project-Specific Amendments *348*

11.1.1 Revised (or Wider) Definitions *348*

11.1.2 Additional Measures to Assist the Employer *348*

11.2 Amendments with a Practical Benefit *349*

11.3 Increased Liability for Contractor's Design *350*

11.4 Reduced Employer's Risk/Increased Employer's Benefit *352*

11.5   Reduced Contractor Rights (Not Design Related) *353*
11.6   Increased Benefits to the Contractor *355*
11.7   Clarifications *356*

12   Possible Future Development in Design and Build *359*
12.1   Continuing Trend of Reducing the Extent of Contractor Design *359*
12.1.1   JCT Design and Build Contracts *359*
12.1.2   FIDIC Yellow Book Contracts *360*
12.1.3   NEC4 Contracts *360*
12.2   Novation of Design Consultants *361*
12.3   Detailed Pricing Documents *362*
12.4   Detailed Procedures for Testing and Commissioning *363*
12.5   Collateral Warranties (FIDIC and NEC4) *365*

Appendix A   JCT Design and Build Contracts Reconciliation *367*

Appendix B   FIDIC Yellow Book Reconciliation *381*

Appendix C   NEC Engineering and Construction Contract (ECC) Reconciliation *387*

Index *395*

List of Figures

Figure 5.1   Contracts for construction and engineering projects *74*

Figure 8.1   Pre-commencement stages under the FIDIC Yellow Book 2017 *137*

Figure 8.2   FIDIC Yellow Book 2017 interim payment procedure (including HGCRA 1996 amendments) *170*

Figure 8.3   NEC4 Compensation Event Procedure *196*

Figure 9.1   FIDIC Yellow Book 2017 (tests on completion) *266*

Figure 9.2   FIDIC Yellow Book 2017, Clause 10.1, taking over procedure *274*

Figure 9.3   JCT DB 2016, procedure for the employer to deduct liquidated damages *281*

Figure 9.4   JCT DB 2016, procedure for submitting the Final Statement *300*

Figure 9.5   JCT DB 2016, Final Statement and Final Payment procedure *302*

Figure 9.6   FIDIC Yellow Book 2017, procedure for the Final Payment Certificate *303*

Figure 9.7   FIDIC Yellow Book 2017, clause 3.7, procedure for agreement or determination *318*

Figure 10.1   Novated consultant's liability transfer *345*

List of Tables

Table 1.1   Project stages in the 2020 Royal Institute of British Architects' *Plan of Work*   *2*

Table 2.1   Specific information contained in a construction contract   *9*

Table 2.2   Categories of risk for the employer and the contractor   *15*

Table 3.1   Employer's requirements for JCT DB 2016, FIDIC Yellow Book 2017, and NEC4   *29*

Table 3.2   Contents of Employer's Requirements (NRM2 and CESMM4)   *31*

Table 3.3   Contractor's proposals for JCT DB 2016, FIDIC Yellow Book 2017, and NEC4   *35*

Table 3.4   Contractor's proposals – common format and contents   *36*

Table 3.5   Comparison of JCT and FIDIC design review procedures   *38*

Table 3.6   'Contract price' for JCT DB 2016, FIDIC Yellow Book 2017, and NEC4 (main option clauses A–E)   *42*

Table 3.7   Pricing document for JCT DB 2016, FIDIC Yellow Book 2017, and NEC4   *42*

Table 3.8   NRM2: Fixed and time-related charges   *46*

Table 3.9   NRM2 template for the contract sum analysis   *49*

Table 3.10   Provisional sum procedures   *50*

Table 4.1   Roles in the RIBA Professional Services Contract (PSC)   *61*

Table 4.2   Contract administration duties for JCT DB 2016, FIDIC Yellow Book 2017, and NEC4   *70*

Table 6.1   Subcontracts for JCT DB 2016, FIDIC Yellow Book 2017, and NEC4   *99*

Table 6.2   Subcontracting criteria under JCT DB 2016, FIDIC Yellow Book 2017, and NEC4   *104*

Table 6.3   Instructions for 'Named' subcontractors under JCT DB 2016 and JCT ICD 2016   *110*

Table 7.1   Bonds and guarantees under JCT DB 2016, FIDIC Yellow Book 2017, and NEC4   *125*

Table 8.1   JCT DB 2016 entries for completion by sections   *140*

Table 8.2   Summary of sectional completion provisions   *143*

Table 8.3   Summary of milestone/key date provisions   *144*

Table 8.4   Summary of acceleration procedures   *155*

Table 8.5   Requirements of the contractor's programme (FIDIC and NEC4)   *159*

Table 8.6   Requirements of revising the contractor's programme (FIDIC and NEC4)   *160*

Table 8.7   Reasons for rejecting the contractor's programme – and procedure (FIDIC and NEC4)   *161*

Table 8.8   JCT DB 2016 'Gross Valuation' under Alternatives A and B   *167*

Table 8.9   NEC4 – 'Price for Work Done to Date' (Main Option Clauses A–E)   *176*

Table 8.10   Conditions of Engagement complying with HGCRA 1996   *178*

Table 8.11   NEC4 Compensation Events cross-references   *189*

Table 8.12   Applicable definitions of 'Defined Cost'   *206*

Table 8.13   Explanation of actual and forecast Defined Cost   *206*

Table 8.14   Summary of NEC4 Main Option Clauses' assessment of compensation events   *210*

Table 8.15   Critical Path Analysis (CPA) methods   *217*

Table 8.16   JCT DB 2016 Relevant Events (clause 2.26) and Relevant Matters (clause 4.21)   *219*

Table 8.17   Matters under clause 8.5(b) which entitle the contractor to an extension of time/EoT   *226*

Table 8.18   Summary of testing and defects procedures.   *241*

Table 9.1   Grounds for termination   *246*

Table 9.2   Consequences of termination under JCT DB 2016   *251*

Table 9.3   Consequences of termination under FIDIC Yellow Book 2017   *257*

Table 9.4   NEC4 – Clause 93 (A4): application of 'fee percentage' on termination   *262*

Table 9.5   Summary of NEC4's termination reasons, procedures, and methods of calculating the amount due on termination   *263*

Table 9.6   Summary of obligations/procedures on 'Completion'   *277*

Table 9.7   Summary of damages for late completion   *286*

Table 9.8   Summary of rectification provisions after 'completion'   *297*

Table 9.9   NEC4: Elements of the 'contractor's share' calculation   *309*

Table 9.10   NEC4 clause 53.3 Options W1, W2, and W3   *313*

Table 9.11   Dispute resolution methods   *315*

Table 9.12   FIDIC Yellow Book 2017, clause 3.7.3(a) determination or agreement subclauses   *317*

Table 9.13   Summary of dispute resolution methods   *329*

Table 9.14   Summary of methods of concluding the contract   *334*

List of Cases

Atkins Ltd v Secretary of State for Transport [2013] EWHC 139 (TCC) *181*

Balfour Beatty Construction Ltd v Grove Developments Ltd [2016] EWCA Civ 990 *354*

Blyth & Blyth Ltd v Carillion Construction Ltd [2001] ScotCS 90 (18 April 2001) *344*

Bovis Lend Lease Ltd v Braehead Glasgow Ltd [2000] EWHC Technology 118 *280*

Bramall & Ogden Ltd v Sheffield City Council (1983) 29 BLR 73 *280*

Cavendish Square Holding BV v Talal El Makdessi [2015] UKSC 67 *279*

City Inn Ltd v Shepherd Construction Ltd [2008] BLR 269 *218*

Costain Ltd v Bechtel [2005] EWHC 1018 *69, 218, 260*

Costain Ltd v Tarmac Holdings Ltd [2017] EWHC 319 (TCC) *16*

Courtney & Fairbairn Ltd v Tolaini Bros (Hotels) Ltd [1975] 1 WLR 297 *82*

Dairy Containers Ltd v Tasman Orient Line CV [2005] 1 WLR 215 *192, 195*

Darlington Borough Council v Wiltshier Northern Ltd [1995] 1 WLR 68 *118*

Dunlop Pneumatic Tyre Co Ltd v New Garage & Motor Co Ltd [1915] AC 79 HL *279*

Emoson Eastern (in receivership) v EME Developments Ltd (1991) 55 BLR 114 *270*

English Industrial Estates Corporation v Kier Construction Ltd (1991) 56 BLR 93 *180*

Grieves & Co (Contractors) Ltd v Baynham Meikle & Partners [1975] 1 WLR 1095 *58*

Hampshire County Council v Stanley Hugh Leach & Others (1990) 8 Const LJ 174 *39*

Henia Investments Inc. v Beck Interiors Ltd [2015] EWHC 2433 (TCC) *171*

HW Nevill (Sunblest) v William Press (1981) 20 BLR 78 *270*

Independent Broadcasting Authority v EMI Electronics Ltd and BICC Construction Ltd (1980) 14 BLR 1 ('IBA v EMI & BICC') *58*

J Crosby & Sons Ltd v Portland UDC (1967) 5 BLR 21 *181*

J Jarvis & Sons Ltd v Westminster Corp (1970) 1 WLR *269*

John Barker Construction Ltd v London Portman Hotel Ltd (1996) 83 BLR 31 *218*

Kaye Ltd v Hosier & Dickinson (1972) 1 WLR 146 *270*

London Borough of Merton v Stanley Hugh Leach Ltd (1986) 32 BLR 51 *16*

Luxor (Eastbourne) Ltd v Cooper (1941) AC 108 HL *16*

PC Harrington Contractors Ltd v Co-Partnership Developments Ltd (1998) 88
  BLR 44   *168*

Peak Construction (Liverpool) Ltd v McKinney Foundations Ltd (1970) 1 BLR 111   *280*

Percy Bilton v Greater London Council [1982] 1 WLR 794 HL   *110, 280*

Piggott Foundations Ltd v Shepherd Construction Ltd (1993) 67 BLR 48   *279*

Robinson v Harman (1848) 154 ER 163   *226*

Rotherham MBC v Frank Haslam Milan & Co Ltd [1996] 78 BLR 1   *39*

Strachan & Henshaw Ltd v Stein Industrie (UK) Ltd and GEC Alsthom Ltd (1998) 87
  BLR 52   *195*

Sunlife Europe Properties Ltd v Tiger Aspect Holdings [2013] EWHC 1161 (TCC)   *180*

Sutcliffe v Thackrah [1974] AC 727   *218*

Taylor Woodrow Holdings Ltd v Barnes & Elliott Ltd [2004] EWHC 3319 (TCC)   *139, 281*

Temloc Ltd v Errill Properties Ltd (1987) 39 BLR 30   *279*

Vinci Construction UK Ltd v Beumer Group UK Ltd [2017] EWHC 2196 (TCC)   *280*

Walter Lilly & Co Ltd v Mackay & Another (No. 2) [2012] EWHC 1773 (TCC)   *270*

Weldon Plant Ltd v Commission for the New Towns [2000] EWHC 496   *199, 200*

West Faulkner Associates v London Borough of Newham (1994) 71 BLR 1   *146*

Preface

For the 10 years leading up to 2020, I worked in both private practice and latterly as a commercial director for a building contractor. When working in both types of organisation, it quickly became apparent that 'design and build' seemed to be the most popular means of procuring buildings and large-scale plant and civil engineering projects. Of course, this evidence was initially anecdotal and taken as a selective sample from the types of contract with which the particular multi-disciplinary consultancy I worked for (albeit, a large, global one) was involved. Firmer evidence of the popularity of 'design and build' was provided in the form of the RIBA's Construction Contracts and Law Reports, which were published on four occasions between 2012 and 2018. The RIBA's reports showed a gradual decrease in 'traditional' procurement (where the employer is responsible for design), seemingly at the expense of a growing 'design and build' market. The RIBA's 2022 report indicates that these trends have continued.

The possible effects of these trends on the construction and engineering industries are likely to be widespread. In my experience, industry usually leads the way, and academia follows. However, 'industry' is not a uniform entity, and within it misunderstandings and mistakes can be rife. The concept of 'design and build' ought to be a simple one, in which the contractor is responsible for design. At one end of the spectrum, many contractors, employers, and consultants still treat 'design and build' as an offshoot of 'traditional' procurement, and on relatively simple projects experience difficulties over the extent of design responsibility. At the spectrum's other end, the increasingly complex nature of construction and engineering projects often requires different parts of the works to be designed by different parties, such as the contractor's specialist sub-contractors, the contractor, and the employer. Difficulties in integrating the various designs may arise, not only in terms of physically joining the contractor's and employer's designs, but also in terms of who bears responsibility for doing so.

Ordinarily, industry relies upon academia to provide it with graduates that it goes on to train to become the likes of engineers, project managers, architects, quantity surveyors, and so on – often to become chartered professionals. However, despite industry's increasing use of 'design and build', it is surprising how few UK universities teach dedicated 'design and build' modules, for what (at time of publication) could be the most popular means of procuring construction and engineering works. This should not be taken as a criticism of university courses for the following reasons. First, there are only a handful of what could be described as 'design and build' textbooks to rely upon, and many of these were published decades ago. Second, in order to effectively teach students about 'design and build', the courses will almost certainly have to teach students how 'design and build' was originally intended to operate – with the contractor carrying out the design and being responsible for it. 'Design and build''s original intention is often very different to the way it operates today. The design is produced by the employer and passed to the contractor, whose only design input is effectively limited to a design coordination role. Nevertheless, the contractor remains responsible for the entirety of the project's design. There may be variants of this method, with the employer's design consultants being passed on to (or 'novated') to the contractor, or the contractor working alongside the employer and design consultants from a much earlier (pre-contract) stage of the project.

I have attempted to write this book in the general order that a construction or engineering project is carried out. Following the sequence of the RIBA's 'Plan of Work 2020' (which is often followed by the Institution of Civil Engineers' 'Civil Engineering Procedure'), I have attempted to set out the practical steps under three of the most popular forms of building contract: namely, the JCT Design and Build 2016 contract, FIDIC's Conditions of Contract for Plant and Design-Build 2017 (the Yellow Book) and NEC4. Whilst this book necessarily examines the various procedures under these contracts, it is not intended to be a 'commentary' on those forms, as far more comprehensive guides to each of those contracts (and their predecessors) are widely available. The book also considers subcontracting and the engagement of consultants (in both supervisory and designing roles).

I would like to express my thanks and appreciation to Dr Paul Sayer, along with all the copyediting staff at Wiley for their forbearance, encouragement, and support.

1

Introduction

The concept of engaging a single organisation to carry out both the design and construction for a project is nothing new. In various subsets within the engineering and construction sectors, the concept is regularly adopted, used to varying extents, and continues to evolve to suit the prevailing market conditions.

This book examines the methods of what we now know as 'design and build' by considering how works are procured under three of the UK's standard forms of building contract. The book begins with a general introduction to 'design and build' and construction contracts in Chapters 1 and 2. Part One (consisting of Chapters 3–9) looks at how these contracts are intended to function in their unamended form throughout each stage of a project, from bidding to the project's conclusion. Part Two (comprising Chapters 10–12) examines how those procedures have changed in recent years, who brought about these changes, and the effects these changes have had on employers, contractors, and consultants (such as designers and those with responsibility to certify matters under the building contract).

1.1   Some Types of 'Design and Build'

The terminology for engaging a single entity to design and construct the works can be varied and depends upon (i) the nature of the works themselves and (ii) the extent to which the employing party is involved. For projects within what can be described as the 'construction sector' (such as constructing commercial buildings and residential construction by a non-housebuilder), this type of procurement is usually called 'design and build'.

For projects of a more 'engineering' nature, such as civil engineering infrastructure or process plant projects, terms such as 'engineering, procurement, and construction' (EPC), 'design, build, operate' (DBO), and 'turnkey' may be used. The EPC and DBO terms usually apply where the employing party retains an element of control over the design. However, 'turnkey' is often used where the employer has little involvement with the project until its completion, when he or she takes possession of the finished works to 'turn the key' to begin their operation.

Design and Build Contracts, First Edition. Guy Higginbottom.
© 2024 John Wiley & Sons Ltd. Published 2024 by John Wiley & Sons Ltd.

1.2 The Nature of Design and Build

Any project, whether its nature is construction or engineering, can be procured in a number of ways. All projects begin with a concept, and depending upon their nature, the 'initial brief' will vary in form, length, and detail. The initial brief may consist of no more than an idea, and it may never even be written down. All projects must be designed. The extent of the design can range from a general idea of what to build to a design that's fully developed, specified, and coordinated for each of its constituent elements.

To understand the nature of 'design and build' as a concept, it is useful to compare the project stages at which the employer engages the design team and the contractor under more 'traditional' procurement. 'Traditional' procurement is often used to describe projects for which the employer firstly obtains the design and then engages the contractor to construct that design but without being responsible for it. The Royal Institute of British Architects (RIBA) publishes its *Plan of Work*, which divides a project into distinct stages. Its most recent version was published in 2020 and divides the work into eight stages as shown in Table 1.1.

For each stage, RIBA's 2020 plan of work describes the common activities under the following headings:

o Stage Outcome
o Core Tasks
o Core Statutory Processes
o Procurement Route
o Information Exchanges

It seems appropriate to refer to the 2020 plan of work for two reasons. Firstly, the 2020 version is substantially the same as its predecessor (from 2013), and it is likely that the building contracts referred to in this book were drafted with the plan of work in mind. Secondly, the Institute of Civil Engineers' *Civil Engineering Procedure* (7th ed., 2015) contains key references to RIBA's 2020 *Plan of Work*, so it has obvious applications to both construction and engineering projects.

Initially (through RIBA Stages 0, 1, and 2), the employer develops the project's general requirements – often with a design team comprising an architect, engineer, and possibly a cost consultant (such as a quantity surveyor). Subsequently, and following the latter

Table 1.1 Project stages in the 2020 Royal Institute of British Architects' *Plan of Work*.

General description	Project stage(s)
o Initial brief	'Strategic Definition' and 'Preparation and Briefing' (Stages 0 and 1)
o Design	'Concept Design', 'Spatial Co-ordination' and 'Technical Design' (Stages 2, 3, and 4)
o Construction	'Manufacturing and Construction' and 'Handover' (Stages 5 and 6)
o Operation	'Use' (Stage 7)

design stages (RIBA Stages 3 and 4), a contractor is selected; this is done either following negotiations, such as a pre-qualification process, or in a competitive tender. The contractor's involvement usually starts at this point and will normally be formalised by entering into a building contract just before the start of RIBA Stage 5 (Construction). The employer remains responsible for the project's design and bears the risk if the design is incorrect, uncoordinated, or causes delay if it is not provided to the contractor on time. The employer firstly will have to assess and pay the contractor for any rework and overruns and must then take action against the offending member (or members) of the design team.

For design and build projects, the contractor's involvement may begin at any point between RIBA Stages 0 through 5. Depending upon their detail, the employer's requirements may be finalised at any point between RIBA Stage 1 (Preparation and Brief) and Stage 5. The contractor reviews the project's requirements (normally called the 'employer's requirements') and then develops proposals for carrying out the works. The contractor's proposals normally include any contractor's design, the specification for the works, a method statement, and health and safety information. The extent of design provided by the contractor will depend upon the extent of design provided by the employer's requirements. This may simply be 'filling in the gaps' in cases where the employer's requirements are quite prescriptive, but it may include a more detailed design if the employer's requirements are fairly brief.

Allied to the proposals, the contractor will normally set out a price for the works in a separate document. Typically, this provides far less detail than in bills of quantities, with the price simply being broken down into building elements (such as substructure, superstructure, etc.). Importantly, and unlike traditional procurement, the contractor assumes responsibility for the adequacy, accuracy, completeness, and consistency of the design. If it has any shortcomings, including an undue delay in its development or coordination, then the contractor will have to bear the cost of putting this right. The main advantage to the employer is that he no longer carries the design risk, as this is now held by the contractor. For example, if the design contains errors, isn't properly coordinated, or takes longer to develop, this remains the contractor's problem. The employer only has to manage dealing with the contractor (and not a design team), which is called 'single point responsibility'.

In the past, many contractors marketed their design and build expertise by claiming it added 'buildability' to a project. This term implied that the contractor would optimise the design so it became more practicable and would avoid design that was unnecessarily complicated and difficult to construct. Whilst the concept of 'buildability' may have originated from a desire for more efficient construction, it also provided the contractor with opportunities to increase profits by adopting a cheaper design. If the employer's requirements did not set out a precise specification or standard, contractors were often able to reduce their build costs by using cheaper materials and products, whilst still complying with the contract and remaining entitled to the same contract sum. Many employers soon countered this by ensuring their requirements contained specific quality standards. They also sought continuity of design from pre-contract (RIBA Stages 0–4) into construction (RIBA Stages 5 and 6), by transferring the employment of its pre-contract design team to the contractor for the construction phase. This concept is called novation and remains very popular.

1.3   A Brief History of Design and Build Contracts

Standard forms of building contract in the UK have been around for some time. The Institute of Civil Engineers (ICE) published its first contract in 1945, and the Joint Contracts Tribunal (JCT) published its standard form of contract in 1963 (JCT 1963). Both of these were contracts for 'traditional procurement', and since then both have been subject to many revisions. ICE produced a further six editions (up to the seventh edition in 1999) before becoming the Infrastructure Conditions of Contract in 2014, when ICE began to solely concentrate on publishing its suite of New Engineering Contracts (NEC). JCT 1963 was revised in 1980, and this was subject to around 18 amendments up to 1998, before JCT adopted its modern format (nine sections of clauses) in 2005. JCT's standard building contracts underwent further revisions in 2011 and 2016.

Design and build contracts, as we now know them, did not appear until 1981 when JCT published the 'with Contractor's Design' contract (CD 1981). This followed the format of the 1980 edition of JCT's standard building contract. CD 1981 was revised in 1998, and again (adopting the nine-section format of the standard building contract) in 2005, 2011, and 2016. This has become one of the most popular contracts for procuring construction projects in the UK.

Since 1943, government and public sector procurement used the CCC/Works/1 contract, until the General Conditions of Contract for Building and Civil Engineering (GC/Works/1) were published in 1959. In 1977 and 1989, second and third editions followed, and soon after, a 'GC/Works/1 (Edition 3) Single Stage Design and Build Version' was published in July 1993. Ultimately, three versions called GC/Works/1 (1998) were published in 1998, which included the single-stage design and build version. However, since then, use of the NEC contracts replaced GC/Works/1 (1998) as the preferred route for public procurement.

For engineering projects (such as civil and process engineering), contracts published by ICE and FIDIC are very popular. ICE published the first edition of its 'Design and Construct' contract in 1992. A second edition followed in 2001 – each edition being based upon the sixth and seventh editions of what was then known as the ICE Conditions of Contract. ICE did not publish any subsequent versions, as it chose to concentrate on the NEC suite of contracts. The Association for Consulting Engineers took over publication, and in 2011, issued a revised 'Design and Construct' contract – re-badged as the Infrastructure Conditions of Contract.

The NEC contracts were first published in 1993 and contained provisions for a contractor's design. These provisions remained largely unchanged in NEC2 (published in 1995) and NEC3 (issued in 2007 and revised in 2013) and, although slightly expanded, remain in the current NEC4 contracts that were published in June 2017. NEC contracts have been adopted for use by public bodies, for many high-profile projects such as the 2012 Olympic Games and Crossrail. They have largely replaced the previous suite of government contracts – GC/Works/1.

The newest version of a design and build contract was developed by FIDIC (the Federation Internationale des Ingenieurs-Conseils) when it published its *Conditions of Contract for Plant Design-Build* (Yellow Book) in 1999, with a second edition following in 2017.

The JCT design and build contract provided the most comprehensive procedures for a project with the contractor's design. The contract documents consist of the 'Employer's

Requirements' and the 'Contractor's Proposals' (this terminology is also adopted by the FIDIC Yellow Book), and the price for the works (Contract Sum) is shown in the 'Contract Sum Analysis'. The conditions also deal with design matters such as the contractor's liability for design, resolving discrepancies, divergences and inadequacies, the employer's approvals, design changes and their valuation, along with the rights of the employer (and third parties) to retain and use the contractor's design. Almost every version of JCT's standard forms of contract now contain some kind of design and build provisions. These are either provided in 'with design' versions of contracts (such as the Intermediate or Minor Works contracts) where the entire works will be contractor designed, or as options where the contractor is only designing part of the works under the JCT Standard Form of Building Contract.

The FIDIC Yellow Book contains some procedures analogous to those in the JCT design and build contract, but NEC's design and build procedures are far more comprehensive. NEC contracts do invite some adaptations by providing for 'Additional Conditions of Contract' (the Z clauses), and it is fairly common to find design and build procedures added here. Owing to the relative decline in use of the ICE and GC/Works/1 contracts, many of their specific provisions are now inserted as Z clauses. Indeed, some Z clause amendments have also been seen to introduce JCT-style design and build provisions into NEC contracts.

1.4 Recent Developments in Design and Build

On five occasions since 2012, RIBA Enterprises' NBS Limited has published its *National Construction Contracts and Law Survey*, with the latest edition arriving in July of 2022. These surveys have shown a number of trends in the UK, the following two of which are particularly pertinent:

o Between 2012 and 2018, the use of design and build contracts increased from an average of 30% to around 40%, whereas the use of traditional procurement declined from 60% to just over 40%.

o Between 2015 and 2018, the use of JCT contracts rose sharply (from below 60%) to 70%, whereas the use of NEC contracts has dropped by almost 15% (decreasing from 53% to 39%). Whilst there was a slight increase in FIDIC contracts in 2015 (to around 18%), their usage has returned to around 10%.

o From 2018 to 2022, the use of JCT contracts (both design and build and traditional procurement including a CDP) increased slightly to 71%. The usage of NEC contracts fell by 8% (from 39% to 31%) and the use of FIDIC contracts fell by 4% (from 10% to 6%).

These trends may be partially explained by the changing economic climate over the last 10 years. When UK construction felt the effects of the global financial crisis around 2010, there was a decline in privately funded development. Many contractors and consultancy firms adapted by taking on more work in the public sector. This may explain the growth in NEC usage from 2012 to 2015. Private sector developers and funders became more risk-averse, so this may explain the increase in design and build procurement, seemingly at the expense of 'traditional' procurement, and the private sector's preferred vehicle remains the JCT design and build contract. Not only does this offer the employer 'single point responsibility' and the most comprehensive procedures, but by introducing bespoke amendments, employers can reduce their risk even further. The relative reduction of design carried out by the contractor and the transfer of risk away from the employer have often led to design and build being called 'risk and build' and 'design and dump'! However, these developments do not come free of charge. Commercially astute contractors will include added risk contingencies within their pricing, and depending upon the prevailing market, contractors may be prepared to reduce their price in return for the employers accepting a greater share of risk.

Perhaps owing to the Covid-19 crisis from 2020 to 2021, RIBA did not commit to a further survey until July 2022. Their 2022 inclusion of CDP elements within the published figures for 'traditional' procurement is unfortunate, as it does not allow a consistent comparison with the data from 2012 to 2018. It therefore remains to be seen if design and build has overtaken traditional procurement to a significant degree. Furthermore, with the publication in 2017 of NEC4 and FIDIC's second edition contracts, it is difficult to predict whether the increasing trend in using JCT contracts will be maintained. However, given that JCT design and build procedures are often adopted even if the JCT contract itself is not, it seems likely that JCT will remain an influential force in design and build procurement – for both construction and engineering projects.

1.5 How to Use This Book

This book attempts to bridge the gap between how design and build is meant to be operated under the standard forms of contract and how it is now commonly operated in practice. It may be more accurate to suggest that there are two bridges to overcome. Firstly, as a concept, design and build is not widely taught in universities at the undergraduate level, as most construction courses focus on traditional procurement. In the event that design and build overtakes its traditional counterpart, it is increasingly likely that employers, consultants, and contractors will not be able to select candidates who are familiar with what may become the UK's most popular construction procurement method.

Secondly, the way design and build is widely practiced is very different from how it may be taught if the student purely studies the standard forms of contract and how they were originally intended to operate. For example, neither JCT, NEC4, nor FIDIC contracts provide for the novation of designers – a common and widespread practice. In another example, and as briefly described above, a contract may state 'design and build' on the front cover, but there may be little design for the contractor to undertake. He or she will just be 'building', but subject to a very different allocation of risk to that set out in an unamended design and build contract.

This book is not intended to provide purely a clause-by-clause commentary on particular forms of design and build contracts, as there are far more comprehensive texts available to the reader which provide precisely that for the JCT, NEC, and FIDIC forms of design and build contract. This book's principal focus will be on appropriate procedures at each stage of the project from a practical perspective (following the RIBA Plan of Work) and will reference particular contracts and their clauses where applicable.

2

Construction Contracts

2.1 General

All contracts contain a set of reciprocal obligations for each party to perform. In its simplest form, a construction contract provides that the employer pays the contractor who, in return for payment, constructs (and in this book's context, also designs) the works. The contract should create certainty so that each party knows the extent of its obligations to the other. All construction contracts should include (what lawyers refer to as) the following 'essential terms':

o Scope – what the employer requires the contractor to design and build.
o Time – when the contractor can start and when the works must be completed.
o Price – how much the employer will pay for the work. This can be either as simple as a single lump-sum or a basis for calculating the price (such as a schedule of rates and prices).

2.2 The Structure of a Construction Contract

Clearly, contracts for modern construction projects contain far more than just the three essential terms. They should have enough conditions to cover statutory provisions (such as interim payments and health and safety), instructing and valuing changes to the scope, testing and rectifying defects, various types of insurance, guarantees, bonds, and collateral warranties, and they should provide a mechanism for resolving disputes. The scope of work may also be very detailed and is normally found in drawings and specifications for each major element of work (such as structural and mechanical engineering, along with architectural design). All of these elements must be effectively pulled together by incorporating them under the umbrella of the construction contract.

Most construction contracts are made up of two main sections:

o One section contains specific information that relates directly to the project.
o The other section contains the conditions of contract, setting out in detail how the parties conduct themselves (see Section 2.2.2). The conditions of contract contain the general

Design and Build Contracts, First Edition. Guy Higginbottom.
© 2024 John Wiley & Sons Ltd. Published 2024 by John Wiley & Sons Ltd.

conditions and any optional clauses that the parties choose and are drafted with a 'one size fits all' approach, so that they can be applied to any project. These are often called the 'standard conditions', as found in contracts published by the JCT, NEC, and FIDIC (see Section 2.2.3).

2.2.1   The General Nature of Standard Forms of Design and Build Contract

This book examines typical procedures found in the three design and build contracts; namely JCT DB 2016, FIDIC Yellow Book 2017, and NEC4. However, whilst many of the procedures explored in this book operate within the terms of these contracts, the contracts themselves can operate very differently. JCT and FIDIC contracts share many features and in many respects operate in similar ways, whereas the NEC4 contract is intended to operate differently. Later in this book, these similarities and differences are examined, compared, and discussed in more detail, but it is useful to summarise the general differences at the outset:

o   The general principle of JCT and FIDIC contracts is that the parties eventually obtain the right result. Both contracts provide:
   – Confirmation that payments made during the course of carrying out the works are not conclusive[1] and can be revised; and
   – What can be described as a 'final account' procedure,[2] which effectively concludes the contract after the works are completed and defects have been rectified.
o   The NEC4 contract, on the other hand, provides that the parties obtain quick answers to their questions whilst carrying out the works, even though they might be incorrect. Previous NEC versions didn't include a 'final account'; however, a 'final assessment' procedure[3] was introduced with NEC4. Whilst initially, this appeared to contradict the NEC ethos, it is currently understood that this new procedure will not permit wholesale reassessments, so its application is relatively limited.

2.2.2   The Specific Information

The specific information sets out the essential terms in greater detail, along with other important matters.

Depending upon the form of contract, the specific information (see Table 2.1) is usually found in the section setting out what the parties have agreed to. This also contains the part of the contract that the parties sign, which for JCT DB 2016 is the 'Attestation' and for FIDIC Yellow Book 2017 is the 'Contract Agreement'.

NEC4 adopts a different format. Firstly, it doesn't have an agreement section as such. It sets out and incorporates all the specific information, along with the conditions of contract, within a section called the 'Contract Data'. Secondly, NEC claims that the lack of a signature page is deliberate, leaving the method of execution up to the parties, and

---

1 JCT DB 2016 clause 1.9 (Effects of payment other than payment of Final Statement), and FIDIC clause 14.6.3 (Correction or mediation).
2 Refer to Section 9.6.1 for JCT DB 2016, and Section 9.6.2 for FIDIC Yellow Book 2017.
3 NEC4, under clause 53 (refer to Section 9.6.3).

Table 2.1 Specific information contained in a construction contract.

o General	– The identity of the employer and contractor, along with their addresses, and any key personnel (NEC4) – The contract conditions – General description of the works – The identity of the supervising officers such as the employer's agent (JCT), project manager and supervisor (NEC4), or engineer (FIDIC) – Any other personnel such as persons engaged to fulfil requirements of the Construction (Design and Management) Regulations 2015 (CDM)
o Scope	– The contract documents – such as the Employer's Requirements and Contractor's Proposals (JCT and FIDIC) or Scope (NEC4) (including the design provided by the contractor in Section 2 of the Contract Data Part Two). The contract documents will contain the necessary drawings and specification for the project, including particular subcontractors or suppliers. – Division of the works into sections (see the 'Time' section of Table 2.1)
o Price and payment	– The Contract Sum (JCT), tendered total of the Prices (NEC4), or Contract Price (FIDIC) – Currency or currencies of the contract – The payment provisions (e.g. stage or 'milestone' payments), or periodic payments (e.g. monthly) – The Contract Sum Analysis (JCT), the Schedule of Rates and Prices (FIDIC), or the *activity schedule* and schedule of cost components (NEC4) – Retention (JCT, NEC4 Option X16), Retention Money (FIDIC) – Fluctuations (JCT) or
o Time	– Date(s) of Possession (JCT), Starting Date (NEC4), or Commencement Date (FIDIC) – Completion Date (JCT, NEC4), Time for Completion (FIDIC) Including any start and completion dates that apply to separate sections of the works – Key Dates (NEC4) – Rectification Period (JCT), *defect correction period* (NEC4), or Defects Notification Period (FIDIC) – may differ for sections – Damages for late completion – Liquidated Damages (JCT), or delay damages (NEC4 Option X7, FIDIC) – may differ for sections
o Insurance	– Contractor's public liability insurance – Employer's insurance – Professional indemnity insurance – Terrorism cover
o Guarantees	– Collateral warranties – Parent company guarantees – Bonds – Third-party rights
o Dispute resolution	– Adjudication: identifying the adjudicator or the adjudicator nominating body, or the adjudication procedure – Means of finally resolving a dispute: arbitration or litigation, including a tribunal (NEC4) or dispute board (FIDIC)
o Miscellaneous	– Framework agreements – BIM protocol – Collaborative working – Performance indicators, benchmarking – Sustainable development – Other employer policies or requirements

as NEC4's use is intended to be international, the formalities of signing a contract in the UK may be different from those for signing a contract outside the UK. This does seem odd, particularly as the FIDIC contract contains a 'signature' page and is widely used outside the UK.

JCT and FIDIC contracts adopt the 'agreement' format. This begins with a general description of the parties and the works, before formally setting out what is known as the 'operative' part of the contract. For JCT contracts, the operative part commences with the words, 'Now it is hereby agreed as follows', and FIDIC adopts similar wording – 'The Employer and Contractor agree as follows'.

The JCT contract includes a section entitled 'Agreement', which is divided into the following subsections:

o 'Recitals' is the introductory section that includes the date of the contract, identifies the parties, and gives a very general description of the works.
o 'Articles of Agreement'
  – This is the operative part of the contract.
  – Articles 1–9 set out the contractor's obligations by incorporating the Conditions[4] and identifying the Contract Sum, the Employer's Agent, relevant CDM personnel, and the methods of dispute resolution.
o 'Contract Particulars' sets out the remaining specific information (from Table 1.1) that is not already described in the Recitals or the Articles of Agreement. This remaining information is likely to consist of the contract documents (such as drawings and specifications provided by the employer and the contractor, and the contractor's pricing documents).
  o 'Attestation' is the part that the parties sign to execute the contract either as a deed or under hand.

The FIDIC contract is similar, as its 'Contract Agreement' contains the following:

o An unnamed introductory section sets out the date of the contract, identifies the parties, and gives a brief description of the works.
o The operative part of the contract then incorporates the specific information (as in Table 1.1) and the conditions of contract by reference to the following documents:
  – The 'Letter of Acceptance',
  – The 'Letter of Tender',
  – Any addenda,
  – The 'Conditions of Contract', which comprise the 'General Conditions'[5] as amended by the 'Particular Conditions',[6]
  – The 'Employer's Requirements',
  – The 'Completed Schedules',
  – The 'Contractors Proposals', and

---

4 JCT DB 2016 – clause 1.1 (Definitions) 'Conditions' means the clauses set out in Sections 1–9, and any applicable schedules.
5 FIDIC Yellow Book 2017 – clause 1.1.43 (Definitions) defines 'General Conditions' means the 'Conditions of Contract' (i.e. clauses 1–21 of the 'Conditions', as defined in clause 1.1.8).
6 FIDIC Yellow Book 2017 – clause 1.1.59 defines the 'Particular Conditions' as including Part A – Contract Data, and Part B – Special Provisions.

- The JV Undertaking (for use where the contractor is a joint venture).
o A signature section is for the parties executing the contract and their witnesses.

Each form of contract has a particular name for the section containing the specific matters. JCT calls this section the 'Contract Particulars', whereas for NEC4 and FIDIC it is called the 'Contract Data'. NEC4 divides the Contract Data into two sections; part one is completed by the employer (which NEC4 calls the 'Client') and part two is completed by the contractor. FIDIC is similar, although the 'Contract Price' is not shown in the Contract Data; rather it is shown in the 'Letter of Acceptance' (one of the documents incorporated by reference and listed in the 'Contract Agreement').

2.2.3 The Conditions of Contract

The second main section of a construction contract contains the general conditions and (in the case of NEC4) any optional clauses. JCT refers to these as the 'Conditions', which include Sections 1–9, along with any applicable 'Schedules' (which contain the optional clauses such as the 'Supplemental Provisions' in Schedule 2). NEC4 contains the 'Core Clauses' in Sections 1–9, along with the 'Main Option Clauses' (Options W1, W2, and W3), and the 'Secondary Option Clauses' (Options X1 to X21, Y(UK)1 to Y(UK)3, and Z). The 'Contract Particulars' (JCT) or the 'Contract Data' (NEC4) set out which options apply. FIDIC contains optional clauses in 'Particular Conditions Part B – Special Provisions'. Similar to NEC4, FIDIC's Special Provisions contain wide-ranging amendments that can be incorporated within the Conditions.

2.2.4 The Documents Comprising the Contract

The construction contract needs to contain all the necessary agreements and information for the works to be designed and constructed, and for the contract to be administered in the manner agreed to by the parties. The JCT and FIDIC contracts achieve this in similar ways, whereas NEC4's method is slightly different.

JCT DB 2016 very clearly provides that the contractor must design and build the works in accordance with the 'Contract Documents'. This obligation is firstly set out in Article 1 (of the Agreement) and again at clause 2.1.1 of the Conditions. The JCT definitions are set out in clause 1.1 of the Contract Documents as follows:

o The Agreement – includes the Recitals, the Articles of Agreement, and the Contract Particulars.
o The Conditions – means all the clauses included in Sections 1–9, and any other clauses in the Schedules (numbers 1–7).
o The Contract Documents – includes 'the Agreement and these Conditions, together with the Employer's Requirements, the Contractor's Proposals, the Contract Sum Analysis, and (where applicable) the BIM Protocol'.

The 'Employer's Requirements', 'Contractor's Proposals', and 'Contract Sum Analysis' are each described against the entries for Article 4 in the 'Contract Particulars' (itself, forming part of the Agreement). Similarly (and if applicable), the BIM Protocol is described against the entry for clause 1.1. in the Contract Particulars. The scope of work is set out in the Employer's Requirements and Contractor's Proposals, and details of the Contract Sum are set out in the Contract Sum Analysis.

FIDIC Yellow Book 2017's approach is similar to that of JCT. In the 'Contract Agreement' (in part 3), the '…Contractor hereby covenants with the Employer to design, execute and complete the Works and remedy any defects therein, in conformity with the provisions of the Contract'. The similarities with JCT continue, as this obligation is repeated at clause 4.1 of the Conditions. FIDIC's definition (in clause 1.1.9) of the Contract is as follows:

> 'Contract' means the Contract Agreement, the Letter of Acceptance, the Letter of Tender, any addenda referred to in the Contract Agreement, these Conditions, the Employer's Requirements, the Schedules, the Contractor's Proposals, the JV Undertaking (if applicable), and the further documents (if any) which are listed in the Contract Agreement or in the Letter of Acceptance.

NEC4 is slightly less straightforward. Unlike JCT and FIDIC, it does not contain a succinct definition of the 'contract'. Instead, the parties must refer to the following:

o Clause 10.1 provides that 'The Parties…shall act as stated in this contract'.
o Clauses 20 and 21 provide that the contractor respectively constructs and designs the work in accordance with the 'Scope'.
o 'Scope' is defined in clause 11.2(16) as being contained in the 'Contract Data'.
o The introductions to both Parts 1 and 2 of the Contract Data state as follows: 'Completion of the data in full, according to the Options chosen, is essential to create a complete contract'.
o The Contract Data incorporates the following documents (assuming no Secondary Option clauses are selected):
  – 'Scope' and 'Site Information' are in Contract Data Part 1.
  – The 'scope for [the contractor's] design', and the priced 'activity schedule', the bill of quantities, and date for the schedule of cost components (including the data for its shorter version) are in Contract Data Part 2.

When read together, it should be clear that the 'Contract Documents' in NEC4 are the 'Core Clauses' (along with any 'Main Option Clauses' and 'Secondary Option Clauses' that are selected), 'Contract Data Parts One and Two', the 'Scope', the 'Site Information', the 'scope for the contractor's design', and the priced 'activity schedule'.

Clause 12.4 of NEC4 provides that the contract is an 'entire agreement' between the client and contractor. This means that all statements made by either party before entering into the contract have no bearing on the contract itself. Thus it is all the more important for both parties to ensure that the Scope, the Site Information, and all parts of the Contract Data capture the full extent of their agreements.

2.2.5 The Relationship Between the Conditions of Contract and the Contract Documents

Construction contracts are often voluminous documents that, by necessity, have to include many 'contract documents' covering all elements of the project's scope, as well as the specific information that the parties have agreed to and that sets out how the contract is to be performed. For any contract to be successful, it must provide clear directions as to which party carries out a particular obligation, when this obligation should be carried out, and in some instances, the manner in which it should be carried out.

As described in Section 2.2.4 of this book, the JCT, FIDIC, and NEC4 contracts all contain a hierarchy that states that the contract agreement and conditions all govern the contract documents. This has the effect that nothing set out in the contract documents (predominantly, in the employer's requirements or contractor's proposals) will override the contract conditions. JCT and FIDIC contracts provide for this hierarchy in similar ways:

- o JCT DB 2016 – Clause 1.3 sets out that the Agreement and the 'Conditions' override the Contract Documents (or any Framework Agreement). At first this may seem a little cyclical and confusing because by definition, the Conditions form part of the Contract Documents. It therefore appears that for the purposes of interpretation, the Conditions sit above the Employer's Requirements, the Contractor's Proposals, and the Contract Sum Analysis. The JCT DB 2016 Conditions do allow the following matters to be set out in the Employer's Requirements:
  - (Clause 3.12) – the Employer's Requirements (and/or the Contractor's Proposals) can set out that where work complies with the contract, the costs for opening up, testing, and making good may not be added to the Contract Sum.
  - (Clause 4.7.4) – specifies information to accompany the Contractor's Payment Applications.
  - (Clause 6.5.1) – says that insurance is required.
  - (Schedule 1) – specifies the format of the Contractor's Design Documents and the manner in which they should be submitted.
  - (Schedule 2) – identifies any Named Subcontract Work.
  - (Schedule 4) – specifies tests for the Code of Practice.
- o FIDIC Yellow Book 2017 contains a simple hierarchy at clause 1.5 (Priority of Documents), which states that any 'Particular Conditions' (at clause 1.5(d) and 1.5(e) in the hierarchy), along with the 'General Conditions' (at clause 1.5[f]), take precedence over the 'Employer's Requirements' and the 'Contractor's Proposals' (further down the hierarchy at clauses 1.5(g) and 1.5(i), respectively).
- o NEC4 doesn't provide anything like the clear guidance set out in the JCT or FIDIC contracts in terms of a priority of documents. From Section 2.2.4 in this chapter, it appears that the contractor's first obligation is to act in accordance with the contract (clause 10.1), and then design and build the works in accordance with the Scope (clauses 20 and 21). If there appears to be an inconsistency or ambiguity between any of the contract documents (which seemingly include the contract conditions), then this is a matter for the project manager to resolve (clause 17.1). If discrepancies between contract documents are of a technical nature (e.g. conflicting specification and drawings) then the project manager would be the ideal person for its resolution. However, if there are conflicts between the contract conditions (such as the Core Clauses) and the Scope, which affect the operation of the contract, then this is likely to be a matter for which the parties should obtain legal advice.

The parties could agree to the order of precedence, but this would have to be set out in the Z clauses.

2.3 Fairness as a Concept

Dictionary definitions of fairness include 'reasonable', 'just', 'in agreement with the rules', and perhaps the most appropriate definition in the context of a construction contract is 'equitable'. This implies a balance, with each party not only acknowledging its obligations to the other, but also being satisfied that the bargain struck is a fair one – where reciprocal obligations essentially match in terms of value. Of course, 'value' is not always measured in terms of money, bricks and mortar, or the simple calculation of 'function divided by cost'. Each party may value the bargain of a construction contract in different ways. An employer may feel there is added value from the prestige of a new building, having a development in a prime location, or completing the building ahead of a competitor. A contractor may value the enhanced reputation of successfully completing the building, may have been awarded the project ahead of a competitor, or may be keen to win the work in order to reach that year's turnover target. Making the 'contractual seesaw' balance seems relatively simple when all the facts and circumstances are known in advance.

Unfortunately, construction projects rarely provide all of the facts and circumstances from the outset. Market conditions may fluctuate, and the full extent of risks may not be discovered until the works are underway, or even when completed. Balancing the figurative see-saw can become a complex task – particularly for the contractor, who will have far more matters to consider than the employer, as Table 2.2 shows.

The contractor's difficulty is often compounded by the nature of contracting itself, as it's often a high-turnover business with relatively low margins. This equates to carrying out a lot of work whilst managing many different trades, all with a low margin for error. At the risk of making an overly simplistic comparison, all that the employer may have to worry about is the budget.

However, this perceived imbalance of risk should not be unmanageable. Contractors can assess the risks and make the appropriate allowances within their tender offers. These allowances are in terms of money and time. Commercially astute contractors may couch their tender in terms of a 'base offer', calculated by pricing and programming the works, evenly sharing the risks between the employer and contractor. The contractor then assesses the actual risk allocation as a priced list of add-ons to accompany the base offer. This can be very useful for both parties. The contractor can be confident that his tender (comprising the base offer and add-ons) hasn't missed any major risks. The employer can see precisely how much his proposed allocation of risk for the works will cost, and how long it will take. If the parties are engaged in negotiations, or a two-stage tender, this can provide a useful agenda for 'value engineering', particularly if the parties are also discussing alternative designs and construction methods.

In the contractual context, and irrespective of any disparity between the extent of each party's risks and obligations, the see-saw becomes balanced upon agreeing to the contract. The opportunities to change the agreed upon contract will be few. Afterwards, a perception of unfairness may arise if one of the parties feels 'caught' by a term of the contract, particularly if something occurs that puts him at a commercial disadvantage. Two relevant examples are the employer finding himself obliged to pay the contractor a sum he didn't expect to have to pay, and the contractor discovering that completing the works will be more difficult or more involved that he previously thought and will cost him far more than he'll recover through the contract.

Table 2.2 Categories of risk for the employer and the contractor.

Risk category	Employer	Contractor
Procurement and management	'Single point responsibility' of just procuring and managing the contractor Performance bond	'Multi-point responsibility' – many different parties to manage: – Subcontractors – Suppliers – Designers
Contractual	Means of transferring risk to the contractor	Time-bar clauses and contractual 'hurdles' restricting entitlement. Can the main contract risks be transferred to the subcontractors?
Price	Is the price (including contingencies) within budget? Reducing the risk of price increases?	Is the price sufficient to cover the cost of the design, the physical works, overheads, and return a profit? What contingencies should be included? What ground conditions, price fluctuations, additional work are required to achieve the design?
Scope	Is the scope sufficiently detailed? If so, there should be no changes/variations.	Contractor remains responsible for designing and constructing the works. Can the full scope of work be ascertained from the drawings and specifications? Is the design complete, or will allowances for additional design and physical work be required? Is the design sufficiently coordinated? Is the design adequate? Contractor is responsible for the quantities. Whether the contractor is to be made responsible for: – Ground conditions – Condition of any existing buildings (e.g. refurb)
Quality	Quality standards established in the contract. Contractor remains responsible.	Can each subcontractor/supplier achieve the required quality? Any allowance for rectification or rework?
Time	Can the contractor realistically complete on time? Are the delay damages sufficient protection if the contractor is late?	What 'time risks' are there? – Approval by statutory bodies – Key subcontractor failure to complete on time How much will it cost for each week of an overrun? – Employer's delay damages, plus – Contractor's preliminaries – Subcontractor costs – Effect on cash flow or borrowing costs Proving entitlement to extensions of time/disruption is not straightforward or cheap.

2.3.1 Assistance from the Contract Terms?

Can the contract terms alleviate a perception of unfairness? NEC4 (at clause 10.2) provides what Sir Rupert Jackson refers to[7] as one of the more 'aspirational provisions': 'The Parties, the Project Manager and the Supervisor act in a spirit of mutual trust and co-operation'. This aspirational language is often found in 'partnering' contracts, which were heavily promoted, along with NEC contracts from the mid 1990's.[8] Sir Rupert questioned what, if any, effect such aspirational provisions may have on the 'black letter terms of the contract' and, citing *Costain Ltd v Tarmac Holdings Ltd,*[9] concluded that they have little effect. The reasoning in *Costain* was predominantly based upon a number of implied terms[10] that were already introduced:

o A degree of cooperation, as follows:
  – The employer will not hinder or prevent the contractor from carrying out its obligations in accordance with the terms of the contract and from executing the works in a regular and orderly manner.
  – The employer will take any steps reasonably necessary to enable the contractor to discharge its obligations to execute the works in a regular and orderly manner.
  – The employer has a general duty to provide the necessary cooperation.[11]
o A party is not obliged to set aside its own self-interest.[12]

Unless expressly excluded, these terms will be implied in any construction contract in the UK, be it based upon the JCT, FIDIC, or NEC terms. Whilst many commentators have praised NEC's risk-sharing model, it does come with a sting in the tail. NEC removes the contractor's entitlement to additional payment or an extension of time if it fails to give the relevant notice to the employer within eight weeks. The perception of unfairness will arise if the contractor is clearly entitled to significantly more money and time, only to have this snatched away because of being a day late in sending a particular piece of paper! Accepting a risk is far easier than accepting its consequences.

2.3.2 Understanding the Contract and its Implications

Two vital ingredients to ensuring a project's success is that both parties understand the contract and its implications. Experienced parties will also have the foresight to anticipate particular risks and make reasoned, commercial decisions on their effects without resorting to a worst-case scenario. A party that doesn't have the experience, either in terms of the particular contract or of the nature of work to be undertaken, should seek advice. There are many experienced professionals who can assist, with the best being able to apply their specialist knowledge to the practicalities of construction. Parties should not be afraid of

---

7 Jackson, Sir R. (2018). Does good faith have any role in construction contracts? Society of Construction Law.

8 Latham, Sir Michael (1994). *Constructing the Team*. The Stationery Office. Paragraphs 5.26 and 6.47.

9 EWHC 319 (TCC), (2017).

10 *London Borough of Merton v Leach,* 32 BLR 51, (1986).

11 *Keating on Construction Contracts,* (10th ed.), at 3-063 citing *Luxor (Eastbourne) Ltd v Cooper,* AC 108 HL, (1941).

12 *Costain* at 124.

obtaining advice from a different sector of the industry, particularly if it shares a number of characteristics or procedures with the parties' current project. Looking beyond the usual horizon can lead to innovation and an improvement of existing ideas. As construction and engineering projects not only share the 'essential terms' (summarised in Section 2.1), but also many similar procurement procedures, these are a major reason why this book considers design and build under the JCT DB 2016, FIDIC Yellow Book 2017, and NEC4 forms of contract. Incorporating a policy or procedure from outside the usual contract framework may significantly benefit a particular project.

Part I

3

Design and Build as Originally Intended

Section 1.2 of this book briefly summarised the features of early design and build contract practice. The concept of 'single point responsibility' was attractive to the employer, and contractors were keen to market their 'buildability' expertise by providing a building with an efficient design. For some years after JCT published the CD81 contract, design and build was successfully used to procure buildings intended for a simple function for which a simpler design would suffice. Buildings such as factory and warehouse units were procured in this way, and employers were happy to make contractors responsible for designs that focused on the functional rather than the aesthetic. The employer would tell the contractor *what* to build, and the contractor would work up a design that showed *how* it would be built.

3.1   The Parties' Primary Obligations Under a Design and Build Contract

In the simplest of terms, the employer's obligation is to pay the contractor, who is obliged to design and build the project for the employer. Owing to the sheer number of technical tasks that the contractor must perform in satisfying its primary obligations in terms of design and construction, the majority of obligations under a design and build contract have to be performed by the contractor. However, that is not to say that the employer has the single obligation of payment. Many of the employer's obligations involve administering the contract. In practice, most of the employer's obligations are carried out by the employer's representative (see Sections 3.2 and 4.2.8).

3.1.1   The Employer's Primary Obligations

The employer's obligations under a design and build contract focus on facilitating and controlling the works. The majority of the employer's facilitating obligations take place between RIBA (Royal Institute of British Architects) Stage 4 (Technical Design), Stage 5

Design and Build Contracts, First Edition. Guy Higginbottom.
© 2024 John Wiley & Sons Ltd. Published 2024 by John Wiley & Sons Ltd.

(Manufacturing and Construction), and Stage 6 (Handover). Aside from administering any residual maintenance or rectification duties that the contractor may have, it will be uncommon for the employer to have any major duties for the contractor from RIBA Stage 7 (Use). The employer's primary facilitating duties may include:

o Payments to the contractor (including issuing the relevant notices);
o Providing information to the contractor such as site survey, topographical information, permits, licences, and contract documents;
o Pre-construction information under the CDM (Construction Design and Management) Regulations 2015;
o Access and possession of the site;
o Insuring the works;
o Appointing the employer's representative(s); and
o Supplying the contractor with materials and/or plant.

The employer's obligations to control the works may include:

o Issuing instructions (often via the employer's representative);
o Managing progress, such as extending the period for completion or instructing acceleration;
o Controlling quality of works with inspections, tests, and instructing remedial works;
o Reviewing and approving the contractor's design documents;
o Valuation of variations (changes);
o Determining when the works are complete; and
o Final account or concluding the contract.

3.1.2 The Contractor's Primary Obligations

The contractor's primary obligations (to design and construct the works) will take place during RIBA Stage 4 (Technical Design) and Stage 5 (Manufacturing and Construction). It is very common for the contractor to have some obligations at RIBA Stage 6 (Handover), as the JCT DB 2016, FIDIC Yellow Book 2017, and NEC4 contracts all oblige the contractor to rectify defects for a period after the works are complete. Unless the particular contract contains some maintenance obligations that go beyond a defects rectification period, it is unlikely that the contractor will have many obligations to the employer during RIBA Stage 7 (Use).

The contractor's primary obligation is described in the three design and build contracts as follows:

In JCT DB 2016, clause 2.1.1 sets out the contractor's primary obligation in terms of both carrying out the physical works and completing the design.

o The physical works must be carried out:
  – In a 'proper and workmanlike manner';
  – In compliance with the Contract Documents (i.e. the Employer's Requirements, the Contractor's Proposals, and the Contract Sum Analysis);
  – In compliance with the CDM Regulations (i.e. the Construction Phase Plan); and

- The Statutory Requirements – UK legislation including the Building Act 1984, The Building Regulations 2010,[1] the Building (Amendment) Regulations 2018, the Sustainable and Secure Buildings Act 2004, and the Defective Premises Act 1972.

o The design must be completed, which includes:
  - Providing a design for the entire works (i.e. filling in any 'gaps' left in the Employer's Requirements);
  - Specifying the standards of any goods, materials, and workmanship that are not described in the Employer's Requirements or the Contractor's Proposals; and
  - Clause 2.17 provides that the standard of design is equivalent to that of a competent architect or other professional designer (refer to Section 4.2.1 of this book).

FIDIC Yellow Book 2017 describes the contractor's principal obligation in overall similar terms: Clause 4.1 (Contractor's General Obligations) deals with executing the physical works, and clause 5.1 (General Design Obligations) deals with the design.

FIDIC clause 4.1 provides that the contractor must execute the physical work:

o In accordance with the contract;
o So it is fit for:
  - The purpose defined in the Employer's Requirements; or
  - The works' ordinary purpose (if not set out in the Employer's Requirements); and
o So it satisfies:
  - The Employer's Requirements, the Contractor's Proposals and the Schedules; or
  - Any work that is implied by the contract; and
  - All work necessary to complete the works that is not mentioned in the contract.

It seems clear that like clause 2.1.1 of JCT DB 2016, clause 4.1 of FIDIC Yellow Book 2017 provides that the contractor's principal obligation extends to *completing* the physical work, as the contractor must effectively 'fill in any gaps' by completing all work that is necessary but perhaps not specifically described in the contract.

For the contractor's design, FIDIC clause 5.1 provides that:

o The contractor is responsible for the design of the Works.
  - By referring to the 'design of the Works', it appears that clause 5.1 implies that the contractor becomes responsible for *all* of the Works' design, which may include any preliminary design carried out by the employer. However, this is subject to the operation of clause 1.9 (Errors in the Employer's Requirements).
o The design must be carried out by properly qualified and competent designers who have the appropriate experience.
o The design must comply with the Employer's Requirements (see above regarding clause 1.9).

---

1 To achieve the standards set out in the Building Regulations 2010 ('Building Regs'), the UK Secretary of State has approved a set of documents ('the Approved Documents') that provide guidance for complying with the Building Regs. The Approved Documents state: 'It is the responsibility of those carrying out building works to meet the requirements of the Building Regulations 2010. Although it is ultimately for the courts to determine whether those requirements have been met, the approved documents provide practical guidance on potential ways to achieve compliance with the requirements of the regulations in England.'

Under NEC4, the contractor's principal obligations to provide the physical works are contained in clause 20.1 (Providing the Works) and its design obligations are set out in clauses 21.1 and 21.2 (The Contractor's Design). Both clause 20.1 and 21.1 provide that the works and its design must be provided in accordance with the Scope. In particular, clause 21.2 clearly implies that if the contractor's design is not in accordance with the Scope, then it is highly unlikely that the project manager will accept it.

The importance of the employer's requirements (for JCT DB 2016 and FIDIC Yellow Book 2017) and the scope (NEC4) cannot be overstated (see Section 3.4 in this book). These documents ought to precisely specify what the contractor is to provide. Ideally, they should not only specify particular materials, goods, and equipment (either by product name or class), and performance specifications, but should also provide particular standards of workmanship (e.g. including specified tolerances or industry standards). It is therefore advisable for the employer's requirements to set out what elements of the work are of prime importance to the employer. For engineering projects, the works may have to achieve a certain level of performance, whereas for construction projects, there may be a particular standard of finish that must be achieved. Expressing these important elements in definite, empirical terms leaves less margin for error or uncertainty.

Conversely, there may be elements of the work that are less important to the employer. For an engineering project, the standard of finish for a plant room is likely to be less important than the performance of the plant it contains. For the design and specification of such less important elements, the employer may leave more or all of the design to the contractor.

Attempting to set out the employer's expectations is problematic as it is almost a completely subjective task, so precisely worded employer's requirements should leave as little room for interpretation as possible. For employers, the rule of 'minimal performance' should be avoided, where the contractor's work complies with the contract but is less than the employer may have expected. The 'minimal performance' rule is particularly relevant for design and build projects (see Section 8.8, Variations). In cases where the employer's requirements allow the contractor considerable discretion in how to achieve a particular standard of work, the contractor will usually choose the cheapest option, which may not be to the employer's liking. If the cheapest option nevertheless satisfies the employer's requirements, the only remedy to the employer is to instruct the contractor to provide work to a higher standard. Under JCT DB 2016, FIDIC Yellow Book 2017, and NEC4, such an instruction would entitle the contractor to additional payment for carrying out the physical works, and possibly an extension of time, loss, and expense.

3.2 The Employer's Representative

Continuing with the theme of 'single point responsibility', employers may seek to rely upon the contractor's account of how the works are progressing. The employer will receive some guidance from his representative, but as the design team are working for the contractor, it is unlikely they will be able to report on matters of progress, commercial issues, and quality directly to the employer. If they did, they would risk breaching the terms of their appointments with the contractor, and more generally, creating a conflict of interest (for more details, see Section 10.2.2). It is therefore sensible for the employer to engage his own representative to advise him in these respects, and to act on his behalf.

If infrastructure or property forms a significant part of the employer's business, the employer is likely to have an in-house team of specialist advisors. Typically however, these advisors are unlikely to act on the employer's behalf under the terms of the building contract, so most employer's representatives will be appointed from external companies. Under all of the design and build contracts considered by this book, the employer's representative has to carry out many important duties, and failure to do so can have serious and expensive consequences for the employer. Therefore, many employers prefer an externally appointed representative, as they can be held to account and pay (via their professional indemnity insurance) for any failures for which they are found liable. These days, it is perhaps only the occasional public body or local authority that may have an 'in house' employer's representative.

The employer's representative's duties vary under each of the design and build contracts considered below:

o JCT DB 2016. The employer's representative is called the 'Employer's Agent' (EA). The EA does not have any duties that are distinct from the employer, as Article 3 essentially provides that the EA acts for the employer under the conditions of contract. This relationship between the EA and the employer is more aligned to the traditional role of an agent (hence the title EA is particularly apt) who acts on behalf of the employer (or principal), albeit Article 3 seems to limit the EA's authority to acting under the terms of the contract. The EA's role and details are provided in Article 3.

o FIDIC Yellow Book 2017. Under the FIDIC contract, the employer's representative is the engineer, whose role is defined in clause 1.1.35 and whose details are set out in the Contract Data. Unlike in JCT DB 2016, FIDIC provides distinct duties for the engineer, who acts as an independent professional person, and not as agent of the employer. The engineer may call upon a number of assistants; each called the 'Engineer's Representative' (see clause 3.3), provided that the engineer notifies the employer and contractor of each assistant's duties and authority.

o NEC4. The project manager is the employer's representative under NEC4 and acts in a manner similar to the engineer (FIDIC), as an independent professional and not as agent of the employer. The details are set out in Contract Data Part One and the role is described in clause 14. The project manager is be assisted by the supervisor (clause 14), particularly when inspecting and reporting defects.

It seems that under FIDIC or NEC4, the employer may be able to call upon more persons to assist him than he can under JCT DB 2016, as the EA appears to be a single person. The employer may require the engineer (FIDIC) or project manager (NEC4) to carry out the bulk of the contract administration, and delegate other duties to an engineer's representative (FIDIC) or to the supervisor (NEC4). These other duties may include giving opinions over the quality of work (carrying out a 'clerk of works' role) or quantity surveying and valuation functions. If the project involves a significant input from a particular discipline (such as a structural or building services engineer), then it is likely that a dedicated engineer's representative or supervisor will be appointed.

In many cases however, the employer's representative is not an individual, but a company – such as a multidisciplinary consultancy, whose staff can cover all the disciplines such as project management, contract administration, managing the quality of work and testing, and commercial issues.

The roles and duties of the employer's representatives (and their assistants) under JCT DB 2016, FIDIC Yellow Book 2017, and NEC4 are explored in greater detail in Chapters 4 and 10.

3.3 The Relationship Between the Employer's Requirements and the Contractor's Proposals

This chapter considers design and build contracts being operated in the way they were originally intended. As such they are vehicles for procuring a project where little or no design is carried out by the employer, so the majority or all of the design is produced by the contractor. In this context, the words 'employer's requirements' simply mean the employer's instructions to the contractor, describing what the employer wants the contractor to design and then build. Likewise, the words 'contractor's proposals' should be interpreted in the same way, meaning the contractor's explanation of how he intends to interpret and implement the employer's instructions. Neither term should be inferred to have the meaning ascribed by JCT DB 2016 or FIDIC Yellow Book 2017 unless the terms are capitalised (see Introduction Section 1.5 in this book).

This chapter will not consider any modifications that change the manner in which a design and build contract may be operated.

Understanding the relationship between the employer's requirements and the contractor's proposals should be straightforward. The employer's requirements instruct the contractor *what* works to provide, and the contractor's proposals inform the employer *how* the contractor intends to provide them. It seems a fairly simple concept that the employer's requirements necessarily take precedence over the contractor's proposals. If the contractor's proposals do not fulfil the employer's requirements, then the contractor will not provide the building, structure, or plant that the employer wants.

3.3.1 FIDIC Yellow Book 2017

FIDIC Yellow Book 2017 clearly and simply spells out this relationship in clause 1.5 (Priority of Documents) – that the 'Employer's Requirements' (at subclause [g]) take precedence over the 'Contractor's Proposal' (shown later at subclause [i]).

3.3.2 JCT DB 2016

JCT DB 2016 adopts the same approach, but unfortunately, it does not express the precedence of the employer's requirements with anything like the simplicity and clarity found in the FIDIC Yellow Book 2017. However, several important clues are found in JCT DB 2016, namely:

o The Third Recital confirms that the employer has reviewed the Contractor's Proposals and is satisfied that they appear to meet the Employer's Requirements. The Third Recital's footnote adds some clarity that in the event that the Contractor's Proposals are divergent from the Employer's Requirement in some way, that the employer may accept the divergence and should amend the Employer's Requirements accordingly before executing the contract.

o Clauses 2.2.1 and 2.2.2 stipulate that materials, goods, and workmanship must comply with the specific descriptions in the Employer's Requirements, and if they are not as 'specifically described', only then will descriptions in the Contractor's Proposals apply.

o Clause 5.1 of JCT DB 2016 defines a 'Change' (or variation) as being a change in the Employer's Requirements. There is no provision for the employer to instruct a change in the Contractor's Proposals. This is simply because any deficiency in the Contractor's Proposals will mean that the contractor will fail to meet the Employer's Requirements and will be in breach of the contract. There is no need to instruct a change in the Contractor's Proposals, as the criteria with which the contractor must comply is already stated in the Employer's Requirements.

o Clause 2.14.2 of JCT DB 2016 provides that if there is a discrepancy within the Employer's Requirements (e.g. a drawing shows a 300-mm-thick concrete slab, but the specification only describes a slab that is 200 mm thick), then its resolution is a 'Change' under clause 5.1, unless the Contractor's Proposals resolve the discrepancy (in this example, by specifying the concrete slab's thickness).

o Clause 2.12 deals with inadequacies in the Employer's Requirements in the same way.

3.3.3   NEC4

NEC4 (clause 20.1) states that the contractor not only must provide the 'Works' in accordance with the Scope, but must design the parts of the Works that the Scope says the contractor must design, and design the designated works in accordance with the Scope (clause 21.2).[2] The clause 11.2(16) definition states that the Scope is the description and specification for the works,[3] and is contained or referred to in the Contract Data. However, the Contract Data includes two references to the Scope:

o The first reference is in Contract Data Part One and refers to the documents provided by the Client, which state what works the contractor must design (clause 21.1). From this, it seems that (what this book refers to as) the 'Part One Scope' (or the relevant parts of it) is perhaps analogous to the 'employer's requirements' (as found in JCT DB 2016 and the FIDIC Yellow Book 2017).

o The second reference to Scope is in Contract Data Part Two, which refers to the scope provided by the contractor for its design. This also seems to fit with the clause 11.2(16) definition of Scope, albeit it describes or specifies the work that the contractor must design. Seemingly, (what this book calls) the 'Part Two Scope' has more similarities to the Contractor's Proposals found in JCT DB 2016 and FIDIC Yellow Book 2017.

However, given that the Scope in its overall sense[4] includes both the Part One Scope and the Part Two Scope, the following difficulties may arise, particular if the client's Part One

---

2 If it is not in accordance with the Scope, the project manager may not accept the contractor's design.
3 It seems to be common sense that 'description' includes drawings or graphical descriptions of the works and not just text describing the works.
4 Applying the clause 11.2(16) definition.

Scope does not set out detailed requirements for what the contractor is to provide as his Part Two Scope:

o As NEC clearly points out that there is no 'hierarchy of documents',[5] it may be too difficult (from a contractual perspective) to establish a precedence of the Part One Scope over the Part Two Scope as both are contained in the Contract Data and therefore satisfy the clause 11.2(16) definition of the Scope.

o Clause 63.10 of NEC4 refers to an instruction changing the Scope[6] in order to resolve an ambiguity or inconsistency, which becomes a Compensation Event[7] under clause 60.1(1). It is clear that the Scope in clause 63.10 means the Part One Scope, and the Part Two Scope, as assessing the Compensation Event, is interpreted against the party who provided the Scope. Therefore, the interpretation is against the client if the Part One Scope contains the ambiguity or inconsistency, and it is against the contractor if either are found in the Part Two Scope.

o On above basis, once the Part Two Scope is entered into Contract Data Part Two, the contractor's design becomes part of the Scope (within its clause 11.2(16) meaning). This may affect the project manager's ability to reject any design submitted by the contractor (under clause 21.2). As the Scope contains both the Part One Scope and the Part Two Scope and there is no express mechanism for either (i) establishing a hierarchy between the two or (ii) for resolving discrepancies and inconsistencies between them, it is difficult to see how the Scope (which now includes the Part Two Scope) cannot comply with itself.

Some assistance is provided by the exception in clause 60.1(1) which states that any change to the Part Two Scope that the contractor requests, or which is made in order to comply with the Part One Scope, is not a compensation event. However, the clause 60.1(1) exception is likely to best assist the parties when the Part One Scope clearly specifies what the Part Two Scope should contain.

If the client's Part One Scope is not particularly specific, another way of avoiding discrepancies between it and the contractor's Part Two Scope is simply not to make any entry in Contract Data Part Two for the contractor's design. By doing so, the Scope remains as the employer's instructions to the contractor (i.e. what this book refers to as the Part One Scope), and the contractor's design (the Part Two Scope) will be the particulars of its design submitted for the project manager to accept under clause 21.2, and must therefore be in accordance with the Scope and consequently subservient to it. This way, the project manager will have no difficulty in rejecting any of the contractor's design if it does not comply with the Scope. This approach would be familiar to design and practitioners less familiar with NEC4, as the Scope is clearly similar to the Employer's Requirements, and the contractor's design (under clause 21.1) is similar to the Contractor's Proposal(s) under JCT DB 2016 and FIDIC Yellow Book 2017.

---

5 Mitchell and Trebes, *Managing Reality – Book 3, Managing the Contract*, (3rd ed.), 2018, p. 42.
6 NEC4 – A 'Compensation Event' to change the Scope.
7 'Compensation events' include the NEC4 equivalent of a 'Change' (JCT DB 2016) or a 'Variation' (FIDIC Yellow Book 2017).

NEC4 shares the following procedural similarities with JCT DB 2016, but they seem to be carried out after the contract has been concluded (whereas JCT DB 2016 envisages pre-contract):

- o NEC4 clause 21.2 provides that the project manager must accept the contractor's design before the contractor can proceed with any of its designed work. This may place a considerable initial burden on the project manager if a significant amount of the contractor's design is set out in the Part Two Scope. To protect the project manager and avoid creating unnecessary delays and compensation events,[8] it is advisable for Contract Data Part One to include a longer 'period for reply' for the project manager to approve the contractor's Part Two Scope.
- o Clause 45 of NEC4 permits the client to change (reduce) the Scope so either a defect does not have to be corrected or, if a defect is accepted, the Scope is changed so the previously defective work now complies with the Scope. This permits an acceptance of the contractor's design in a similar way envisaged by the footnote to the Third Recital in JCT DB 2016.

3.4 The Employer's Requirements

The 'Employer's Requirements' is the document in which the employer tells the contractor what he wants built. The employer's requirements will vary according to the nature and complexity of a project. In addition to instructing the contractor what to design and build, they may also contain instructions telling the contractor how the employer wants the works to be carried out, along with his requirements for how the contractor administers his part of the contract. The format for the employer's requirements is not prescribed in any of the main design and build contracts (JCT, FIDIC, NEC4), so this is left to the employer and his advisors. The employer's requirements often contain a number of documents, so a common format is a clearly referenced general or introductory document to which any other documents are appended. By the contract documents including a clear reference to the employer's requirements (that also includes all other appended documents), it can be easily identified in the 'specific information' for each particular contract (see Section 2.2.2), as shown in Table 3.1.

This is far more convenient than having to append a schedule of the constituent documents on a separate sheet, as these can often become detached from the contract and lost.

Table 3.1 Employer's requirements for JCT DB 2016, FIDIC Yellow Book 2017, and NEC4.

Contract	Employer's requirements	Reference in the contract documents
JCT DB 2016	'Employer's Requirements'	Article 4, Contract particulars
FIDIC Yellow Book 2017	'Employer's Requirements'	Entry 2(e), Contract agreement
NEC4	'Scope'	Section 1, Contract Data Part One

---

8 For example, failing to reply within the relevant period – clause 60.1(6).

3.4.1 The Employer's Requirements – Essential Content

The employer's requirements must provide sufficient information to enable the contractor to produce a design from which he can construct the works. It's unlikely that the employer would want the contractor to carry out the 'Initial Brief' stage (refer to Section 1.2), so the employer's requirements would ordinarily meet the following stages of RIBA's Plan of Work:

o Stages 0 (Strategic Definition) and 1 (Preparation and Briefing). The employer should have already completed all feasibility studies along with the project brief, necessary site information, and a project budget.
o Stage 2 (Concept Design). As a minimum, this should include the outline specification. Depending upon the extent of the contractor's design, architectural concepts may be part of the contractor's design.
o Stage 3 (Spatial Co-ordination). The architectural concept may be developed by the employer or left with the contractor. A description of the spatial coordination may suffice, if it's sufficient to allow the contractor to complete this as part of the design.

It is unlikely that the employer's requirements would contain any documents for Stage 4 (Technical Design) as this (in keeping with this chapter's example), along with any remaining design from Stages 2 and 3, would be carried out by the contractor.

3.4.2 The Employer's Requirements – Common Format and Contents

The employer's requirements often follow the 'preliminaries' or 'general' section of one of the methods or rules of measurement. For many years, the employer's requirements for construction projects adopted the 'Preliminaries/General Conditions' section from the RICS (Royal Institution of Chartered Surveyors) 'Standard Method of Measurement – 7th Edition' (SMM7). In 2012, the RICS revised SMM7, which became the second volume of its 'New Rules of Measurement', entitled 'NRM2: Detailed measurement for building works'. NRM2's 'Preliminaries (main contract) – Part A: Information and requirements' section broadly follows the sequence of its predecessor (SMM7) but is significantly expanded to reflect the requirements of modern projects.

NRM2's use of the words 'Employer's Requirements' has the same meaning as in Section 3.1 of this book (i.e. it means the employer's instructions to the contractor), so it should not be mistaken for having the same meaning as defined in JCT DB 2016 or FIDIC's Yellow Book 2017. This is because NRM2 contains the measurement rules for compiling a bill of quantities (for example, to be used with traditional forms of contract such as the JCT Standard Building Contract 2016 SBC/Q – refer to Section 3.6.1). NEC4 does not define 'employer's requirements' (refer to Section 2.2.4) as the extent of the contractor's design must be contained within the Scope (clause 21.1).

The ICE's Civil Engineering Standard Method of Measurement (CESMM4) includes similar sections to NRM2's 'Preliminaries (main contract)' in its 'Class A: General Items' section, but in far less detail. CESMM4 cross references are shown in square brackets in Table 3.2. If NRM2's format is followed, the employer's requirements will contain the sections shown in the table.

Table 3.2 Contents of Employer's Requirements (NRM2 and CESMM4).

NMR2 Section	Likely contents
1.1 Project particulars	o Project name, description, and location o Contact details for employer and his representatives 'Employer's Agent' – JCT DB 2016, 'Project Manager' – NEC4, or 'Engineer' – FIDIC

1.2 Drawings	o In the context of this chapter, the contractor is carrying out all or most of the design, so only general information is likely to be shown on the drawings, which may include:
	  – Site plans – both existing and proposed
	  – General arrangement
	  – Elevations
	  – Possibly a general internal layout
	  – Standard details to be incorporated. For a 'construction' project, this could include typical layouts for a hotel room or an office. For an 'engineering' project, it may include standard details for highways or rail works.
	o Specification of goods, materials, and workmanship
	o Performance specification
	o Pre-construction information

1.3 The site and existing buildings	This may include descriptions of: o The site (if not already described in Preliminaries Section 1.1) o Any existing services o Site information such as the health and safety file, and visiting details
1.4 Description of the works	This is likely to be included in Preliminaries Section 1.1 unless there is work carried out by others.

1.5 The contract conditions [CESMM4, Class A – 1]	This section should include:
	o The form of contract
	o As much of the Specific Information (see Section 2.2.2) that is currently known, which would subsequently be included in:
	  – The Contract Particulars (JCT DB 2016);
	  – The Contract Agreement (FIDIC Yellow Book);
	  – Contract Data – Part One (NEC4) including which of the Main Option Clauses (A, B, C, D, E, or F) and Secondary Option Clauses applies; or
	o Collateral warranty requirements (e.g. for funder, purchaser, operator, or tenant).
	(For the purpose of this chapter, no amendments to the contract conditions are considered).

NRM refers to 'Employer's requirements' in its Preliminaries Sections 1.6 to 1.12	*See note regarding NRM's use of 'Employer's requirements' (above)*
1.6 Employer's requirements: provision, content, and use of documents	Preliminaries – Section 1.6 is very applicable to design and build contracts, as it covers documents provided by both the employer and the contractor. This is particularly important for NEC4 projects because the extent of the contractor's design must be set out in the Scope. o It should include any documents included in the employer's requirements that are not set out in Preliminaries Section 1.2.

(Continued)

Table 3.2 (Continued)

NMR2 Section	Likely contents

	o Importantly, it should specify any desired format for documents submitted by the contractor, such as:
	  – The Contractor's Proposals (JCT and FIDIC)
	  – The Contract Sum Analysis (JCT)
	  – The Completed Schedules (FIDIC)
	  – Information to be provided with Contract Data Part Two (NEC4)
	o Design information
	  – Required types of drawings (plans, sections, details)
	  – Drawings and information required for each discipline – such as architectural, structural engineering, building services, etc.
	  – Product and technical information (for constructing the works, along with handover, operating, and maintaining)
	o Any relevant definitions for products and standards
	o Procedures for design submission and approval
	o Details for the parties' communications. Note: For NEC4 projects, this information should be set out in Preliminaries, Section 1.5 – The contract conditions.

1.7 Employer's requirements: management of the works [CESMM4, Class A – 3.7]	o Site management including specific supervision requirements o Monitoring and reporting (e.g. programme, contractor's reports for site meetings, photographs, cash flow forecasts) NRM2 also provides for additional requirements for dealing with 'compensation events' (NEC4) and notifying extensions of time (JCT, FIDIC). Employers should avoid attempting to modify the contract conditions through the contract documents, as the latter are often subservient to and are overridden by the former.
1.8 Employer's requirements: quality standard and quality control [CESMM4, Class A, 2.5, 2.6]	o Applicable standards of workmanship and materials (including samples, tolerances, and setting out) o Inspection and testing procedures o Samples o Certification of building services, BREEAM, and air testing
1.9 Employer's requirements: security, safety, and protection [CESMM4, Class A – 3.2]	o Pre-construction information o Format for the health and safety file o Site-specific and general security arrangements o Identification and management of hazards – such as existing services, hazardous materials (such as asbestos) o Protection against nuisance, noise, pollution, particular substances, moisture, and site operations (e.g. waste disposal and burning) o Protection of existing site features, both externally (e.g. wildlife, fauna, services, paths and roads) and internally
1.10 Employer's requirements: specific limitations on method, sequence, and timing	o Giving details of completing the works in sections (see also Section 1.5 – The contract conditions), or any phasing of the works o NRM2 does mention the 'method' of working, but ordinarily this is at the contractor's discretion with certain exceptions (e.g. health and safety, logistical and working hours restrictions, etc.) o Details of any handover procedures required by the employer

Table 3.2 (Continued)

NMR2 Section	Likely contents
1.11 Employer's requirements: site accommodation/services/facilities/temporary work [CESMM4, Class A – 2.1, 3.2, 3.5, 3.6]	o General facilities such as the contractor's site accommodation (offices, meeting rooms, sanitary facilities, canteen, telephones, IT, parking, etc.) o Services for the works – including lighting, water, power, storage of equipment, materials, and use of existing land o Site access – temporary roads, hardstandings, paths; maintaining existing roads, access ways

1.12 Employer's requirements for completion	o Handover or completion procedures
	o Specifying the format and contents of documents:
	  – Operation and maintenance (O&M) manuals
	  – Health and safety file
	o Any post-completion maintenance obligations (likely to be for projects under FIDIC Yellow Book 2017)

3.4.3 Responsibilities Retained by the Employer

This chapter considers that most if not all of the physical works' design is carried out by the contractor. However, the employer may retain some responsibility over design, in particular, the consistency of the employer's requirements and their compliance with statutory regulations or procedures. This is essentially because the employer is the master of his own requirements and is responsible for (often with the aid of professional advice) the employer's requirements (including any subsequent instructions changing them) conveying a clear, adequate, and consistent set of instructions to the contractor. The nature of this responsibility does not mean that a failure to provide adequate, clear, or consistent employer's requirements results in the employer breaching the contract. Instead, the employer accepts that in order to resolve any lack of clarity, inconsistency, or inadequacy, the contractor may not only be entitled to an additional payment for carrying out additional (or changed) works but also an extension of time along with any additional costs it attracts. Such obligations to make additional payments to the contractor and give him additional time are in opposition to one of the main reasons for the employer selecting design and build – that being a reduced likelihood of variations and having to pay extra for the works.

Essentially, each party is responsible for the documents it prepares, so the employer is responsible for the employer's requirements (including instructions varying the works), and the contractor is responsible for his own proposals. From Section 3.1, there should be no discrepancy, inadequacy, or divergence between the employer's requirements and the contractor's proposals, as the former should prevail over the latter.

For JCT DB 2016, clause 2.11 clearly states that the employer remains responsible for the Employer's Requirements in terms of their contents' consistency, the adequacy of their design, and obviously, the site boundary. The only exception comes under clause 2.15, where the contractor is expected to know how to resolve a divergence between the Employer's Requirements, and any 'Statutory Requirements' current at the contract's 'Base Date' (often set as the tender return date). It is expected that any divergence, inadequacy, or discrepancy in the Employer's Requirements is firstly to be resolved by the Contractor's Proposals. However, in the event that any of these is not resolved:

o If such a divergence is created because the Statutory Requirements are changed after the Base Date, then this is usually a 'Change' (or variation) entitling the contractor to additional payment and/or time. This includes making necessary changes to the Contractor's Proposals, unless the Employer's Requirements provide otherwise (clause 2.15.2.2).
o Save as above, the Employer's Requirements shall be changed ('corrected, altered, or modified') to resolve the discrepancy, divergence, or inadequacy.

FIDIC Yellow Book 2017 at clause 1.9 (Errors in the Employer's Requirements) initially adopts a similar approach to JCT DB 2016, insofar as it recognises the possibilities of 'errors' (defined as an 'error, fault, or defect') in the Employer's Requirements and also compensating the contractor in terms of time and money for the measures required for their rectifying those errors. FIDIC takes a broader view than JCT DB 2016 regarding the site, as the employer remains responsible for errors in documents describing the site's nature and characteristics (see clauses 2.5, 4.7.2, and 5.1), and not just a discrepancy with the site boundary (as in JCT). However, FIDIC (at clause 1.9) places more onus, along with a time limit (the default period is 42 days from the Commencement Date), on the contractor to discover any errors. Upon discovery, the contractor notifies the employer's representative (the engineer), who then agrees or determines (under clause 3.7 – see Section 9.7.2.1) the following:

o If the Employer's Requirements contain an error;
o If, exercising due care, an experienced contractor would have discovered the error; either during the tendering period, or within the contractual time limits (default periods are 28 days under clause 4.7.2, or 42 days under clause 1.9); and
o What (if anything) the contractor must do to rectify the error (the measures).

The employer's representative's decision under the second point is pivotal. If he determines that an experienced contractor would have discovered the error, then the contractor will not be entitled to a variation (under clause 13.3.1) or additional payment and/or time for carrying out the necessary measures required to rectify the error.

3.5 The Contractor's Proposals

The 'Contractor's Proposals' is the document that the contractor prepares to tell the employer how he is going to build what the employer wants, so by providing the design, he explains how he intends to construct the works. The contractor's proposals often begin with the contractor's design and include specification to the extent not already given in the employer's requirements. JCT DB 2016, FIDIC Yellow Book 2017, and NEC4 use different terminology for the contractor's proposals as shown in Table 3.3.

The definition of 'Scope' in NEC4 clause 11.2(16) is very wide and appears to include the Scope provided by the client (as identified in Part 1 of Contract Data Part One), and the Scope provided by the Contractor for its design (in Part 2 of Contract Data Part Two – refer to Section 3.3.3).

Table 3.3 Contractor's proposals for JCT DB 2016, FIDIC Yellow Book 2017, and NEC4.

Contract	Contractor's proposals	Reference in contract documents
JCT DB 2016	'Contractor's Proposals'	Article 4, contract particulars
FIDIC Yellow Book 2017	'Contractor's Proposal'	Entry 2(g) of the contract agreement
NEC4	'Scope provided by the contractor for its design' 'Particulars of its design'	Section 2 of Contract Data Part Two Submitted to the project manager under clause 21.2

3.5.1 Contractor's Proposals – Common Format and Contents

The format for the contractor's proposals is often stipulated in the employer's requirements (e.g. if adopting NRM2's format, this should generally follow the 'Preliminaries', Section 1.6). The contractor's proposals are likely to contain the information shown in the following subsections.

3.5.2 Contractor's Proposals – Review of Contractor's Design

The contractor normally has to provide specific documents with his 'contractor's proposals'. The form and content of some of these documents may be prescribed by law[9] or will be as set out in the employer's requirements.

The form and content of the contractor's design documents will typically comprise the kinds of information described in Table 3.4. JCT DB 2016, FIDIC Yellow Book 2017, and NEC4 all contain procedures by which the employer (or its representative) can review the contractor's design, as follows:

o JCT DB 2016: Schedule 1 (Design Submission Procedures). 'Contractor's Design Documents' are provided to the employer in accordance with the Design Submission Procedure unless they are already contained in the Contractor's Proposals, and the information required by clause 2.8 (Construction Information).

o FIDIC Yellow Book 2017: clause 5.2 (Contractor's Documents). Unlike JCT DB 2016, clause 5.2.2 states that either the Employer's Requirements or contract Conditions[10] specify which of the Contractor's Documents must be submitted for review. Whilst clause 5.2.2 principally states that documents for review are set out in the Employer's Requirements, it adds that any documents necessary to satisfy statutory approvals, permits, licences, and the like must also be submitted for review.

o NEC4: clause 21 (The Contractor's Design) and clause 23 (Design of Equipment). It seems that under clause 21.2, the project manager must review and accept all of the contractor's design, initially including the Scope provided by the contractor for the design[11] and any

---

9 For example, under the CDM Regulations.

10 'Contractor's Documents' are defined in clauses 1.1.14, 5.2(a), 5.2(b), 5.2(c), and 5.2.2 of FIDIC Yellow Book 2017.

11 As set out in NEC4, section 2 of the Contract Data Part Two (The Contractor's Main Responsibilities).

Table 3.4 Contractor's proposals – common format and contents.

Section	Likely contents
General company information	o Brief company history o Relevant experience of similar projects o Outline of financial information (e.g. published company accounts) o Banking details

Contractor's design	o Details of the contractor's designer
	  – External company or in-house?
	  – Relevant experience/CVs
	o Drawings – typically include:
	  – Site plan
	  – Existing site features (e.g. for an 'engineering' project showing ground conditions, existing services, etc.)
	  – General arrangements for each level, such as floor and roof plans or layouts
	  – Plant and equipment – layouts and schematics
	  – Elevations
	  – Sections
	  – Covering architectural, structural engineering, building services engineering, landscaping
	  – Details of specific construction elements (e.g. lifts, pools, plant rooms), including those by specialist subcontractors
	o Calculations
	o Specification and schedules
	  – Plant, equipment, and product details, samples
	  – Quality assurance information, quality plan
	  – Schedules of finishes, drainage, and components
	o Proposed subcontractors or suppliers for major work elements
	o Proposals for alternative products, goods, or subcontractor works (each should be cross-referenced to the pricing document)
	o Identifying and proposals for resolving any inadequacies, divergences, or discrepancies in the employer's requirements
	o BREEAM and air test documents

Carrying out the works	o Method statement
	  – General and (if known) specific methods of work
	  – Protection of the works, stored materials, or goods
	  – Protection of surrounding site, existing site features, and the environment
	  – Cross reference to 'logistics' (see above)
	o Lifting and access plans (e.g. cranes, hoists, scaffolding, etc.)
	o Programme, including:
	  – General construction (or master) programme
	  – Intended critical path of the works
	  – Design programme
	  – Procurement programme
	  – Any sections, phasing, or key dates
	o BIM information

Contractual details	As stated in Section 3.2.2, under '1.5 The contract conditions', this chapter does not consider any amendments to the contract conditions.
	o Insurance details:
	  – Employer's liability/public liability
	  – Professional indemnity

Table 3.4 (Continued)

Section	Likely contents

	o Confirming agreement to wording for any nonstandard bonds and/or collateral warranties
	o Any example pro forma contractual documents, such as:
	  – Interim Applications (JCT DB 2016)
	  – Early Warning notices (NEC4)

Management of the works	o Personnel information:
	  – CVs and contact details for site staff/supervision or technical staff
	  – Organogram of staff including directors, visiting staff, or those not fully based on site
	o Site accommodation facilities (including any dedicated facilities for the employer and staff)
	o Security information
	  – Security measures
	  – Means of site access
	  – Entry requirements
	o Logistics information
	  – Access routes
	  – Times/dates/hours for routine deliveries and special deliveries
	  – Delivery areas
	  – Storage areas
	  – Off-site areas

Health and safety and CDM information	o Construction phase plan o Health and safety file o H&S and CDM contacts o Risk and COSHH assessments o Protection measures (if not included in the method statement) o Site emergency plan (fire assembly points, first aid stations)

Other documents	o Typical documentation, such as:
	  – Example O&M manuals
	  – Example maintenance programme (for an engineering project)
	  – Plant operation procedures
	  – Example handover procedures and checklists

subsequent design provided by the contractor under clause 21.2. However, it does not appear that the contractor has to submit *all* of the design at once, as clause 21.3 provides that parts of the design can submitted for approval on the basis that each part can be 'fully assessed'. Clause 21.3 implies (and possibly encourages) that the contractor may submit the design for particular elements of work progressively, in advance of carrying out the relevant element of work on site.

The design review procedures in JCT DB 2016 and FIDIC Yellow Book 2017 contain clear procedures and timescales, whereas NEC4's procedure is less detailed. Clause 1.1 of JCT DB 2016 defines the 'Contractor's Design Documents' being prepared by the contractor to cover materials, workmanship, and goods in the form of drawings, specifications, and other information (such as details and other documents) along with any other information to be provided under the applicable BIM (Building Information Modelling) protocols. The definitions in the FIDIC Yellow Book 2017 (Contractor's Documents) and NEC4

(The Contractor's Design) are similar in that the contractor has to provide the documents either specified in the Employer's Requirements (FIDIC) or in the Scope (NEC4). FIDIC adds that the contractor's documents must also ensure that the contractor complies with the applicable regulations (i.e. the design satisfies clause 1.13 – Compliance with Laws) and, quite practically, includes 'as-built' and O&M information. JCT DB 2016 contains a similar obligation under clause 2.37 (as-built drawings), but the contractor's compliance may be at the employer's discretion.

A comparison of the JCT and FIDIC design review procedures is shown in Table 3.5.

The procedure under NEC4 is less detailed. It appears that the project manager must either accept or reject the 'Contractor's Design' (clause 21.2) within the 'period for reply' (clause 13.3), although a longer period for replying under clause 21.2 can be stated as an exception in Contract Data Part One. The reasons for not accepting the contractor's design

Table 3.5 Comparison of JCT and FIDIC design review procedures.

Procedure	JCT DB 2016	FIDIC Yellow Book 2017
Review period	14 d (Schedule 1, para 2)	21 d (cl 5.2.2)
Design acceptance	'Marked A' (Schedule 1, para 5.1)	Engineer's notice of no objection (cl 5.2.2 [a])
Acceptance subject to employer's comments	'Marked B' (Schedule 1, para 5.2)	No procedure
Design rejected	'Marked C' (Schedule 1, para 5.3)	Engineer's notice that the contractor's document fails (cl 5.2.2 [b])
Acceptance by default if no response from employer or engineer	Schedule 1, para 3	Clause 5.2.2
Contractor can work to approved design	'Marked A' (Schedule 1, para 5.1) 'Marked B' (Schedule 1, para 5.2)	Clause 5.2.2 (a)
Contractor must not proceed and re-submit for review	'Marked C' (Schedule 1, paras 5.3 and 5.6)	Clause 5.2.3
Entitlement to additional time or money	(Schedule 1, para 7) Only if the contractor's design document complies with the contract, so the employer's comments (under Schedule 1, para 5.2 or 5.3) are a 'Change' (under cl 5), which may entitle the contractor to additional payment and an extension of time (cl 2.26.1) and loss and/or expense (cl 4.21.1).	No extension of time for re-submitting the contractor's documents for review (cl 5.2.2 [iii]). A contractor who disagrees with the engineer's notice of failure (cl 5.2.2 [b]) can seek a 'determination' (under cl 3.7 – see Section 9.7.2.1) or seek resolution under cl 21 (disputes and arbitration).

are that it does not comply with the Scope or the applicable law and, also in the case of equipment design (clause 23.1), the equipment must comply with the design that the project manager has already accepted.

Similar to JCT ('Marked C') and FIDIC, the contractor must not proceed with the designed work until the project manager has accepted it.

The contractor should bear in mind that even though the supervising officer has approved the contractor's design, such approval does not mean that the contractor is no longer responsible for his design,[12] nor will it remove the contractor's liability for latent defects in materials if the supervising officer has approved those.[13]

3.5.3 Contractor's Proposals – Programme (Documents for Managing the Works)

The requirements for the contractor to provide a programme vary significantly across the three contracts (see Section 8.6). JCT DB 2016 does not even mention a 'programme' unless paragraph 7 of the Supplemental Provisions applies, and even then, it is just a fleeting description. This lack of requiring a programme is surprising for such a modern building contract, particularly when contrasted with the far more comprehensive requirements of FIDIC and NEC4 and recently published guidance.[14] The only exception under JCT would be if additional programming requirements were set out in the employer's requirements, or if they were imposed by a particular BIM protocol.

FIDIC's Yellow Book 2017 arguably requires the contractor to provide the most detailed programme. It is the only contract whose conditions specify that programming software must be used (clause 8.3) and that activities are logic-linked to a specified level of detail and show the critical path (clause 8.3[g]). Furthermore, FIDIC requires the contractor to provide a supplemental report that includes site resources, method statements, and any proposed measures to mitigate delays (clause 8.3[k]).

Both FIDIC and NEC4 require that the programme is accepted by the engineer (FIDIC) or project manager (NEC4), and that the contractor provides revised programmes which are also subject to acceptance in the same way (FIDIC clause 8.3[f] and NEC clause 32.1). It is perhaps a little strange that for all the detail FIDIC requires the contractor to show on and with its programme, that the programme itself is not used as a tool for assessing extensions of time (under clause 8.5) in the same way that NEC4 uses the programme (or the Accepted Programme) to assess the delay to the completion date under clause 63.5. In general terms, the contractor must show the following information on the programme under clause 8.3 of FIDIC Yellow Book 2017 or clause 31 of NEC4:

o Start date, completion date, time for completion (FIDIC), and the planned completion date (NEC4).

---

12 *Hampshire CC v Stanley Hugh Leach,* 8 Const LJ 174, (1990).
13 *Rotherham MBC v Frank Haslam Milan & Co Ltd,* 78 BLR 1, [1996].
14 Society of Construction Law, *Delay and Disruption Protocol,* 2nd ed., (February 2017), and the Chartered Institute of Building's *Guide to Good Practice in the Management of Time in Major Projects,* 2nd ed., (2018).

o The order and timing of the works (including any work undertaken by nominated subcontractors – FIDIC).

o Access dates on which the contractor needs access to parts of the works.

o Dates on which the employer/client must provide something (such as materials, plant, or goods).

o The periods for the employer's representative to accept or review contractor's documents.

    – In the event that all or a significant proportion of the contractor's design is completed prior to entering into the contract (referred to in part 2 of Contract Data Part Two[15]), the contractor should either allow an appropriate period within its programme for the project manager to approve that design under clause 21.2, or submit parts of the design for acceptance under clause 21.3.[16] In any event, such a period should account for the 'period(s) for reply' in Contract Data Part One.

o Testing and inspections (FIDIC) or the same if they are stated as a 'procedure' in the contract (NEC4).

o Any remedial works or rectification of defects.

o Other durations such as recognised holidays (FIDIC) or float or time-risk allowances (NEC4).

Interestingly, aside from requesting information that is relevant to Key Dates (NEC4), neither FIDIC's nor NEC4's programmes require much in the way of procurement information. Employers are likely to keep a close eye on the procurement of key subcontract packages, particularly those whose works sit on the critical path.

There is tendency for the employer's requirements under JCT DB 2016 contracts to include many of the programme criteria found in FIDIC and/or NEC4. This in itself is a good idea as it provides the employer with an accurate tool for monitoring progress and generally managing time. However, the requirements and contents for programmes should be proportionate to the project's complexity and time constraints, so very detailed programmes may not be required for projects where the scope of work is relatively straightforward or where managing time (for the employer) is less important.

3.5.4 Contractor's Proposals – Documents for Managing the Works (Cost)

The contractor's proposals are unlikely to contain much in the way of contemporaneous cost information for the employer, as (in most cases) they will be generated before the project gets underway. A cash flow forecast is often required, and this should align with the contractor's programme and pricing document (see Section 3.6). Depending upon the employer's funding arrangements, a cash flow forecast may take on considerable importance if it provides an accurate indication when the employer will reach particular levels of expenditure that trigger the drawdown of money from the funder.

The contractor may provide examples of other documents, such as vesting certificates for equipment or goods, or a template for interim payment applications (if no format is prescribed by the employer's requirements).

---

15 The Scope for the contractor's design.
16 Refer to Section 3.5.2.

3.6 The Pricing Document

For the purposes of this chapter, it should be assumed that the Pricing Document (and the Contractor's Proposals) fully comply with the Employer's Requirements. However, continuing with Part One's theme that all or the majority of the project's design will be undertaken by the contractor, this section will also provide for the contractor offering priced alternatives to those set out in the Employer's Requirements.

JCT DB 2016, FIDIC Yellow Book 2017, and NEC4 use different terms for the 'contract price'. NEC4 does not express the total of the 'Prices' as a single figure in the same way as JCT DB 2016 sets out the 'Contract Sum', or FIDIC Yellow Book 2017 sets out the 'Contract Price'. This is because NEC4 includes five Main Option Clauses (A to E[17]), each of which provides a different way of calculating the 'total of the Prices'. NEC4 refers to the 'tendered total of the Prices'[18] for Main Option Clauses A, B, C, or D. However, this 'tendered total of the Prices' is not used when assessing the amount due for interim payments[19] or variations,[20] and simply appears to show a record of the contractor's tendered price. The contract price for each contract is summarised Table 3.6.

Each contract also uses different terminology for the pricing document itself. JCT DB 2016 and FIDIC Yellow Book 2017 adopt the 'Contract Sum' and 'Contract Price', respectively, whereas for NEC4, the prices often have to be calculated by reference to the pricing document relevant to each Main Option Clause. A summary of the different pricing documents is shown in Table 3.7.

3.6.1 Pricing Document – Common Format and General Contents

If the employer requires the contractor to submit the contract price in a particular format or to express it to a particular level of detail, his stipulations should be set out in the employer's requirements. If the employer adopts NRM2's format, his requirements as to price and detail of pricing should be described in NRM2's Section '1.6 – Employer's Requirements: Provision, content, and use of documents' (see Table 3.2).

It is uncommon for design and build contracts to adopt a bill of quantities as its Pricing Document, but the following choices remain available to employers who would like the contractor to show the contract price in detail or to ensure that it is priced in accordance with a particular set of rules. For building projects, this is likely to be NRM2 (or possibly its predecessor, SMM7), and for engineering projects, CESMM4 is more likely to be adopted.

o JCT DB 2016 no longer contains an option for a bill of quantities, so there is no reference to any methods of measurement in the contract conditions.

In the event that the employer wanted to use a bill of quantities to describe the Works in the Employer's Requirements, he could adopt paragraph 3 (Bills of Quantities) from

---

17 NEC4 does contain a Main Option Clause – Option F (Management Contract) but that is not considered here.

18 NEC4 – Contract Data Part Two (Data Provided by the Contractor), part 5 'Payment'.

19 NEC4 – For interim payments, the 'Price for Work Done' is defined in clause 11.2(2) – Option A, 11.2(3) – Option B, clause 11.2(31) – Option C, clause 11.2(31) – Option D, and clause 11.2(31) – Option E.

20 NEC4 – 'Compensation events' are variations in NEC4, and their assessment is expressed as a 'change to the Prices' in clause 63.1.

Table 3.6 'Contract price' for JCT DB 2016, FIDIC Yellow Book 2017, and NEC4 (main option clauses A–E).

Contract	Contract price	Contract documents' reference to calculation of contract price
JCT DB 2016	'Contract Sum'	Article 2, defined in clause 1.1
FIDIC Yellow Book 2017	'Contract Price'	Defined in clauses 1.1.12 Clause 14.1 provides it is the 'Accepted Contract Amount[a]' set out in the Letter of Acceptance[b]
NEC4	'The Prices'	Option A – Priced Activity with Activity Schedule: total of the lump sum prices on the Activity Schedule[c] Option B – Priced Contract with Bill of Quantities[d]: lump sums calculated by multiplying the BoQ rate by the quantity Option C – Target Contract with Activity Schedule[e]: total of the lump sum prices on the Activity Schedule Option D – Target Contract with Bill of Quantities[f]: lump sums calculated by multiplying the BoQ rate by the quantity Option E – Cost Reimbursable Contract: Defined Cost plus Fee[g]

a) FIDIC Yellow Book 2017, clause 1.1.1 defines the 'Accepted Contract Amount'.
b) FIDIC Yellow Book 2017, clause 1.1.50 defines the 'Letter of Acceptance'.
c) NEC4 (Option A) – clause 11.2(21) and clause 55.1 define the 'Activity Schedule', and clause 11.2(32) defines 'the Prices'.
d) NEC4 (Option B) – clause 11.2(22) and clause 56.1 define 'Bill of Quantities', and clause 11.2(33) defines 'the Prices'.
e) NEC4 (Option C) – clause 11.2(21) and clause 55.2 define the 'Activity Schedule', and clause 11.2(32) defines 'the Prices'.
f) NEC4 (Option D) – clause 11.2(22) and clause 56.1 define 'Bill of Quantities', and clause 11.2(33) defines 'the Prices'.
g) NEC4 (Option E) – clause 52.1 defines 'Defined Cost', clause 11.2(10) defines the 'Fee', and clause 11.2(34) defines 'the Prices'.

Table 3.7 Pricing document for JCT DB 2016, FIDIC Yellow Book 2017, and NEC4.

Contract	Pricing document	Reference in contract documents
JCT DB 2016	Contract sum analysis	Article 4, Contract particulars
FIDIC Yellow Book 2017	Schedule of rates and prices[a]	Forms part of the 'Schedules' in 2(f) of the contract agreement[b]
NEC4	Varies	Section 5, Contract Data Part Two Data for the short schedule of cost components (only used with Options A and B) Data for the schedule of cost components (only used with Options C, D, or E)

a) FIDIC Yellow Book 2017 – clause 1.1.75 defines the 'Schedule of Rates and Prices', which is one of the 'Schedules' (defined in clause 1.1.72), which will be attached to the Letter of Tender (forming part of the 'Contract' – as defined in clause 1.1.9).
b) FIDIC Yellow Book 2017 – clause 1.1.10 defines the 'Contract Agreement', which forms part of the 'Contract' defined in clause 1.1.9 (refer to Section 2.2.4).

'Supplemental Provisions' found in the previous version (JCT DB 2011).[21] In that case, the Employer's Requirements should identify the method of measurement to be used.

o FIDIC Yellow Book 2017 is intended to be based upon a lump sum, but encourages 'detailed price break-downs, including quantities'[22] – which are likely to be a quantified schedule of rates or bill of quantities, and called the 'Schedule of Rates and Prices'.[3] In the event that the employer elects not to solely rely upon the engineer's expertise in valuing work carried out, or variations,[23] FIDIC permits measurement of the works if the contractor is to be paid on the basis of the quantity of work done. Clause 14.1 (The Contract Price) states that the 'provisions for measurement and valuation' should be set out in the 'Particular Conditions', however, Part A (Contract Data) of the Particular Conditions does not include any reference to measurement and valuation provisions in this regard. Therefore, the employer may select (or the Parties may agree to use) the 'Example Sub-clause for Measurement' (clause 14.1) found in Part B (Special Provisions) of the Particular Conditions, which then requires:

– The Contract to specify which parts of the Works are to be measured; and
– The Schedule of Rates and Prices to be in sufficient detail to allow the measurement and valuation of the Works which the Contract states are to be measured.

The method of measurement in the 'Example Sub-clause for Measurement' is relatively simple, using the 'net actual quantities of those parts, notwithstanding local practice', but the parties may agree on another (possibly more detailed) method of measurement.

o NEC4's Main Options A to E each contain differing pricing options, any of which may be used where the contractor is to provide the design. The relevant pricing document will depend upon which Main Option is selected as follows:

– If Option A[24] or C[25] is used – 'activity schedule'.
– If Option B[26] or D[27] is used – 'bill of quantities' (and the 'method of measurement' is set out in Contract Data Part One[28]).
– If Option A, B, C, or D is used – 'The tendered total of the Prices'.
– If Option F[29] is used – each 'activity' and its corresponding 'price'.
– Data for the Schedule of Cost Components (only used with Options C, D, or E[30]).
– Data for the Short Schedule of Cost Components (only used with Options A or B).

If the employer's requirements contain no such stipulations, then for a construction project, contractors often adopt a similar format to what NRM2 refers to as a 'Template

---

21 JCT DB 2011. The 'Supplemental Provisions' are contained in Schedule 2. The entry in the Contract Particulars (against Seventh Recital and Part 1 of Schedule 2) selects if paragraph 3 (Bills of Quantities) applies.
22 FIDIC Yellow Book 2017, Particular Conditions Part B (Special Provisions), introductory text to 'Sub-Clause 14.1 – The Contract Price'.
23 Regarding 'variations', FIDIC Yellow Book 2017, clause 13.3.1 provides that unless a 'Schedule of Rates and Prices' is included in the contract, the value of a variation shall be 'Cost Plus Profit' (as defined in clause 1.1.20). Also refer to Section 8.8.4.
24 NEC4 Option A: Priced contract with Activity Schedule.
25 NEC4 Option C: Target contract with Activity Schedule.
26 NEC4 Option B: Priced contract with bill of quantities.
27 NEC4 Option D: Target contract with bill of quantities.
28 NEC4, Contract Data Part One, 'method of measurement' is entered in Section 6 'Compensation Events' if Option B or D is used.
29 NEC4 Option F: Management contract.
30 NEC4 Option E: Cost reimbursable contract.

for pricing summary for elemental bill of quantities (condensed),[31]' which typically breaks down the contract price into the following construction elements:

o Substructure
o Superstructure
o Internal finishes
o Fixtures and fittings
o Building services (sometimes the mechanical and electrical elements are shown separately)
o External works
o Risks and contingencies
o Provisional sums (note that the contractor cannot introduce a provisional sum in JCT DB 2016, it has to be set out in the Employer's Requirements)
o Any credits (e.g. for excavated material, or scrap sold for the employer's benefit)
o Dayworks (see Section 8.9.1)
o Contractor's preliminaries (which should include any design fees)
o Overheads and profit

An example format[32] for contract sum analysis is set out in Table 3.9.

For engineering projects, contractors may adopt CESMM4's 'Work classification',[33] which separates the contract price into the following elements:

o Class A: general items	o Class N: miscellaneous metalwork
o Class B: ground investigation	o Class O: timber
o Class C: geotechnical and other specialist processes	o Class P: piles
o Class D: demolition and site clearance	o Class Q: piling ancillaries
o Class E: earthworks	o Class R: roads and pavings
o Class F: in situ concrete	o Class S: rail track
o Class G: concrete ancillaries	o Class T: tunnels
o Class H: precast concrete	o Class U: brickwork, blockwork, and masonry
o Class I: pipework – pipes	o Class V: painting
o Class J: pipework – fittings and valves	o Class W: waterproofing
o Class K: pipework – manholes and pipework ancillaries	o Class X: miscellaneous work
o Class L: pipework – supports and protection, ancillaries to laying and excavation	o Class Y: sewer and water main renovation and ancillary work
o Class M: structural metalwork	o Class Z: simple building works incidental to civil engineering works.
	o Grand summary[a]

a) CESMM4, Section 8 'Work Classification', p. 8.

---

31 NRM2 – Appendix D, p. 272.
32 The example in Table 3.9 also contains pricing alternatives, but generally follows the structure of NRM2's 'Template for pricing summary for elemental bill of quantities (condensed)'.
33 CESMM4, Section 8 'Work Classification' p. 8.

Engineering projects for particular industries (such as for highways, rail, power generation, water treatment, nuclear power, etc.) may have their own methods of measurement based upon the applicable work breakdown structure.

In the author's experience, many contractors have been reticent to show the contract sum in significant detail. There may be many reasons for this, and perhaps the risk of competitors learning your 'commercial hand' is the most important. There are other risks, such as the employer (or his advisors) misunderstanding or misusing information that is clearly commercially sensitive to the contractor or that may become sensitive in the future – such as the contractor submitting claims for an extension of time or loss and/or expense. However, contractors should temper this reticence as firstly, the pricing document should contain sufficient information from which to value any variations (see Section 8.9); and secondly, a failure to provide sufficient pricing detail may render their tender noncompliant.

In general, contractors provide less pricing detail for design and build contracts than they would provide under a traditional contract, which may require a detailed bill of quantities or schedule of rates. In a construction project (particularly, those procured using a JCT design and build contract), the pricing document will often consist of summary prices against each of the elements shown in NRM2's 'condensed' template. Given that a design and build contract presents the employer with relatively less risk than its traditional counterpart, the employer may not be too concerned with the individual prices for each element of the works, and instead is just focused on the total (see Section 3.6.5). This often enables contractors to 'move' money around within the pricing document. For example, the contractor may see an opportunity to enhance (or 'front load') his cash flow, if he increases the value of the initial activities (such as any values for setting up site, or for the substructure works), by a commensurate reduction in value of activities that occur later in the programme. This has the following advantages:

o The project will commence with an improved 'cash positive' position for the contractor.
o Activities such as groundworks involve a number of unknowns (such as unforeseen ground conditions[34]), so the contractor may firstly allocate all or a substantial part of his 'risk or contingency' sums to groundworks; and secondly, may transfer any margin he predicts he will earn on later activities (such as building services or finishes) to the earlier activities, to protect his cash flow in the event that the initial activities make a loss.

If the employer's requirements instruct the contractor to provide his price in more or specific detail, then this may follow NRM2's 'Template for pricing summary for elemental bill of quantities (expanded)[35]' format. This adopts the headings in the 'condensed' version of the template (see the above example), but it expands these by showing additional pricing detail against subheadings.

Owing to the relatively less intricate nature of engineering works, CESMM4 sets out the contract price in less detail than NRM2, but shares many of its core characteristics.

---

34 JCT DB 2016 puts the risk of ground conditions firmly with the contractor. Despite FIDIC Yellow Book 2017 and NEC4 placing some risk associated with ground conditions with the employer, the contractor still has to demonstrate that he could not have reasonably foreseen any adverse conditions in the ground.
35 NRM2 – Appendix E, p. 273.

3.6.2   Pricing Document – Contractor's Preliminaries

It is not uncommon for the contractor's preliminaries to simply be shown as a single figure in the pricing document (see NRM2's condensed example in Table 3.9). However, this can cause difficulties when the contractor is applying for interim payments. The only practical ways to value preliminaries when they are expressed as a single lump sum are:

o To apply a simple percentage completion (in the same way as is applied to sums for each element of the physical works as it progresses); or
o To value the entire sum for preliminaries on a time-related basis, dividing the total by the contract duration's total number of weeks, then multiplying it by the elapsed number of weeks at the valuation date.

It is unlikely that the contractor will incur preliminary costs in the even, linear way that the above methods would indicate. He is likely to expend sums on design fees, haulage, accommodation deposits and rental, statutory authorities, services, and power supplies for the site before commencing work. This may adversely affect cash flow in the short-term, unless he separates his preliminaries costs into 'fixed charges' and 'time-related charges'. NRM2's Appendix B shows a condensed pricing schedule template for the main contractor's preliminaries. Using this as a guide, Table 3.8 shows which categories of the main contractor's preliminaries typically fall into the categories of fixed and time-related charges (numbers in brackets are references to NRM2's 'components').

For construction projects with a relatively high value or engineering projects that are publicly funded, the employer's requirements may specify a particular level of detail for the main contractor's preliminaries costs. NRM2's 'Appendix C: Template for preliminaries

Table 3.8   NRM2: Fixed and time-related charges.

Fixed charges	Time-related charges
(1.1.3 and 1.2.10) Completion and post-completion requirements	(1.1.1) Site accommodation
(1.2.2) Site establishment	(1.1.2) Site records
(1.2.3) Temporary services	(1.2.1) Management and staff
(1.2.5) Safety and environmental protection	(1.2.6) Control and protection
(1.2.8) Temporary works o Some scaffolding or shoring is charged on a fixed-charged basis	(1.2.7) Mechanical plant
(1.2.12) Fees and charges o Likely to be some lump sums, particularly for design work already carried out (RIBA stages 0 to 3) and further lump sums for design work during RIBA stages 4 to 6	(1.2.8) Temporary works o Hire costs are time-related
(1.2.14) Insurance, bonds, guarantees and warranties	(1.2.9) Site records
	(1.2.11) Cleaning
	(1.2.12) Fees and charges o May be some time-related charges (e.g. electricity, gas, traffic management charges)
	(1.2.13) Site services

(main contract) pricing schedule (expanded)' is often used as a guide by contractors for showing their preliminaries in more detail, and many contractors provide more detail than is set out in NRM2's expanded preliminaries template. In addition to separating each cost component into either the fixed or time-related charge categories, contractors will often show how a particular charge is calculated.

Taking the example of '(1.2.1) Management and Staff', contractors often provide the following additional information:

o Separate costs for each individual discipline of staff member (e.g. project manager, site manager, quantity surveyor, planner, document controller, etc.)
o The calculation of cost for each discipline of staff member, comprising:
   – The number of staff for each discipline.
   – The number of weeks that a member of staff is required. This is often split into pre-contract, construction, and post completion stages.
   – The percentage of each week that the staff will be engaged on the project. For site-based staff (such as a site manage, or engineer) this is normally 100%, but some staff (such as a quantity surveyor or planner) may be part-time (e.g. 50%), as they may be engaged on more than one project.
   – The weekly rate for each staff discipline.

Similar details are often given for preliminaries costs in the '(1.2.2.1) Site accommodation', '(1.2.7) Mechanical plant', and '(1.2.8) Temporary works' categories – with contractors often expressing these costs as time-related charges.

If the contractor is following the format of NRM2's templates, he will need to include design fees under '(1.2.12) Fees and charges'. Most contractors will show design fees as 'fixed charges' for each of the RIBA stages. This is sensible, as it is likely that by the time the contract is awarded, the design may have progressed to RIBA Stage 3 (Spatial Co-ordination) or into Stage 4 (Technical Design). Even if the contractor has not engaged the designer for the early RIBA stages on a speculative basis (i.e. the designer will only be entitled to fees if the contractor wins the project), the contractor will want to recover any design fees which he has either expended or has become liable for as soon as possible. The design for Stage 4 is often shown as a fixed charge, but the designer's work for Stages 5 (Manufacturing and Construction) and 6 (Handover) may be of a more periodic nature; for example, a number of days per week allowing for a site visit, design coordination, inspections, and quality monitoring once the bulk of the design work is completed. In such cases, showing design fees for Stages 5 and 6 as time-related charges may be appropriate.

3.6.3 Pricing Document – Pricing Alternatives

The contractor may choose to include cross references to parts of other documents (such as the employer's requirements – including particular drawings or sections of the specification, or the contractor's proposals) within the pricing document. This adds clarity to show the employer what is included within the contractor's price, and what pricing adjustments may be available if the employer wants to consider any alternatives proposed by the contractor.

Some contractors will offer a 'base' proposal and price, which should naturally satisfy the employer's requirements. However, the contractor may want to show further pricing alternatives, including adjustments to the base price where appropriate. Further pricing detail can be used if the pricing document is adopting either of NRM2's condensed or expanded templates (see Section 3.4.1 of this book), as the following examples describe:

o If the contractor considers that the work presents an additional risk, he may provide more detail in the 'risks and contingencies' section. For example, if the works require the contractor to take responsibility for ground conditions or the condition of an existing structure, the contractor may show an increased price (or premium) for taking that risk. If this premium puts the contractor's price in excess of the employer's budget, the employer may want to share some of the risks, in return for the contractor reducing the risk allowance in his price.

o The contractor may identify a premium within the base price for an element of the works. To continue the example of the contractor being responsible for the groundworks, the contractor may show a fixed price that is calculated on a 'worst case scenario' basis. Clearly, this will be a high figure, and if the employer accept this price, this is the price he will pay. However, if the contractor suggests that the groundworks becomes a 'Provisional Sum' (see Section 3.6.4) in the employer's requirements, then the employer can consider the risk of unnecessarily paying a premium (if the groundworks are easier to carry out than the contractor thought when submitting his price) against paying the actual cost of the groundworks.

o The contractor may offer a number of alternatives to those specified in the employer's requirements – such as different methods, products, goods, and materials. These would normally be shown as a series of 'adds' and 'omits' where the base price is adjusted upwards or downwards accordingly.

Following the condensed NRM2 template example, the alternatives described above could be set out as shown in Table 3.9.

3.6.4 Provisional Sums

A 'provisional sum' is often used to provide a budget within the contract sum for an uncertain element of work. The uncertainty could arise for any of the following reasons where the work in question:

o Has not been fully designed (e.g. the employer may not have decided upon the precise specification for an item of work).

o May not ultimately be required (e.g. stabilisation of ground or a structure).

o Its extent (or quantity) is unknown (e.g. whilst the employer may definitely require some refurbishment work to be carried out, its extent may depend upon the results of a subsequent survey or report).

The contractor does not carry out any provisional sum work (including its design) unless the employer instructs the provisional sum to be expended. The employer has no obligation to instruct the provisional sum's expenditure, and it is not uncommon for provisional sums not to be instructed. If the employer does instruct its expenditure, then the amount of money

Table 3.9 NRM2 template for the contract sum analysis

Element	Sub-Total	Total
(All as per the Employer's Requirements)		
Substructure		**£1,446,908**
Superstructure		**£3,101,364**
Internal Finishes		**£3,353,038**
Fixtures and fittings		**£471,874**
Mechanical	£3,059,441	
Electrical	£1,662,458	
Lifts	£228,030	
BWIC Services	£242,026	
Building Services		**£5,191,955**
External works		**£283,219**
Risks and contingencies		**£700,001**
Provisional sums		**£1,487,934**
Dayworks		**£0.00**
Preliminaries	£2,426,085	
Insurances	£163,500	
Design Fees	£264,445	
Bond	£62,582	
Contactor's Preliminaries		**£2,916,612**
Overheads and profit		**£1,083,803**
"Base" Price (Tender Sum)		**£20,036,709**
Alternatives (as per Contractor's Proposals):		
1) Reduced risk for sharing ground conditions	(£200,000)	
		£19,836,709
2) Provisional Sum (exc Groundworks):	£1,487,934	
Add Groundworks	£348,000	
Total Provisional Sums	£1,835,934	
Total Fixed Price:	£18,200,775	
Total Price (Tender Alternative 2)		£20,036,709

(Continued)

Table 3.9 (Continued)

Element	Sub-Total	Total
3) Alternatives proposed by Contractor		
(a) Trad build in lieu of Prefab bath Pods:		
Omit Bathroom Pods	(£480,000)	
Add bathrooms traditional construction	£280,000	
Add 4wks additional Preliminaries	£96,000	
Total Price (Tender Alternative 3a)	(£104,000)	
		£19,932,709
(b) Alternative air-conditioning spec:		
Omit specified air conditioning units	(£634,000)	
Add alternative air conditioning units	£518,000	
Add additional ceiling works	£30,000	
Add additional M&E design	£4,200	
Add 1 week Preliminaries (commissioning)	£24,000	
Total Price (Tender Alternative 3b)	(£57,800)	
		£19,978,909

expressed as the provisional sum is omitted from the contract sum and is replaced with the sum of money calculated in accordance with the contract. Depending upon the 'valuation rules' in a particular contract (see Section 8.9), the calculation may be as simple as totting up the actual costs of the work and adding the contractor's overheads and profit.

The definitions of provisional sums and their instructed expenditure being treated as a variation under the following terms of design and build contracts are described in Table 3.10.

Table 3.10 Provisional sum procedures.

Contract	Clause reference
JCT DB 2016	o Clause 1.1 – Contained in the employer's requirements o Clause 3.11 – Employer instructing expenditure of provisional sums o Clause 5.1 – If the provisional sum qualifies as a 'Change' under that clause o Clause 5.2 – Valuation of Changes and provisional sum work
FIDIC Yellow Book 2017	o Clause 1.1.68 – Specified by the employer in the contract o Clause 1.1.88 – 'Variation' defined as an instruction under clause 13 ('Variations and Adjustments') that changes the works o Clause 13.3.1 – The engineer instructs the contractor to carry out a variation o Clause 13.4 – The engineer's instructions to the contractor to expend a provisional sum
NEC4	o An instruction changing the Scope is a compensation event under clause 60.1(1).

By including a provisional sum, the employer accepts all or a significant cost risk for that work, as he cannot hold the contractor to a fixed price, and consequently the contractor faces considerably less commercial risk. The use of provisional sums can therefore benefit both parties, as the employer will not face the risk of paying a high (or 'worst case scenario') price for work for which the actual cost may be much less, and the contractor is not held to a fixed price for work the scope and extent of which cannot be priced with certainty.

As provisional sum expenditure is likely to be a variation under JCT DB 2016, FIDIC Yellow Book 2017, and NEC4, then in addition to bearing the cost risk, the employer will bear the time risk as well. A variation (be it a 'Change' under JCT DB 2016, a 'Variation' under FIDIC Yellow Book, and a 'Compensation Event' under NEC4) under all three forms of contract may entitle the contractor to an extension of time. It is generally accepted that because of provisional sums' temporary nature, the contractor does not include any time allowance within his programme for carrying out provisional sum work. Neither should there be any allowance included within the contractual period for carrying out the works (calculated from the contracts' start and completion dates).

The absence of any time allowance (in either the programme or period for completion) for provisional sum work can often be contentious. It is fairly common for the employer's requirements to try and distinguish between types of provisional sum, often describing them as 'defined' and 'undefined'. If these distinctions are set out in the employer's requirements, the employer will deem that the contractor has included sufficient allowances for carrying out the defined provisional sum work in both its programme and in pricing its preliminaries.

The difference between defined and undefined provisional sums has been set out in various methods of measurement for building works. It was previously described in Rule 10.3 of SMM7, and from 2012 is now described in a similar form in Section 2.9 of NRM2. However, whilst some employer's requirements may adopt the format and content of NRM2 (see Section 3.4.2), methods of measurement often have limited application to design and build contracts, because bills of quantities are rarely adopted as the contract's pricing document (see Section 3.4.1 in this book).

If bills of quantities are adopted as the pricing document and NRM2 is the stated method of measurement, then the employer's requirements' description for defined provisional sums will have to contain sufficient information for the contractor to price the work, as if he was offering a fixed price. This information should include the following:

o Descriptions of the type of work, how it is to be constructed, and from what it is to be constructed.
o The location(s) at which it is to be fixed, how it is to be fixed (including any other work that is to be fixed to it), and any details of any specific limitations must be given.
o Sufficiently accurate quantities.

If the provisional sum's description is lacking in respect of any of the above, then the provisional sum will be undefined, and the contractor will not be deemed to have made any allowance for its execution, either within his programme or in pricing its preliminaries.

3.6.5 Employer's General Analysis of the Pricing Document

The degree of analysis the employer may apply to the contractor's proposals and pricing documents will vary depending upon the employer's needs for the project. For building projects of a simple nature, the employer may only be interested in the tender or contract sum, or very little analysis may be undertaken as long as the contractor's price is within the employer's budget. For more complex projects of either a building or engineering nature, the employer may need a much more in-depth analysis (explored in further detail in Chapter 5). He may need the contractor to provide far more pricing detail than in NRM2's condensed format, so the employer's requirements should specify that the pricing document is to be provided in a more detailed form – similar to NRM2's expanded format (refer to Section 3.4.12).

In addition to the pricing document itself, the employer may ask the contractor to provide further information such as a cash flow forecast. Ideally, this should link the pricing document to the contractor's programmes (see 'Carrying out the works' in Table 3.4 in Section 3.5.1), and show the following detail:

o Value earned each month
   – For all forecast (or programmed) activities
   – For key activities (such as high-value works or activities on the critical path)
   – Showing funding 'milestones' (used where the employer can draw down funding in prescribed intervals or amounts)
o Cumulative value shown progressively each month (the 'S' curve)
o Identifying expenditure in advance of work executed on site:
   – Costs for manufacturing equipment off site
   – Design costs
   – Deposits for equipment or materials
   – Statutory (or services) undertakers

The employer and his consultants should not assume that the contractor's tender price and any breakdown of it has been calculated in a consistent, linear manner; that being each element of the contractor's price precisely represents the following:

o The actual elemental costs of labour, plant, materials, and subcontract prices;
o A consistent element of risk or contingency; and
o Consistent additions (percentages) for overheads and profit.

As described in Section 3.6.1, contractors often 'move' money around within their overall bid, so their prices for individual elements of the works may not precisely reflect the costs of the physical work, along with consistently applied additions for risk, contingency, overheads, and profit. In addition to the example of contractors 'front-loading' their cash flow, they may make arbitrary commercial adjustments to the prices for individual building elements, or to the tender sum as a whole. It is not uncommon for a contractor's estimating team to accurately arrive at a robust and sustainable tender sum, only for a commercial reduction to be applied in order to make the contractor's tendered price more appealing, to increase the chances of winning the job.

Whilst consultants (quantity surveyors in particular) often expect to see such a linear and consistent calculation of the contractor's price, this is not always the case. A consultant should not make undue assumptions or over-analyse a contractor's tender or build-up to its tender or contract price. Often the contractor's explanation of or account for his pricing is the most accurate, irrespective of how inconsistent or illogical a cost consultant suspects it may be.

An experienced quantity surveyor or cost consultant (i.e. one who is familiar with the prevailing, local market conditions) who carries out a thorough analysis of the contractor's tendered price, tender programme, and cash flow forecast should be able to identify any pricing 'anomalies' within the contractor's tender. He or she is likely to employ cost information obtained from similar projects such as:

o The RICS' Building Cost Information Service (BCIS), making the appropriate indices adjustments for the current project's location and year.
o Tenders received for other previous similar projects:
  – From other contractors, or
  – From the tendering contractor.

Any suspected pricing anomalies should be brought to the employer's attention in the tender reporting, so the employer can obtain any clarification from the contractor when considering his bid before negotiating or letting the contract.

4

Consultants and the Employer's Representative

When undertaking a design and build contract as it was originally intended (i.e. where the contractor produces and is responsible for the majority of the design without novated consultants), consultants would act for the employer and for the contractor in differing roles. For the employer, various consultants may be appointed for the project's early stages; typically RIBA (Royal Institute of British Architects) Stage 0 (Strategic Definition), Stage 1 (Preparation and Briefing), Stage 2 (Concept Design), and Stage 3 (Spatial Coordination). The employer may either retain one of those (such as an architect, engineer, project manager, or quantity surveyor), or appoint another to act as his representative during the following project stages. Typically, the employer's representative will be engaged prior to procuring the works and will be retained during RIBA Stage 4 (Technical Design), Stage 5 (Manufacturing and Construction), Stage 6 (Handover), and Stage 7 (Use). The employer's representative's role usually ends when the building contract is concluded with the contractor.

For the contractor, one or more design consultants would be appointed during the tender period (between RIBA Stages 4 and 5), and certainly during and possibly beyond the construction stages (RIBA Stages 5, 6, and 7). Unlike when working for the employer, a consultant will not adopt a role that involves supervising the project, although a design consultant's input into quality and design coordination matters has obvious benefits. Some contractors already have their managers or directors in place and carry out the commercial and administrative functions under the main and subcontracts themselves, so it is unlikely they will engage a quantity surveyor. However, contractors will, of course, require design input. Continuing the example of looking at design and build contracts as originally intended, the contractor is likely to need a much greater input from the design consultants than they provided to the employer. For the contractor, consultants are likely to be appointed during the procurement period – in this example, sitting somewhere between RIBA Stage 3 (Spatial Coordination) and Stage 4 (Technical Design). The level of design input that the contractor requires for delivering a successful project can be far greater and may require many more design consultants than the employer engaged during the early project stages. In addition to carrying out the primary design (such as structural or civil engineering for engineering projects or architectural input for building projects), the contractor may need to appoint specialists consultants to complete individual parts of the overall

Design and Build Contracts, First Edition. Guy Higginbottom.
© 2024 John Wiley & Sons Ltd. Published 2024 by John Wiley & Sons Ltd.

design. Producing the individual design elements is often only part of the task; their respective designs must also be coordinated so the building can be constructed efficiently. This latter aspect formed a significant part of the 'buildability' service that was marketed by contractors (see Section 1.2 in Chapter 1).

4.1 Means of Appointing Consultants

Part One of this book considers design and build as originally intended, and without any amendments to the standard forms of design and build contract. The same principle is applied to the conditions of engagement for consultants in Part One. Part Two of this book considers the common amendments to standard forms of building contract and consultants' terms of engagements (along with bespoke terms).

Unlike contractors, who are providing the physical works, a consultant usually just provides a service. Therefore, aside from drawings, documents (such as specifications and calculations), and other intellectual property, there is nothing physical for the consultant to produce, so the ordinary standard to which consultants must perform their services is one of exercising reasonable skill and care. There are no goods or materials to be utilised and no standard of undertaking physical works to be applied. The standard forms for engaging consultants reflect this.

There are a number of standard forms for engaging consultants. Their form and content varies according to the nature of the project for which the design is required, or their suitability for providing the design under a particular form of building contract. The standard terms of appointment considered next only apply to designing consultants (such as engineers or architects) and those acting as the employer's representative. They will not consider cost consultants such as quantity surveyors. The four conditions of engagement (considered here) that are widely adopted are as follows:

RIBA Standard Professional Services Contract 2020
FIDIC Client / Consultant Model Services Agreement, 5th ed., (2017)
Association of Consulting Engineers (ACE) Professional Services Agreement 2017
NEC4 Professional Services Contract (June 2017 with Jan 2019 amendments)

The RICS (Royal Institute of Chartered Surveyors) also publishes forms of consultants' appointments (both in standard and short versions), along with its Employer's Agent Services (May 2022 versions).[1] However, the principal focus for this book will be on the RIBA, FIDIC, and NEC4 conditions of appointment.

4.1.1 Conditions of Engagement – Typical Format and Contents

The RIBA, FIDIC, and ACE conditions are, in general, similar documents. Whilst the precise clauses and contents differ, they appear typical of consultant appointment documents. The NEC4 Professional Services Contract (PSC) differs significantly, being almost

---

1 The RICS' versions of the consultant's appointments are the terms of appointment, to which specific services (such as those for a quantity surveyor, project manager, or principal designer) can be incorporated.

an abridged version of the NEC4 Engineering and Construction Contract (ECC) but adapted for appointing a consultant. This is perhaps unsurprising, given the obvious and intentional continuity running through the NEC4 suite of contracts. For this reason, the NEC4 PSC will largely be considered separately from the other three.

In general terms, the RIBA, FIDIC, and ACE conditions have similar contents that provide for the following:

o General
  – Company and contact details for the client and the consultant.
  – Project details, including the site address, works to be constructed, and form of building contract to be used.
o Services provided by the consultant
  – Standard for performing the services.
  – Scope of services for the consultant to provide (set out in a schedule appended to the relevant appointment).
  – Programme for providing the services.
  – Format of information and building information modelling (RIBA and ACE).
o Fees payable to the consultant (and the payment basis)
o Liability
  – Professional indemnity insurance
  – Liability and any limitations thereto
o Termination
o Dispute resolution

The NEC PSC contract is set out in the same format as the ECC as follows:

o Core clauses
  – 1 General
  – 2 The consultant's main responsibilities
  – 3 Time
  – 4 Quality management
  – 5 Payment
  – 6 Compensation events
  – 7 Rights to material
  – 8 Liabilities and insurance
  – 9 Termination
o Main option clauses: A (Priced Contract with Activity Schedule), C (Target Contract), and E (Cost Reimbursable Contract).
o Resolving and avoiding disputes: Options W1 and W2.
o Secondary option clauses: X1 (Price adjustment for inflation), X2 (Changes in the law), X3 (Multiple currencies), X4 (Ultimate holding company guarantee), X5 (Sectional completion), X6, (Bonus for early completion), X7 (Delay damages), X8 (Undertaking to others), X10 (Transfer of rights), X11 (Termination by the client), X12 (Multiparty collaboration), X13 (Performance bond), X18 (Limitation of liability), X20 (Key performance indicators), Y(UK)1 (Project bank account), Y(UK)2 (HGCRA 1996), Y(UK)3 CRoTP Act 1999, Z (Additional conditions of contract).

o Schedule of cost components and short schedule of cost components.
o Contract data: Part One (data provided by the client) and Part Two (data provided by the consultant).

4.2 Obligations and Services

4.2.1 Applicable Standard for Performing the Services

Ordinarily, a design consultant has to exercise reasonable skill and care[2] when carrying out his or her services. Occasionally, and under special circumstances, the design consultant may have to exercise the higher standard of ensuring that the design meets a fitness for purpose obligation, but this is rare, and it is thought that clear and express terms in the appointment would have to be used.[3] The difference between the two standards (the comparison is most appropriate when considering design services) can be simply explained as follows:

o Reasonable skill and care. As long as the design consultant can demonstrate having exercised reasonable skill and care when providing the services, he will not be liable if the design fails or is defective.
o Fitness for purpose. Irrespective of the standards exercised by the design consultant, he will be liable if the design fails or is defective.

The four conditions of engagement considered by this chapter all require the consultant to exercise the standard of reasonable skill and care and diligence[4] when providing a design or carrying out other services as follows:

o RIBA Professional Services Contract (RIBA PSC) – clause 3.1.
o FIDIC Model Services Agreement (FIDIC MSA) – clause 3.3.1.
o ACE Professional Services Agreement (ACE PSA) – clauses 2.1(i) and 2.1(ii).
o NEC4 Professional Services Contract (NEC4 PSC) – clause 20.2, albeit this refers to the consultant using the skill and care normally used by professionals carrying out similar services.[5]

4.2.2 Typical Duties Owed by the Designer to the Employer

For most projects, the design at various RIBA stages (or their equivalent) will be carried out by an architect or an engineer. Whilst the term 'architect' has a specific meaning and usage,[6] the term 'engineer' does not. However, it is generally thought that the duties of an engineer carrying out design are analogous to those of an architect. The architect's or engineer's specific duties normally consist of a set of obligations in the appointment's conditions, along

---

2 s13 of the Supply of Goods and Services Act 1982.
3 See *IBA v EMI and BICC,* 14 BLR 1, (1980) and *Grieves & Co v Baynham Meikle,* 1 WLR 1095, [1975].
4 'Diligence' is only mentioned in the RIBA, FIDIC, and ACE conditions, not in the NEC4 PSC.
5 This is understood to be equivalent to the standards required by the RIBA, FIDIC, and ACE conditions when used in the UK.
6 Unless a person is registered in the Register of Architects (see s20(1) of the Architects Act 1997 as amended).

with the particular services to be performed (usually set out in an appendix or an attached schedule).

Without formal conditions of engagement (either one of the four above or a bespoke agreement), the design consultant may have the following duties to the client:

o General responsibilities:
- The architect's appointment cannot be delegated to another (either individual or firm), so in that sense, the appointment is personal (applying to an individual or firm). Even if the work is delegated to someone whom the architect knows is reasonably competent, that will be no defence.
- To perform their duties in accordance with the contract with their employer (client), and/or under the building contract.
o Liaison
- Generally consult with and advise the client.
- If the method of working is pivotal to successfully constructing the design, the architect or engineer may have to liaise with the contractor and any specialist subcontractors.
- Follow statutory obligations to liaise with the contractor under the relevant parts of the CDM Regulations 2015.
o Procurement advice
- Depending upon the circumstances, advise the client as to the most effective procurement route.
- If required, send out tenders and advise whether appointing a quantity surveyor is necessary for their evaluation.
o Adequacy of design
- Examine the site including its surroundings and subsoil.
- Prepare a design that is 'buildable' by the contractor.
- Consult with and/or engage an appropriate engineer if (i) there are questions as to the 'buildability' of the architect's design, and (ii) it either is or becomes apparent that an inadequacy in the architect's design cannot be resolved by the architect alone. For (ii) the architect may have to engage the engineer (or an engineer may have to engage an additional, specialist engineer) at his own expense unless the client agrees otherwise.
- Prepare sketch plans and a specification and, if necessary, expand upon, modify, or amend the plans and specifications and issue the contractor with such further drawings, specifications, and instructions as are necessary.
- If the architect or engineer wants to change the design from that previously approved by the client, it will be necessary to explain the implications of the proposed change to the client and seek further instructions from the client.
- Exercise reasonable skill and care when recommending particular contractors, specialist subcontractors or use of materials. The architect or engineer should reconsider any recommendations if they become aware of circumstances that may adversely affect the work in question.
o Budgetary
- Provide an estimate of cost if requested by the client.
- If the client instructs the architect or engineer that the project's budget has a specific financial limit, the design must meet the budget.

- If the client's budget is limited or constrained, the architect or engineer should advise the client of any deficiencies in cheaper materials or work that were necessarily specified in order to keep to the client's budget.
- If the project cannot be designed to meet the client's budget, the architect or engineer must discuss this with the client. If the design significantly exceeds the budget, it may be of no use to the client, and the architect or engineer may not be entitled to payment.
  o Supervisory
    - Ensure that the contractor and other consultants give effect to the architect's or engineer's design and follow instructions.
    - Supervise the work at appropriate times (e.g. inspect critical work before it is covered up).

The extent to which the above obligations may apply may depend upon the circumstances in which the consultant is engaged. However, employers will benefit from having a specific scope of services for the consultant to perform and incorporate this in the consultants' terms of appointment (see Sections 4.2.3 and 4.2.4).

4.2.3 Scope of Services

As is often the case with the scope of a contractor's work, the scope of services that a consultant is to provide will depend upon the nature, complexity, characteristics, and circumstances of each project. In the context of this chapter, the employer will engage a consultant or a team of consultants to carry out the initial design necessary to meet the employer's requirements (RIBA Stages 0–3, possibly to part of Stage 4). Thereafter, the employer will engage or possibly retain a consultant to act as the employer's representative. Conversely, the contractor will require design consultants to develop the design contained in the employer's requirements, to a standard and extent from which it can construct and complete the works (from RIBA Stage 3, or part of Stage 4, to RIBA Stage 7). There will be some inevitable overlap of design services provided to the employer and to the contractor between RIBA Stages 3 and 4, and this will depend upon how much spatial coordination (RIBA Stage 3) or technical design (RIBA Stage 4) is contained within the employer's requirements or tender documents.

Therefore, for each of the RIBA stages, the general scope of a consultant's or consultants' duties can be determined, with the precise scope to be determined by the employer. The employer who is not (what is often referred to as) a 'professional client' (i.e. one whose core business does not involve the procurement of construction or infrastructure projects) will require further advice and guidance as to the consultants' scope of services. Unfortunately, the FIDIC, ACE, and NEC4 conditions provide little assistance, as the scope of services is wholly left to the employer to complete.

However, the RIBA PSC is enormously helpful in this regard. In much the same way as NRM2 provides guidance as to the contents of the employer's requirements (see Section 3.3.2), the contractor's proposals (see Section 3.4.1), and the pricing document (see Section 3.5), the RIBA PSC provides the following:

  o The nature of the individual consultant's appointment, by specifying what roles the consultant is engaged to fulfil, for example:
    - A designer (such as an architect or consultant)
    - The lead designer

- The project leader
- The contract administrator
o The roles set out in the RIBA PSC are essentially as follows:
  - Design services – architect/engineer/consultant (see Section 4.2.5). This principal function is to develop the design – from the initial feasibility stages through to developing detailed specifications and drawings. The initial stages also require the architect/consultant to assist in developing the project programme and budgets, along with other services (e.g. town planning).
  - Lead designer (see Section 4.2.6). This is a specific coordination service, whereby the lead designer is responsible for managing the coordination of design produced by all disciplines of consultant (e.g. architect, structural engineer, interior designer, building services engineer, landscaping designer, etc.). It is likely that the lead designer will also be carrying out the architect/consultant or designing engineer role and be preparing the majority of the project's design.
  - Project lead (see Section 4.2.7). This is predominantly a project management function, whereby the project lead essentially ensures that all the other consultants are fulfilling their functions. The project lead's services may not require any design or coordination to be carried out – it may be a purely managerial role.
  - Contract administrator (see Section 4.2.8). This role is as the 'employer's representative', as the principal function is to administer the building contract between the employer and the contractor. In the context of this book, the contract administrator will act as the employer's agent (JCT DB 2016), the engineer (FIDIC), or the project manager (NEC4).
o The 'services' are shown as examples, specific to:
  - Each of the RIBA Stages of Work (from Stage 0 to Stage 7); and
  - The express services for each role are included at each RIBA stage as summarised[7] in Table 4.1.

Table 4.1 Roles in the RIBA Professional Services Contract (PSC).

RIBA stage	Architect/ consultant	Lead designer	Project lead	Contract administrator
Stage 0 – Strategic definition	✓a)	_a)	✓a)	_
Stage 1 – Preparation and briefing	✓a)	_a)	✓a)	_
Stage 2 – Concept design	✓a)	✓a)	✓a)	_
Stage 3 – Spatial coordination	✓a)	✓a)	✓a)	_
Stage 4 – Technical design	✓a)	✓a)	✓a)	✓a)
Stage 5 – Manufacturing and construction	✓a)	✓a)	✓a)	✓a)
Stage 6 – Handover	✓a)	✓a)	✓a)	✓a)
Stage 7 – Use	_b)	_b)	_b)	_b)

a) No express, additional (or other) services specified, as the RIBA PSC envisages that such services would be provided under a separate appointment.
b) Allows option for additional (or other) services to be specified.

---

7 Express services shown only.

4.2.4 Incorporation of Services

All the services that are summarised above, and described in more detail in Sections 4.2.5, 4.2.6, 4.2.7, and 4.2.8, can be easily incorporated into whichever conditions of appointment the employer chooses to adopt as follows:

o RIBA PSC – The Schedule of Services is already incorporated into the RIBA PSC contract. The employer simply 'ticks' which of the roles (see Section 4.2.3) and services applicable to each role apply. The Schedule of Services also allows the employer to add additional services that the consultant must perform.
o FIDIC MSA – Appendix A contains the services to be performed by the consultant. In practice, the services will usually be set out in a separate document, as there isn't enough room on the relevant page to include the services' details. This should be clearly identified as containing the services and incorporated by reference (e.g. writing the name of the services document in Appendix 1 and physically attaching it to the FIDIC MSA).
o ACE PSA – 'Schedule Part 5: The Services' contains the services. Similar to the FIDIC MSA, the services are likely to be contained in a separate document that is often appended to the services contract, as there isn't enough room on the relevant page in the ACE PSA to describe the services in detail.
o NEC4 PSC – The consultant's services are set out in the 'Scope'. The scope is identified in 'Contract Data – Part One, 1 General', within the information to be provided by the employer (or client, to use NEC4 terminology). As with the FIDIC MSA and ACE PSA appointments, the scope is usually a separate document that is incorporated by reference.

4.2.5 Design Services

The role of design consultant (or 'architect/consultant' to use the RIBA terminology) is relatively self-explanatory. The consultant will be providing the design for his or her particular discipline, to the extent required by the appointment. As stated above, this will typically be from Stage 0 until Stages 3–4 if engaged by the employer, or from Stages 3–4 until Stage 7 if engaged by the contractor. Designers engaged by the contractor are likely to take part in the tendering process on the contractor's behalf. The degree of consultants' involvement during tendering (for single-stage and two-stage tendering, along with negotiations) is explored in more detail in Chapter 5.

The services described in the RIBA PSC seem to be more focussed towards the employer's needs than towards those of the contractor. Whilst the services for Stages 4 (Technical Design) to 6 (Handover) include many necessary practical and reporting duties (such as design review, inspections, quality reviews, and providing any further design information that's reasonably required) in order for the works to proceed, they may not adequately provide the contractor with the assistance needed when managing subcontractors or suppliers, particularly those who will be providing design. For a construction project, the following subcontractors are likely to be responsible for the following elements of construction:

o Substructure
  – Piling
  – Foundation design
  – Reinforcement
  – Ground treatments
o Superstructure
  – Concrete frame
  – Steel frame
  – Secondary steelwork
  – Structural framing systems
  – Roofing (including rainwater disposals systems)
  – Cladding systems
o Building services
  – Mechanical and electrical services
  – Fire suppression systems and fire strategy
  – Building management systems
  – Lifts and transport systems
  – Other specialist services
o Finishes
  – Finish elements with a performance specification (e.g. acoustic, insulation, fire protection)

Assuming that the contractor does not already have an in-house resource (such as a design manager or a design coordinator), services such as managing and coordinating subcontractors' design can be provided by a design consultant. The contractor is likely to need advice on the integration of subcontractor designs within the overall design, along with advice on the sufficiency and completeness of the subcontractor's design (i.e. if or to what extent it meets the relevant part of the employer's requirements). If using the RIBA PSC as the basis for the consultant's services, it may be advisable for the contractor to add specific services for the consultant in relation to subcontractors, such as:

o Procurement of subcontracts
  – Developing the design (e.g. drawings, specifications, or performance specification) for subcontract works packages.
  – Reviewing subcontractors' tenders.
  – Advising on the sufficiency of subcontractors' design.
  – Advising the contractor regarding the subcontractors' costs and programme.
o Coordination and integration of subcontractors' design
  – Refer to Section 4.2.6 'Lead Designer.'
o Quality management
  – Conducting inspections.
  – Advising on testing.

- – Reporting quality of subcontractors' work to the contractor.
- – Advising on remedial measures and snagging.
- o Further design
  - – Assisting the contractor and subcontractors with any revised designs (e.g. if a change or variation occurs).
- o Programme and completion advice
  - – Advising on the completeness of subcontractors' work in terms of overall project completion.
  - – Assisting the contractor and subcontractors with completion documents, such as operation and maintenance manuals, as-built drawings, etc.
- o Post-completion
  - – Advising on defects occurring after completion.
  - – Assisting the contractor with concluding subcontracts (subcontractor final accounts).

The contractor can add the above to the services provided under the RIBA, FIDIC, ACE, and NEC4 conditions of engagement.

4.2.6   Lead Designer

When a consultant takes on the role of lead designer, it is generally understood that the role will include additional responsibilities for directing and coordinating design, and often, power to instruct other consultants who are members of the employer's or contractor's design team.

The lead designer's coordination duties are perhaps the most important. This is particularly the case when the lead designer is acting for the contractor. He or she will have to ensure that other consultants' designs are coordinated with the main design discipline. For an engineering project, this is likely to be the main engineering design, which may incorporate separate designs for reinforcement, structural steelwork, ground stabilisation, drainage, and mechanical and electrical plant, etc. For construction projects, the lead designer is normally the architect, who has to incorporate and coordinate designs from the structural engineer and building services consultant; and the lead designer who is working for the contractor (during RIBA Stage 5) may have to incorporate and coordinate the designs produced by specialist subcontractors.

Once the contractor is engaged, he may be working to a tight programme and want to avoid any costs levied by the employer for finishing late (such as liquidated damages). It is therefore very important that the contractor works to a design that is as complete and coordinated as possible, otherwise he may face delays for redesigning the work and for taking down and rebuilding work if the works began without a fully coordinated design. In instances where the lead designer's or consultants' failures have caused a delay, the contractor's losses can be considerable. Not only will the contractor be liable to pay delay damages to the employer, but he will also incur his own costs for running the site for the period of delay. He may look to recover these costs firstly from the lead designer's professional indemnity insurance and, secondly, from the other consultants' professional indemnity insurances.

If using the RIBA PSC, the client (not to be confused with 'Client' under NEC4) enters which RIBA stages will apply to the lead designer by completing the relevant entry in the Role Specifications part of the Schedule of Services. As stated previously (in Section 4.2.4), the lead designer provides an additional service of coordinating the information received from other consultants (described as 'from the Other Client Appointments'[8]). Curiously, the architect (irrespective of whether acting as lead designer) is obliged to collaborate with other consultants (i.e. Other Client Appointments) under clauses 3.2.6 and 4.3 of the RIBA PSC, but there appears to be no reciprocal obligation for Other Client Appointments. The RIBA PSC conditions would benefit from an amendment that obliges a consultant (when engaged under these terms as an Other Client Appointment) to collaborate with the lead designer and comply with any reasonable instructions or directions that the lead designer issues.

FIDIC's MSA form doesn't contain any specific provisions for a designing consultant acting as lead designer, but Appendix A (Scope of Services) encourages the client to 'Specify the responsibility for interface management between the Services[9] and services provided by others (when the provision of services by others is necessary), if not the responsibility of the Client'. The client should therefore include duties as lead designer as part of the services in the FIDIC MSA for the relevant consultant (architect or engineer), and subservient obligations for other consultants to comply with the lead designer's instructions – particularly in relation to providing design information for the lead designer to coordinate (i.e. the 'interface management').

ACE's PSA deals with the lead consultant role in a more straightforward manner. The lead consultant role is defined in clause 1.1, and in part 3.1 of the Schedule, the client confirms that the consultant is taking on that role. It appears that 'Other Consultants'[10] are obliged to comply with the lead consultant's directions if the client so instructs. Under clause 2.3, any consultant (including Other Consultants) engaged under the ACE PSA shall comply with the client's reasonable instructions, and under clause 3.3 (irrespective of whether the consultant is engaged as the lead consultant), the client will assist the consultant and ensure that any other consultants give assistance to the consultant. From the lead consultant's perspective, the operation of clauses 2.3 and 3.3 provide that all other consultants are obliged to assist him, if the client so instructs.

Under NEC4's PSC, the client can simply appoint the consultant to act as the service manager,[11] which then obliges all other consultants (similarly engaged under NEC4 PSC) to comply with his instructions.[12] Otherwise, clause 22 requires the consultant to gain approval[13] from and cooperate with the 'others',[14] which can include any other consultants

---

8 'Other Client Appointments' is a defined term in the RIBA PSC's Contract Conditions.
9 'Services' meaning those carried out by the consultant under the FIDIC MSA, as defined in clause 1.1.23.
10 As defined in clause 1.1 (Definitions and Interpretation), meaning other consultants (save for the consultant appointed under the ACE PSA or the lead consultant).
11 Refer to NEC4 PSC clause 14.
12 NEC PSC clause 24.2.
13 NEC PSC clause 24.1, although the consultants' scope should identify which approvals are to be obtained, and the 'Others' from which approval is to be obtained.
14 See NEC4 PSC definition in clause 11.2(10).

and the contractor. The role of the service manager can be a full-time job, particularly if the consultant is also acting as project manager under the NEC4 ECC building contract, as well as administering several other NEC4 PSC appointments. As the administration role can be quite demanding, and there are many tasks that the service manager has to carry out within specific time periods (refer to the 'period for reply' in clause 13.3), it is perfectly feasible for the service manager to take on a purely coordinating role without carrying out any design himself. If the service manager and project manager are one and the same, the employer (client) is likely to employ a consultancy firm rather than a single individual.

4.2.7 Project Lead

The role of 'Project Lead' is often one of pure management. It is unlikely that the project lead will carry out any design or even design coordination services. Instead, the role effectively requires the project lead to ensure that all other consultants carry out their services in accordance with their conditions of engagement, procedures established by the project lead, and the client's instructions.

The RIBA PSC contains a list of project lead services for each of the RIBA stages. The ACE PSA contains a general clause 2.8, where the consultant acts as agent for the client. The only 'service' specified in clause 2.8 is 'arranging for the performance by other on behalf of the Client'. This seems wide enough to include acting both as project lead (i.e. managing other consultants) and also as contract administrator by managing the contractor (see Section 4.2.7). A more comprehensive list of services can be found in the RICS Project Manager Services 2022 edition (the RICS services are intended to be used with their Standard[15] or Short Forms[16] of Consultant's Appointments). Broadly, both the RIBA and RICS schedules[17] of project lead services comprise:

o General or recurring duties that can be carried out during any of the RIBA stages such as:
  - Chair and attend meetings.
  - Issue instructions to the other consultants.[18]
  - Review and report on consultant's progress and performance against the overall programme and construction budget.
  - Establish and maintain procedures and risk management (risk register).
  - Administer consultants' appointments (e.g. payments, notices, etc.).
o Stage 0 – Strategic Definition
  - Advise the client in terms of consultant selection, responsibilities, and appointment.
  - Provide basic feasibility and strategic brief.
  - Establish project management tools/framework for the project.

---

15 RICS Standard Form of Consultants Appointment (2022 edition).
16 RICS Short Form of Consultants Appointment (2022 edition).
17 The RICS' Project Manager Services also include supplementary services (including for prime cost contracts, further financial and contractual services, BIM, and additional bespoke services).
18 Referred to as the 'Professional Team' in the RICS' Project Manager Services.

o Stage 1 – Preparation and Briefing
  – Provide initial site appraisal, surveys, and report.
  – Liaise with client and other consultants to develop the overall programme and construction budget.
  – Select and appoint consultants (if not already appointed under Stage 0).
  – Give initial procurement advice.
  – Establish a clear client's brief.
o Stage 2 – Concept Design
  – Change control procedure (for the consultants and later for the contractor).
  – Use updated construction information to revise and monitor the overall programme and the construction budget.
o Stage 3 – Spatial Coordination
  – Provide general or recurring duties to Stage 2, with more precise monitoring against the overall programme and construction budget.
o Stage 4 – Technical Design
  – Provide general or recurring duties to Stage 3.
  – Establish scope of the works, including the need for specialists, the content of the building contract, and work required from third parties.
  – Facilitate the coordination of all consultants' designs.
  – Provide procurement advice (including suitable tenderer, assessment of tender submissions including programme and method statements, tender interviews, tender reports, and assist with negotiations).
o Stage 5 – Manufacturing and Construction
  – Carry out general or recurring duties to Stage 4, with greater focus on the project's construction (e.g. regular site meetings, change control, programme, and budget monitoring).
  – Ensure that consultants provide design information to the contractor in a timely way.
  – Obtain client approvals and authorisation.
  – Conduct site inspections and obtain quality and testing information.
o Stage 6 – Handover
  – Carry out general or recurring duties to Stage 5, plus implement handover procedures with the contractor and consultants.
  – Implement any handover plans.
  – Conduct project close-out (ensure that defects are resolved and the contractor's final account and consultants' accounts are concluded).
o Stage 7 – Use

4.2.8 Contract Administration

Contract administration duties are carried out on behalf of the employer (or client). They are self-explanatory: to administer the building contract with the contract on behalf of the employer. As set out previously and in Chapter 3 (Section 3.1), the person engaged by the employer is generally referred to as the 'employer's representative' but will have a different name depending upon the form of building contract, as follows:

o JCT DB 2016 – Employer's Agent
o FIDIC Yellow Book 2017 – Engineer
o NEC4 – Project Manager

The contract administration duties are either carried out in the name of the employer or in the name of the employer's representative on the employer's behalf. The former applies to contract administration duties under JCT DB 2016, as in practice, the majority of the employer's duties can be (and usually are) undertaken by the employer's agent. The remainder could feasibly be carried out by the employer's agent, although the employer may have to confirm that the employer's agent is acting on the employer's behalf and has the appropriate authority. FIDIC and NEC4 are more straightforward as there are express obligations for the engineer and project manager, respectively.

Contract administration duties are of a similar nature under the JCT, FIDIC, and NEC4 contracts. Most of the contract administration duties are carried out during RIBA Stages 5 (Manufacturing and Construction) and 6 (Handover), but it is not uncommon for a consultant to have limited duties from RIBA Stage 4 (Technical Design) onwards.

It is unlikely that a consultant will carry out any contract administration duties on behalf of a contractor, as the contractor usually has their own personnel to undertake the administrative functions under the building contract and under the varioussubcontracts.

The contract administration role is described in each of the professional appointments as follows:

o RIBA PSC – as a specific 'role' in the 'Role Specification' in the Schedule of Services
o FIDIC MSA – in clause 3.9 (Construction Administration) and the corresponding entry in the 'Particular Conditions', with specific services being set out in Appendix A
o ACE PSA – in clause 2.8 (Consultant acting as agent for the client), with specific services being set out in 'Part 5: The Services' of the 'Schedule'

The role is not described in the NEC4 PSA, so all contract administration functions must be set out by the client in the scope (identified in the Contract Data Part One).

The contract administrator's duties for each RIBA stage will normally include:

o Stage 4 – Technical Design
  – Advise the client regarding the building contract.
  – Provide pre-contract advice, such as tender reports, chairing tender interviews, etc.
o Stage 5 – Manufacturing and Construction
  – Administer the building contract (see below), in particular the certification duties such as certifying interim payments, completed stages, sections or the entire works, and extensions of time, along with issuing instructions to the contractor.
o Stage 6 – Handover
  – Manage defects and their rectification.
  – Handle certification (or equivalent) of the contractor's final account.

The employer's representative's duties under Stage 5 will mainly focus on administering the building contract between the employer (client) and the contractor. This is a very important function and can materially affect the outcome of the project for the employer and the contractor. Depending upon the form of building contract, the employer's representative's responsibilities may vary, but generally, he must act fairly and professionally.[19] If the conditions of his appointment do not specify the contract administration services that the employer's representative must carry out, he must ensure that he satisfies all his express obligations under the particular building contract. Determining the scope of services in this way is relatively straightforward for FIDIC and NEC4, as the employer's representative's duties are clearly expressed as being undertaken by the engineer (FIDIC) or by the project manager (NEC4). It is generally thought that the roles and responsibilities of the engineer, project manager, and employer's agent are closely analogous to those of an architect,[20] insofar as any decision making must be fair, unbiased, and hold the balance between the employer and the contractor.

JCT DB 2016 is less straightforward, as the employer's agent is hardly mentioned. The role is identified in Article 3 and clause 1.1, and if the employer's agent is communicating with the contractor, he or she must do so in writing (clause 1.7). Under clause 2.7.3, the contractor should keep copies of the building contract and the Contractor's Design Documents on site and allow access for the employer's agent or any person so authorised to areas on and off site, where the works are being prepared (clause 3.1). For interim payments or the final payment, the employer's agent may give payment notices, pay less notices (clause 4.10.2); and clause 7.4.3 includes the employer's agent within the meaning of 'consultants' in respect of Third-Party Rights and Collateral Warranties. All other duties to the contractor are stated to be performed by the employer.

It therefore appears that under JCT DB 2016, the term 'Employer's Agent' takes on the traditional legal meaning of 'agent'; meaning one who acts on behalf of its principal (the employer). The terms of JCT DB 2016 don't clearly set out the extent of the employer's agent's authority, so depending upon the specific services in his conditions of appointment, he may not be carrying out all the functions that are otherwise expressly allocated to the employer. For example, the employer's agent's duties may only extend to quantity surveying, valuing work, and administering interim payments and concluding the final account. All other matters such as issuing instructions, inspecting the work, and matters of quality, time, and completion may be carried out by the employer. This shouldn't cause any significant problems for the contractor, who will be corresponding and dealing with the employer. However, it is fairly common for the employer's agent to perform all the administrative functions of the employer under the building contract.

In general terms, the contract administration obligations usually carried out by the employer's representative under each of the building contracts are summarised in Table 4.2.

---

19 *Keating on Construction Contracts,* 10th ed., 14-016.
20 *Costain v Bechtel,* EWHC 1018, [2005].

Table 4.2 Contract administration duties for JCT DB 2016, FIDIC Yellow Book 2017, and NEC4.

Functions or duties of the employer's representative	JCT DB 2016[a]	FIDIC Yellow Book 2017	NEC4
Duties and authority	–	3.1, 3.2, 3.6	–
Delegation to issue instructions	–	3.3, 3.4, 4.3(b)	14.2
Issuing instructions for: – Changes (JCT DB 2016) – Variations (FIDIC) – Compensation events (NEC4)	2.1.4, 3.5, 3.8, 3.9, 3.11, 3.15, 5.6, 6.10.4, 6.11.3, Schedule 2 Part 1 – paras 1, 2, 4, 7, 8	3.5, 4.6, 4.7.3, 4.12.3, 4.23, 5.2.2, 5.4, 7.2, 13.1, 13.3, 13.4, 13.5, 13.6	14.3, 16.2, 17.2, 23.1, 60.1(1), 60.1(7), 60.1(10), 65.1, 73.1
Agreement/approval/consent/acceptance	1.7.1, 1.10, 2.2.1, 2.15.1, 2.16.1, 2.21, 3.2, 3.3, 4.6, 4.17, 6.5.2, 7.1, 7.3.1, 7E.2, Schedule 2, para 1.2.2, para 1.3.2 Schedule 3, para A.1	4.3, 4.4, 4.17, 4.19, 5.1, 6.5, 6.12, 7.2, 13.2, 13.3	16.2, 21.2, 23.1, 24.1, 26.2, 26.3, 31.1, 36.1, 40.2, 45.2, 46.1, 46.2, 61.4, 62.3, 65.2, 66.1, 84.1, 86.2, 86.3 Option A: 55.3 Options C, D, E, F: 11.2(21), 50.9, Options C, D: 63.3
Instructions for risk management	–	8.4 (13.3.2)	15.2, 15.4,
Instructions to resolve an ambiguity or discrepancy	2.10, 2.13, 2.15	1.5, 1.9	14.3 (63.10)
Receipt of contractor's notices	1.7, 2.1.1, 2.5.2, 2.10.2, 2.13, 2.14, 2.15, 2.16, 2.20.2, 2.24, 3.4.5, 3.5.2, 3.9.4, 4.9.5, 4.11.1, 4.20.1, 6.11.1, 6.13.1, 6.16, 8.9.1, 8.10.1, 8.11.1, 9.4.1, Schedule 1, para 5.3, Schedule 2, para 1, 3, 10	1.3, 1.5, 1.8, 1.9, 3.5, 4.4, 4.5, 4.7.2, 4.9.1, 4.12.1, 4.16, 4.23, 5.4, 6.5, 7.3, 7.4, 8.12, 9.1, 9.2, 10.1, 10.2, 10.3, 11.2, 11.5, 11.6, 13.1, 13.6, 14.6.3, 14.10(c)(i), 16.1, 16.2, 17.2, 17.3, 18.2, 18.3, 18.5, 18.6 20.2, 21.4.4,	11.2(8), 13.2, 15.1, 17.1, 17.2, 28.1, 31.3, 43.2, 61.3, 64.4, 90.1 Options C, D, E, and F: 11.2(26), 50.9 W1.1, W1.3, W1.4, W2; W2.1, W2.3, W2.4, W3.3, X10.3, X10.4, X11.1
Contractor's non-compliance with instructions	3.6	7.6	27.3
Determinations	–	3.7.2	–
Subcontracting	3.3	4.4, 4.5	26.2, 26.3
Postponement/suspension/resumption of work	2.4, 3.10	8.9, 8.13	19.1, 34.1, 60.1(4)
Commencement	–	8.1	–

Table 4.2 (Continued)

Functions or duties of the employer's representative	JCT DB 2016[a]	FIDIC Yellow Book 2017	NEC4
Programme	–	8.3	32.2
Rate of progress	Schedule 2, para 4	8.7	36.1
Completion/taking over	2.30, 2.37	10.1, 10.2, 10.3, 11.6, 11.9	30.2
Non-completion	2.28	–	–
Testing the works	3.12 Schedule 4,	9.1, 9.2, 9.3, 9.4, 11.6, 12.1	41.2, 41.3, 41.5, 44.3,
Rectifying defects	3.13, 3.14, 2.35	5.8, 7.5, 7.6, 11.1, 11.2, 11.4, 11.8, 17.2	40.3, 43.1
Valuing work	4.2, 4.3, 5.2, 5.3, 5.4, 5.5, 5.7	13.2, 13.4(b), 13.5, 13.7, 15.3	63.1 to 63.4 Option A: 63.12 Option B: 63.12, 63.15, 63.16 Option C: 63.13, 63.14 Option D: 63.13, 63.16
Advance payment	4.6	14.2	X14
Interim payments	4.7, 4.9, 4.10, 4.12, 4.13, 4.14	14.3, 14.5, 14.6, 14.9,	50.1, 50.2, 50.4, 50.6, 51.1
Final payments/account	4.8, 4.9, 4.24	14.11, 14.13	53.1
Extension of time	2.25	8.5, 20.2	64
Loss and expense	4.20	13.6, 20.2	64
Termination	8.4.1, 8.5.1, 8.11.1	15.1, 15.6, 16.3, 18.5	91.2, 91.3, 91.8

a) In practice, the majority of the employer's duties are carried out by the employer's agent under JCT DB 2016.

5

Procurement and Tendering

Owing to the subject of this book, it almost goes without saying that one of the essential procurement steps need not be considered; namely, whether or not the employer (or client) is going to select a design and build option. In terms of the contractual vehicle for procuring the project, and in the context of Part One of this book, it is assumed that the employer will select one of the contracts for the classes of project shown in Figure 5.1.

Of course, the extent of the employer's design and particular form of main contract are just two of the essential matters that the employer should address when procuring a project. This chapter assumes that the employer will desire some element of competition for the project, so any mention of negotiations will be considered as taking place after some type of competitive tender has been concluded. Procurement and tendering usually take place during RIBA Stage 3 (Spatial Co-ordination) to Stage 4 (Technical Design).

In the context of Part One of this book, the contractor will be engaged under an unamended building contract, and is carrying out the majority or a significant element of the project's design, without novated design consultants. The various tendering methods considered in this chapter can all be adopted in this context. It should be borne in mind that there is almost always a 'gap' or 'grey area' of negotiation between the employer or contractor during the period between the completed tender process and the time the building contract is agreed upon. To a degree, it is possible to narrow this gap by the parties setting and agreeing to preconditions, either in the rules of tendering (set by the employer) or in the circumstances under which a bid may be considered or accepted (set by the contractor). However, whether such matters are ultimately become stumbling blocks or are successful will depend upon the parties.

5.1 Procurement

5.1.1 Procurement Criteria – Public Sector to 31 December 2020

Until 31 December 2020, procurement by public bodies in the United Kingdom was subject to the procurement rules of the European Union.[1] These rules were formally applied by the Public Contracts Regulations 2015[2] (the '2015 Regulations'). Pending the UK's

---

1 Directive 2014/14 (the Public Sector Directive) and Directive 2007/66 (the Remedies Directive).
2 SI 2015/102.

Design and Build Contracts, First Edition. Guy Higginbottom.
© 2024 John Wiley & Sons Ltd. Published 2024 by John Wiley & Sons Ltd.

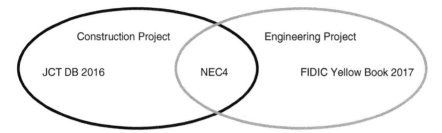

Figure 5.1 Contracts for construction and engineering projects.

withdrawal from the EU ('Brexit'), the 2015 Regulations were amended in 2019[3], largely to accommodate procurement that had already began prior to the end of the 'transition period', expiring on 31 December 2020. As of 1 January 2021, Brexit came into effect, and the UK was no longer subject to the EU's procurement rules but became subject to the World Trade Organisation's (WTO's) Government Procurement Agreement (GPA).

5.1.2 Procurement Criteria – Public Sector from 1 January 2021

At the time of writing this book, the UK government's intended overhaul of public procurement is in its infancy. In December 2020, the government set out its proposals in the Green Paper entitled 'Transforming Public Procurement',[4] which aimed to 'streamline and simplify the complex framework of regulations that currently govern public procurement'.[5] Once the government has considered the responses to the Green Paper's consultation, it should publish draft legislation in a White Paper. Until that legislation comes into force, public procurement remains governed by GPA and the Public Procurement (Amendment) (EU Exit) Regulations 2020[6] ('the 2020 PC Regulations').

   Now that the UK has left the European Union as of 1 January 2021, this book intentionally does not cover procurement in the public sector because it is unclear how long the 2020 PC Regulations will apply and what legislative measures the UK government may impose. However, the government's December 2020 Green Paper submits the following proposals for public consultation:

o Legislating the principles of public procurement; namely, the public good, value for money on the basis of 'most advantageous tender' (including a 10% weighting for social value), transparency,[7] integrity, fair treatment of suppliers[8] and non-discrimination[9];
o Integrating public procurement regulations into a single, uniform framework[10];

---

3 Public Procurement (Amendment, etc.) (EU Exit) Regulations 2019 (SI 2019/560) and the Public Procurement (Amendment, etc.) (EU Exit) (No. 2) Regulations 2019 (SI 2019/623).
4 December 2020 Green Paper. Responses were invited by 10 March 2021 (paragraph 18).
5 December 2020 Green Paper, paragraph 3.
6 SI 2020/1319.
7 December 2020 Green Paper, see also Chapter 6 (Ensuring open and transparent contracting).
8 December 2020 Green Paper, Chapter 8.
9 December 2020 Green Paper, Chapter 1.
10 December 2020 Green Paper, Chapter 2.

o Overhauling the complex and inflexible procurement procedures and replacing them with three simple, modern procedures[11].

o Retaining the current requirements that (i) the contract award criteria must be linked to the 'subject matter of the contract', save for specific exceptions, and (ii) tender evaluation is solely from the point of view of the contracting authority, save only to exceptions defined by a clear set of rules.

o Fast and fair settling of challenges to procurement decisions,[12] and possible exclusion of tenderers on the grounds unacceptable behaviour or previous performance in public contracts[13].

o Forming a new Dynamic Purchasing System,[14] single digital supplier platform,[15] and implementing the Open Contracting Data Standard.[16]

o Providing new framework agreement options.[17]

Many of these principles and requirements of the 2015 Regulations can be thought of as 'good practice' for construction and engineering procurement in both the public and private sectors (described further in Section 5.2).

5.2 Tendering

A general definition of 'tendering' when used in the construction and engineering industries can be expressed as follows:

o It describes some sort of competitive process of selecting an organisation that will carry out works or provide a service; or

o It means an offer or estimate by an organisation to carry out works or provide a service.

These two related definitions can be summarised as being a 'bidding' process and a 'bid', respectively.[18] The competitive element of tendering (in the 'bidding' sense) requires a particular deciding factor to select the successful tenderer. This is often the price for the works,[19] but is not always the case. Other factors may be given equal or greater importance than the offered price. An employer may place particular importance on the following factors, when selecting a tenderer:

o Tenderers' proven ability (for similar projects)
  – Relevant experience
  – Track record

---

11 December 2020 Green Paper, Chapter 3.
12 December 2020 Green Paper, Chapter 7.
13 December 2020 Green Paper, Chapter 4.
14 December 2020 Green Paper, Chapter 5.
15 December 2020 Green Paper, Chapter 6.
16 December 2020 Green Paper, Chapter 6.
17 December 2020 Green Paper.
18 *Oxford Dictionary of Construction, Surveying & Civil Engineering*, 2nd ed., (2020).
19 In the context of this chapter, tendering for 'works' means taking part in a competitive bidding process to provide physical works or services (such as design).

o Achieving the specified quality and performance
o Tenderers' supply chain
  – In particular, access to 'exclusive' suppliers
o Ability to achieve the programmed dates
o Health and safety record
o Tenderers' financial standing
o Tenderers' plans for executing the work
  – Outline method statements
  – Dealing with particular issues
o Other matters, such as
  – Environmental policy and record

In the context of Part One of this book, tendering for a design and build project will invariably include an element of design. The tenderers may need to provide their proposals for:

o Achieving the aesthetic design (predominantly for construction projects).
o Satisfying a performance specification (mainly for engineering projects, but increasingly important for achieving particular certification for building projects).
o Complying with planning requirements and constraints.

Many organisations refer to tendering as 'selective tendering'[20] requiring a shortlist of tenderers to be drawn up (see Section 5.3.1).

It is advisable for employers to set out the method of tender evaluation. If an employer is known for adopting clear, consistent, and transparent methods of evaluating tenders, and applying these methods to appoint the successful contractor, then tenderers should have greater confidence in the employer. Contractors are more likely to commit resources to providing the best and most compliant tender, which clearly benefits the employer.

Tendering for modern construction and engineering projects can be a complicated and expensive enterprise. Contractors often have to spend considerable resources to successfully navigate their way through pre-qualification questionnaires (PQQs) and other selection processes, just to get on a tender shortlist. Then the tendering process itself can be particularly costly. Compiling a bid may require many hours taken up with measuring the works, obtaining prices from subcontractor and suppliers, along with engaging various design consultants to assist with developing the contractor's design. Contractors often engage professional graphic designers to assist their pre-construction and estimating teams in presenting the necessary bidding information in a smart, corporate format. There may also be a significant BIM (Building Information Modelling) requirement to satisfy.

Ordinarily, the contractor's estimating and tendering processes are funded by its overheads (such as its pre-contract or marketing budgets) with the exception of designer's fees, which may be recoverable as part of the contract price, but only if the contractor is awarded the building contract. If the contractor's bid is unsuccessful, he will have to bear and pay all of his tendering costs.

---

20 See the JCT 'Tendering 2017' Practice Note (October 2017).

5.2.1   Tendering Criteria – General Requirements

Depending upon the nature and complexity of the project for which tenders are sought, the employer may send out (or make available), many different documents. However, in general, only the following two classes of document are included:

o Instructions to tenderers, and
o Documents from which the tenderers compile their bid.

The key difference between the two classes of tender documents is that the instructions to tenderers shouldn't be included within the building contract for the successful tenderer. They are simply instructions for how the tendering process proceeds (see Section 5.2.3). The second class of documents contains all the information for the tenderers to consider in order to submit their bid. This invariably includes all the documents comprising the Employer's Requirements for JCT DB 2016 and FIDIC Yellow Book 2017, or the Scope and the Site Information for NEC4 (see Section 5.2.3).

5.2.2   Tendering Criteria – Instructions to Tenderers

The instructions to tenderers are very important as they will set out the parameters and rules governing the tender process. These instructions should include:

o A letter, inviting the tenderer to submit a bid.
  – Brief details of the project, sometimes including its estimated value.
  – Contact details of the employer and the employer's consultants.
  – For completeness, this may confirm that tenderers provide their bids at their own cost.
o The conditions governing the tender process. This is likely to include:
  – Which tendering procedure is to be used – e.g. single-stage, two-stage, or other.
  – The tender timetable – setting out the tender return date, along with a range of dates for interviewing tenderers and an anticipated date for possible contract award.
  – The rules setting out what is and what is not a compliant bid (e.g. any applicable method of measurement, all rates and prices in the pricing document must be shown for individual items of work).
  – Dealing with tender queries.
  – Method of tender submission.
o Instructions for completing the pricing document (e.g. NRM2, CESMM4 – see Section 3.6).
  – Identifying the tenderers' allowances for the main works, provisional sums, preliminaries, overheads, and profit and any design and build risk allowance.
  – Setting out how pricing errors will be resolved.
o The employer's chosen method of tender evaluation, and the 'weighting' or 'scoring' to be applied to the tenderers' bid (i.e. the criteria to apply in determining the successful tenderer)
o The principal documents to be included in the tender return, including descriptions of their format (e.g. native software to be used in their production)[21]:
  – A written (even signed) statement that the tender complies with the employer's tender conditions; and

---

21 See Section 3.4.2 in relation to Section 1.6 of NRM2 (Preliminaries).

- The Contractor's Proposals (JCT DB 2016 and FIDIC Yellow Book 2017) or the Scope for [the Contractor's] Design in Section 2 of Contract Data Part Two (NEC4).
- The pricing document, to be completed by the tenderer – either the Contract Sum Analysis (JCT DB 2016), Schedule of Rates and Prices (FIDIC Yellow Book 2017), or the various pricing documents[22] in NEC4, depending upon which of the Main Option Clauses are selected (refer to Section 3.6).
o Secondary documents to be included in the tender return:
  - Outline programme for the works.
  - Method statements.
  - Evidence of tenderers' financial standing (e.g. parent company guarantee, bonds).
  - Tenderers' track record of similar projects.
  - Organogram of the staff the tenderer proposes to allocate to the project.
  - Details of specialist subcontractors or suppliers for key work elements.
  - Health and safety policies and records.
  - Quality assurance policies.
  - Environmental policies and records.

5.2.3 Information Necessary for the Tenderer to Submit a Bid

This information provided by the employer should be sufficient for the tenderer to provide a complete and compliant bid. It is also the information that will be incorporated into the building contract – forming the Contract Documents (see Section 2.2.4). It should enable him to assess the extent of the design and the quantity of the physical works, and assess their associated risks, following which the tenderer should be able to compile the pricing element of his bid. The essential information should be provided by the following documents (for more details see Section 3.4.2[23]):

o The Employer's Requirements (JCT DB 2016 and FIDIC Yellow Book 2017) or the Scope and the Site Information (NEC4):
  - Project particulars;
  - The site and existing buildings (separately identified as the Site Information in NEC4);
  - Description of the works;
  - Contract conditions;
  - Collateral warranties for the contractor, designer, and subcontractors;
  - Bonds;
  - Drawings and specification;
  - Performance specification;
  - Requirements for managing the works;
  - Quality control requirements;
  - Health and safety, security and protection requirements;

---

22 'Data for the Short Schedule of Cost Components' (Options A and B), 'Data for the Schedule of Cost Components' (Option C, D, or E).
23 Section 2.2.4 gives examples of what may be included in the Employer's Requirements (JCT DB 2016 and FIDIC Yellow Book 2017), or Scope (NEC4) that broadly complies with NRM2 and CESMM4.

- Specific sequence and/or time limitations imposed on the carrying out of the works; and
- Employer's limitations/capacity for the contractor's site accommodation.

5.3 Types of Tender Processes

Tendering for construction and engineering projects can take a number of forms. It is often not a process that can successfully be concluded simply by selecting the lowest price from a number of compliant bids. Design and build contracts complicate the tender process, because they not only require the employer to obtain sufficient information from the tenderer in terms of price, programme, quality, and the tenderers' track records, but also (and in the context of Part One of this book) satisfactory evidence that the tenderers' proposed designs will be acceptable.

The types of tender processes, in particular those involving an element of competition, are often described in terms of their stages – such as 'single-stage' or 'two-stage' tendering. There may also be other processes that sit before or after these tender stages such as pre-qualification procedures taking place beforehand and negotiations taking place after. In general, single-stage tendering can be adopted where the employer's requirements (or Scope) are in sufficient detail to allow tenderers to price the works. Two-stage tendering is often adopted in the following circumstances:

o Where the employer's requirements are not as detailed, and can developed further during both stages of the tendering process; or
o Where different elements of the project are subject to separate bids (or competitions).

5.3.1 Pre-qualification

In its simplest form, pre-qualification can just be the employer asking potential tenderers to express an interest in tendering for an upcoming project. At the other end of the complexity spectrum, it can involve a thorough and wholly competitive selection procedure, from which the employer compiles a list of suitable tenderers. PQQs are a useful way (to use an unkind phrase) of sorting out the 'wheat from the chaff', or to arrive at a 'shortlist'. A typical PQQ for a construction or engineering project may ask the tenderer to provide the following types of information:

o Company information, such as the tenderers' legal name and company number, contact details (including the registered address), parent or holding company.
o Financial and insurance information, such as a copies of the latest accounts, and certificates of insurances.
o Evidence of competency (e.g. track record of similar projects) and BIM usage.
o Professional status, such as disclosing any disqualifications, or (where permissible) any convictions or pending proceedings.
o Health and safety record and policies.
o Environmental record and policies.
o Policies of non-discrimination and diversity.

5.3.2 Single-Stage Tendering

Single-stage tendering is often an out-and-out competitive bidding process for the works, and it will typically commence between RIBA Stage 3 (Spatial Coordination) and Stage 4 (Technical Design) – with the successful tenderer completing the remainder of the design (Stage 4) after the contract has been awarded. RIBA Stages 0 (Strategic Definition), 1 (Preparation and Briefing), and 2 (Concept Design) should have been completed by the employer before embarking upon a single-stage tender. As mentioned above, the employer's requirements are sufficiently detailed and complete to either allow all tenderers to price the works or provide sufficient pricing information for the employer to evaluate the tender and select the successful bid after a single stage of competition. For 'traditional procurement' (i.e. where the contractor carries out little, or no design), single-stage tendering provides about as level a playing field as tenderers can expect. They all receive the same tender information, have the same period in which to compile and submit their bids, and the answers to any queries during the tender period are likely to be relevant to all tenderers' bids. Once the employer evaluates the tenders, it is usually the case that the contract is awarded to the successful tenderer soon after.

Single-stage tendering can be adopted for design and build, but it's unlikely to be suitable for complex projects or if the employer wants to retain a greater degree of control over the design. If the project's design is fairly simple (for example, a warehouse providing 30 000 ft^2 of storage, simple office accommodation, of particular dimensions and constructed from a steel frame with profiled metal roofing and cladding), then the design is likely to be of secondary importance. In such cases, single-stage tendering can be appropriate, as the employer is likely to select the successful tenderer on price, and tenderers' designs are unlikely to vary significantly. This reduces (or even removes) the need for a significant design competition.

5.3.3 Two-Stage Tendering

Giving a precise definition of two-stage tendering is a little more difficult. Often the two stages become a number of stages, with only the initial stage involving direct competition with other tenderers. Given the reduced element of direct competition, it is perhaps a little misleading to describe the two-stage process as tendering which usually concludes with the building contract being awarded. It is often more akin to an extended PQQ exercise than to a bidding competition.

Two-stage tendering is often used where the employer's design hasn't been either completed or sufficiently developed to enable the tenderers to submit a price that is certain. However, many employers find the reduced element of competition (and exposure to increased commercial risk) is outweighed by the benefits of collaborating with the contractor at an earlier stage. After the initial[24] selection stage, the second stage can be prolonged, and in practice, often becomes a number of additional stages before the building contract is awarded. Similar to single-stage tendering, a two-stage tender typically commences

---

24 It seems more appropriate to refer to the 'first stage' as the 'initial stage' as this may avoid the implication that there will only be two stages of tendering before the contract award. In practice, there are often more than two stages.

between RIBA Stages 3 and 4, but the employer and contractor often collaborate to develop far more of the technical design (RIBA Stage 4) before the building contract is awarded than they would under a single-stage tender.

The initial stage usually requires tenderers to submit rudimentary pricing information which can be used to form the basis of a price for the works or a part of them. Such pricing information may include:

o The tenderer's preliminaries price per week.
o Overheads and profit (normally expressed as a percentage).
o A schedule of unit rates and prices.
o Prices for sufficiently designed parts of the project.

The employer then evaluates the initial stage and selects a tenderer who will become what is often described as the 'preferred bidder'. Again, the commonly used terminology can be misleading, as in many cases the competitive bidding part of the process has ended. A more precise description would be 'preferred contractor'.

The preferred contractor will assist the employer in developing the design. This is commonly referred to as 'early contractor involvement' (see Section 5.4). Together, the employer (along with any of his design consultants) and the contractor (and his design consultants) develop the employer's requirements[25] and the contractor's proposals,[26] respectively.

5.3.4 Negotiation

Many employers procure construction and engineering work simply by negotiating with a preferred contractor. Negotiation has been described as a 'single-stage tender with a single contractor',[27] but given the absence of any true competition, it would perhaps be more accurate to describe negotiation (in this context) as a benchmarking exercise. Clearly, the employer will have to operate some kind of selection process to find a preferred contractor and may do this after considering a number of factors. In many cases, the preferred contractor has either worked with the employer before or has been recommended to the employer by another organisation for whom he successfully completed a previous (often similar) project.

Whilst there are a number of public sector and industry guidelines for competitive tendering, any rules for negotiating are few and far between. It is therefore up to the employer and the contractor to set the parameters for negotiations (see Section 5.4).

The negotiation stage can include reaching agreement over many aspects of the project. In terms of the project budget, the employer and his cost consultant often check the contractor's pricing against historical cost data – either from a pricing book or from previous projects. Other aspects of the project, along with the negotiation procedure itself, can be discussed and agreed in a similar way as described below under 'Early Contractor Involvement' (Section 5.4).

---

25 JCT DB 2016 and FIDIC Yellow Book 2017, and the 'Scope' defined in Contract Data Part One of NEC4.
26 JCT DB 2016 and FIDIC Yellow Book 2017, and the 'Scope' defined in Contract Data Part Two of NEC4.
27 RICS Professional Guidance 'Tendering Strategies', 1st ed., (2014).

Negotiation as a procurement option is mostly used in the private sector, because it is not subject to the same regulation and legislation as procurement in the public sector. Negotiation of public sector contracts does take place, but the employer (or contracting authority) often will have to demonstrate that the circumstances of procurement are an exception, and negotiation will still result in the most economically advantageous price for the project.

There is considerable case law stating that there can be no contract to negotiate 'because it is too uncertain to have any binding force'.[28] However, the parties can agree on a procedure which will govern the negotiations.

5.3.5 Mid-tender Events

Issuing of the tender documents by the employer may not be the last event before the tender return date. The employer (or one or more tenderers) may discover errors or inconsistences within the tender documents that may need resolving, in order to reduce any ambiguities and assist with providing the employer with more compliant and consistent tender returns. Tenderers may also withdraw from the tender process, and may be replaced if circumstances are permitting. Typical mid-tender events could include the following:

o Resolution of ambiguities, inadequacies, and inconsistences/correction of errors. Although more likely to be discovered by one of the tenderers, the employer may discover a matter that needs correcting or resolving. The employer should notify all the tenderers of the problem, and issue his solution/correction to all tenderers simultaneously. If the ambiguity, inconsistency, or error is significant, the employer may have to extend the period for tendering to allow the tenderers a reasonable opportunity to deal with it. Under JCT DB 2016 and FIDIC Yellow Book 2017, the contractor may resolve the particular issue in the Contractor's Proposals, but there is no clear obligation for him to do so. If the issue in the Employer's Requirements remains, and providing the relevant conditions are satisfied (FIDIC Yellow Book 2017), the employer may have to resolve the issue at his own cost (see Section 3.4.3). This is also the case for NEC4 if the Scope contained an inadequacy by failing to instruct the contractor to design a particular part of the works. The employer (or client) would have to pay for a compensation event under clause 60.1(1).

o Meetings with the tenderers. It is often essential for tenderers to visit the site to familiarise themselves with the project's likely logistics, so they can make an appropriate allowance in their tender (often in terms of time and money). Employers can find that meeting with tenderers during site visits can provide a useful update on how the tenders are progressing and highlight any difficulties or issues (see above) the tenderers may be experiencing. Tenderers and employers can also find it useful to meet each other's project teams. Employers should ensure that any additional information disclosed to a tenderer during a meeting is disclosed to all other tenderers as soon as possible, in order to keep the tendering process fair. Treating the meeting as an 'interview' is less advisable in the event that different information is disclosed to each tenderer. An employer who intends to hold mid-tender interviews should provide the tenderers with the same list of questions beforehand (ideally with the tender documents).

o One or more tenderers may withdraw from the tendering process before the tender return date. If the employer is satisfied that the competitive element will not be significantly

---

28 *Courtney & Fairbairn Ltd v Tolaini Bros (Hotels) Ltd,* 1 WLR 297, [1975].

compromised (for example, if one out of four tenderers withdraws), he may decide to continue with the remaining tenderers. However, if he feels that the element of competition may be reduced, he may invite replacement tenderers. This can be problematic for two main reasons:

- The employer will almost certainly have to extend the tender period to allow replacement tenderers a reasonable opportunity to properly compile their bids. Whilst this period may be perfectly sufficient, the replacement tenderers may feel that the remaining (original) tenderers have an advantage because they had more time to provide their tender.
- The original tenderers may feel aggrieved because they were clearly the employer's first-choice contractors, and as such the employer should proceed to consider their bids only.

5.4 Early Contractor Involvement

5.4.1 Early Contractor Involvement in Engineering Projects

Early contractor involvement for design and build engineering projects is often an essential part of a project. For a contract to design and ultimately build unique or bespoke items of plant or equipment, the contractor may be engaged from a very early stage to undertake feasibility studies (akin to RIBA Stage 1 – Preparation and Briefing) and produce an initial, 'feasibility' design. The contractor may be a specialist in the design, manufacture, and installation of particular types of plant and equipment, and in such cases, may not need to engage external consultants to advise him during the initial stages, because he'll solely rely on his own expertise and in-house design capability.

The natural first stage would be to engage the contractor to undertake the feasibility stage (to determine if the intended project is viable) and perhaps the concept design. If the studies and feasibility design support the project's viability, the contract can then be extended to develop the concept design (akin to RIBA Stage 2), and then onto the detailed design (RIBA Stage 4). Design work in three stages (RIBA Stages 1, 2, and 4) may be sufficient if the plant or equipment is stand-alone or perhaps requires less integration with other plant. In the event that the project requires plant to be integrated with and serve (or support) existing plant (such as new equipment in a petrochemical plant or power station), then a significant amount of design coordination will be required (RIBA Stage 3 – Spatial Co-ordination). Coordination may have to take place after the majority of the functional detailed design has been completed. The contractor can then proceed with procuring materials and equipment, manufacturing, installation, and testing, and (if required) operation and maintenance (RIBA Stages 5–7).

Engineering projects requiring early contractor involvement are often let progressively. This can be done in a number of different ways.

o The early stages (feasibility and concept[29]) can be let as follows:
  - By engaging the contractor under a 'design only' contract such as the RIBA PSC, FIDIC MSA, ACE PSA, or NEC4 PSC (refer to Section 4.1); or

---

29 RIBA Stages 1 (Preparation and Briefing) and 2 (Concept Design).

- Let under a contract, where the Employer's Requirements (FIDIC Yellow Book 2017) or Scope (NEC4) do not include detailed design,[30] or manufacturing, construction, testing, installation,[31] or operation.[32]

o In the event that the project proceeds to RIBA Stages 4 (Technical Design), 5 (Manufacturing and Construction), 6 (Handover), and 7 (Use):

  - The contractor is engaged under the FIDIC Yellow Book 2017 or NEC4 contract that includes the works required under RIBA Stages 4–7, but within the Employer's Requirements (or Scope), includes the 'feasibility and concept' design studies carried out under the 'design only' contract; or
  - If the feasibility and contract works were let under the FIDIC Yellow Book 2017 or NEC4, then the work required under RIBA Stages 4–7 is added as a Variation instruction (clause 13.3 of FIDIC Yellow Book 2017) or as a Compensation Event (under clause 60.1(1) where the project manager instructs a change in the Scope) under NEC4.

By initially including only the feasibility and concept stages (RIBA Stages 1 and 2) in one contract, the employer will protect himself from any loss of profit/opportunity claims based upon the profit that the contractor may have earned on the subsequent stages (i.e. under RIBA Stages 4–7) had they been included in the contract, and the project does not go ahead.

The option of letting the initial feasibility and concept design under a 'design only' contract, and subsequently incorporating this work (as part of the Employer's Requirements or Scope), offers the most commercial flexibility. If the contractor is satisfied that he can carry out the work under RIBA Stages 4–7 as a fixed price, the parties could opt for either FIDIC Yellow Book 2017, or NEC4 with Main Options A (Priced Contract with Activity Schedule) or Option B (Priced Contract with Activity Schedule). If the contractor could only undertake the work under RIBA Stages 4–7 as a target price or on a cost-reimbursable basis, the parties could use NEC4 with Main Options C (Target Contract with Activity Schedule) or Option E (Cost Reimbursable Contract), respectively.

5.4.2 Early Contractor Involvement in Construction Projects

In its 2020 report 'Fine Margins – Delivering financial sustainability in UK construction', the CBI (Confederation of British Industry) made a number of recommendations, including the early engagement of contractors for design and build projects. Early contractor involvement can provide a number of advantages to both the employer and contractor, the principal advantage being that both parties gain a greater understanding of the project, both from their own and from the other party's perspective. The process of early involvement of contractors shares many aspects with negotiation (see Section 5.3.3) insofar as the employer and contractor are free to agree on the extent and nature of their pre-contract collaboration.

---

30 RIBA Stage 4 (Technical Design) and likely to include RIBA Stage 3 (Spatial Co-ordination) where appropriate.
31 RIBA Stages 5 (Manufacturing and Construction) and 6 (Handover).
32 RIBA Stage 7 (Use).

5.4.3 Early Contractor Involvement (NEC4 Option X22)

NEC4 introduced a new secondary option (X22) that sets out a procedure for early contractor involvement. It is only intended for use with main option clauses 'Option C' (Target Contract with Activity Schedule) and 'Option E' (Cost Reimbursable Contract).

Option X22 divides the work into either two or three sections as follows, which the client must clearly describe in the Scope:

o The entirety of the works is divided into Stage One and Stage Two,[33] so the whole of the works is subject to the early contractor involvement procedure; or
o Only parts of the works are designated as Stage One and Stage Two, with the remaining parts of the works (i.e. the third stage) being described in the Scope in the normal way and not subject to the early contractor involvement procedure in X22.

The client sets the 'Budget[34]' for the works in Stage One and Stage Two, and the incentive mechanism[35] (similar to the 'Contractor's share[36]') encourages the contractor to complete the works for less than the budget, by offering a percentage of the saving (the 'budget incentive'). The contractor then provides the following:

o Stage One – detailed forecasts of the Defined Cost[37] for Stage One at the intervals set out in the Contract Data.
o Stage Two – submits design proposals which include forecasts of their effects on the Project Cost[38] and on the Accepted Programme.[39]
o Stages One and Two – forecasts of the Project Costs[40] (in conjunction with the project manager) at the intervals stated in the Contract Data.
o Approvals and consents from 'Others'.[41]

The project manager considers the contractor's Stage One forecasts and Stage Two design proposals, and either accepts them or rejects them.[42] Following the project manager's acceptance of the contractor's Stage Two design proposal[43] and the client's confirmation that the works are to proceed, the project manager may notify the contractor to proceed to Stage Two.

---

33 X22.1(3) and the corresponding entries in Contract Data Part One against Option X22.
34 X22.1(1) and the corresponding entries in Contract Data Part One against Option X22 for 'Budget' and the 'Budget incentive' percentage – clause X22.7(1).
35 X22.7(1).
36 Option C – clause 54 (The Contractor's share) – see Section 9.6.3.
37 X22.2(1), Option C – clause 11.2(24).
38 X22.1(2), utilising the clause X22.1(4) 'Pricing Information' provided by the Contractor and set out in Contract Data Part Two against Option X22.
39 11.2(1).
40 X22.2(5).
41 X22.3(6), although Option X22 does not specify at what stage the contractor applies for or obtains 'approvals and consents'.
42 The reasons for rejecting (not accepting) forecasts or design submission proposals are that (i) they do not comply with the Scope, (ii) the forecast includes unnecessary work, (iii) the design proposals will cause the client to incur unnecessary cost, and (iv) the project manager is not satisfied that the contractor has properly assessed the prices (or changes to them) – clauses X22.2(2) and X22.3(3).
43 X22.5(1) provides that (i) the contractor must obtain approvals and consents from Others, (ii) Budget changes are agreed to or assessed by the project manager, (iii) the 'total of the Prices' for Stage Two is agreed to by the contractor and the project manager, and (iv) the client confirms that the works can proceed.

Clause X22.5(2) also states that if the notice to proceed to Stage Two is not issued 'for any reason' then the project manager instructs that Stage Two work be omitted from the Scope. Given that the client's consent to proceed is one of the criteria for the project manager issuing the notice to proceed, it appears that Option X22 provides the client with considerable discretion in choosing whether to proceed with the Stage Two works. However, the client can only engage another contractor to carry out the Stage Two works if the contractor has failed to (i) achieve the performance provided in the Scope, and (ii) agree to the Stage Two prices with the project manager.

It is not clear how or even if Option X22 requires the contractor to provide any design for the Stage One works, as it only refers to his design proposals for the works in Stage Two. Presumably, the requirement for any contractor's design for the works in Stage One (and any works falling outside Stages One or Two) could be set out in the Scope and subject to project manager's acceptance under clause 21. It therefore appears that X22's early contractor involvement is more focused on providing monitoring and controlling the budget – total of the prices and project cost – than on obtaining the contractor's design.

It may be difficult for the client to appoint another contractor to carry out the Stage Two works under clause X22.5(3) if the entire works consist of more than the works required for Stages One and Two.[44] Not only would the client have to engage a replacement contractor, which would be likely to cause delay to the project (and entitle the original contractor to a compensation event), but the client would also have to coordinate the works between the new contractor carrying out the Stage Two works and the original contractor completing the works comprising the remainder. In such an event, it is likely that any advantage gained by the X22 early contractor involvement process would be lost.

5.4.4   Advantages of Early Contractor Involvement

The principal advantage of the employer and contractor entering into early, pre-contract dialogue is that both parties gain a greater understanding of the project and each other's positions, particularly when allocating risk. This overarching advantage can be in respect of the project's design and quality, cost, and programme.

o Feasibility/Concept
  – The contractor's involvement may be essential.
  – The contractor may be best placed to interpret data gained from feasibility studies in order to produce the optimum design.
o Design/Quality advantages
  – The Employer's Requirements (or Scope) can benefit from the early input from the contractor, including any design consultants engaged by the contractor, and from the contractor's supply chain.
  – The contractor's specialist subcontractors may be able to offer more efficient or 'state of the art' elements of the works, that the employer's design team may previously not have known.

---

44 The Scope contains works in a remaining 'third stage', in addition to the works required for Stage One and Stage Two.

- Better coordination of the design's key elements when the designers or suppliers of those elements are involved at an earlier pre-contract stage.
o Cost
  - Greater scope for 'value engineering': the earlier the contractor, his designers, and supply chain are involved, in terms of either (i) achieving the same function but at a reduced cost, or (ii) achieving a proportionally greater function, with a lesser cost.
  - Greater transparency of cost if the contractor's supply chain engages with the employer and his cost consultant/quantity surveyor.
  - Opportunity to consider the best pricing model for the contract – e.g. fixed price, target cost, or cost-reimbursable.
o Programme
  - Employer's Requirements can be more precise in terms of programme, including possible completion/handover in sections or stages, or Key Dates[45] for completing particular stages of work.
  - Better understanding of how to programme design development, including the approval/acceptance of the contractor's design by the employer.
  - Early involvement of the contractor's supply chain can better integrate the sequence and timing of specialist subcontractors' works and design input into the contractor's main programme.

Early contractor involvement gives a greater opportunity for the parties to understand and subsequently manage the project's risks. Considering the above matters for design/quality, cost, and programme promotes better understanding of each party's risks. For example, the project's early completion may be subject to particular time constraints imposed on the employer. If the contractor understands this from an early stage, he (along with his design consultants and supply chain) have a greater opportunity to provide the employer with an acceptable programme if the employer discusses this with the contractor, than he would otherwise may have if the employer simply imposed a completion date without further explanation.

The parties should record a clear audit trail of how the project's design develops during the early contractor involvement stage. The employer's initial requirements may change significantly, so that any early designs or specifications are now obsolete and require wholesale revision. An audit trail of design change will provide the parties with a clear idea of how to capture the design development in the contract documents. Attempts at short-cutting this process should be avoided. For example, obsolete initial designs should be revised and brought up to date to reflect the design's development. Employers (and their consultants) should ensure that the version of the Employer's Requirements[46] or Scope[47] incorporated into the building contract reflects the finalised design. It may not be practicable to revise all the individual employer's documents, so the employer and contractor should collaborate to compile a schedule of amendments to the Employer's Requirements (or Scope), which can then be satisfied by the Contractor's Proposals or contractor's design.

---

45 NEC4 – clause 11.2(11).
46 JCT DB 2016 or FIDIC Yellow Book 2017.
47 NEC4.

5.4.5   Disadvantages of Early Contractor Involvement

From the perspective of delivering the project, there are relatively few disadvantages of early contractor involvement. However, in many instances, the early collaboration and dialogue between employer and contractor proceeds without any formal procedure or rules being agreed upon. Without any formal agreement over pre-contract dialogue and negotiation (particularly over price), the whole process can create uncertainty for both the employer and the contractor.

The employer may be concerned that the contractor may withdraw from the early involvement or negotiation at any time. Whilst this may be unlikely where contractors have expended significant efforts during the early involvement/negotiation process, if the contractor receives a better offer from elsewhere, he may decide that the new opportunity is more commercially viable, and either reduce his involvement (by reducing or diverting key staff to another project) or withdraw from the project altogether. The employer may have further concerns if the project is commercially sensitive to him – for example, the employer's business may be adversely affected if a competitor learned of the project. If the contractor withdraws, the employer will want the contractor to treat his participation in the negotiation/early involvement as strictly confidential.

The threat of contractor withdrawal may place the employer at a disadvantage when negotiating a price with the contractor. If the employer faces significant project delays for appointing a replacement contractor, his bargaining position will be weaker; and if the contractor has a reasonable idea of how much replacing him would cost the employer, then he may be in a strong position to increase his price by a similar amount. Reduced competition can present a significant risk to the employer, particularly when the contractor's early involvement has 'crossed the Rubicon' where his withdrawal would place the project in jeopardy.

The reverse situation of 'employer withdrawal' will concern the contractor. He may suspect that his participation in the early involvement/negotiation process will be purely speculative and the employer could ultimately choose not to proceed with the project. There have been many instances where a contractor has carried out significant pre-contract budgetary and design work for a prospective project, only for the employer to cancel the project and sell his interest in the site. As with the above example, the employer's withdrawal is less likely when he has committed significant resources to the project as, like the contractor, he will need the project to proceed to recover the costs of the negotiation/early involvement process and to earn his profit. Another concern the contractor may have is that the employer is simply using the negotiation/early involvement stage as a 'benchmarking' exercise and may ultimately appoint another (possibly cheaper) contractor to undertake the project.

The parties may wish to protect themselves by formally agreeing on conditions for the negotiation/early involvement process as follows:

o Measures to protect the employer:
  – Obtaining a non-disclosure agreement (NDA) or confidentiality agreement from the contractor not to discuss any of the negotiation/early involvement matters without the employer's prior written authorisation, or not to discuss for a defined period.
  – The contractor's agreement to only withdraw from the negotiation/early involvement process after the expiry of a period after giving the employer written notice.

- Tender Security (FIDIC Yellow Book 2017 – see Section 7.4.5).
- Agreement on a method to 'benchmark' or check the contractor's pricing.
o Measures to protect the contractor:
  - The contractor may require an 'exclusivity' agreement from the employer; that the employer will not solicit the involvement of any other contractors unless the parties agree that they will not proceed with the negotiation/early involvement procedure.
  - The contractor may insist on the employer reimbursing his reasonable costs of participating in the negotiation/early involvement process if either (i) the parties agree not to continue with the process, or (ii) the contractor is not awarded the building contract within an agreed-upon period.
  - The contractor retains copyright in all the design material (e.g. drawings, specifications, calculations) he has produced.
  - All the contractor's pricing information will be treated as confidential by the employer.

5.4.6 Pre-construction Services Agreements

The early contractor involvement period may not just involve the employer and the contractor negotiating the terms of the building contract, value engineering, or developing the design. The employer may formally engage the contractor to not only carry out some services, but also some physical works that will either form part of or facilitate the main building contract. In such cases, the parties often enter into a Pre-Construction Services Agreement (PCSA).

A PCSA is essentially a shortform contract that sets out the services and works that the contractor is to provide. JCT publishes two editions of their Pre-Construction Services Agreement (2016): the 'specialist' and 'general contractor' versions. The services[48] include the following:

o Procuring specialist design services,
o Cost advice,
o Programming and sequencing of work,
o Construction advice,
o Construction phase management and communication systems, and
o Any additional services.

Whilst the JCT PCSA 2016 does not specify any 'works',[49] other PCSAs may engage the contractor to undertake some physical works, such as:

o Enabling works, including:
  - Site clearance, demolition, provision of site access;
  - Temporary works (e.g. make a structure safe, or facilitate investigative surveys);
  - Investigations.
o Service diversions
o Site security (e.g. erect fencing/hoarding, provide a site setup for the main contract).

---

48 JCT PCSA (General Contractor) edition 2016.
49 Unless physical works are included under 'Additional Services'.

Some PCSAs (although, not the JCT PCSA 2016) also seek to provide a framework for the negotiation/early contractor involvement stages. They may include provisions where the parties operate an agreed-upon procedure for negotiation (see Section 5.3.3) to assist with their agreeing on the building contract for the main works. Such provisions may include:

o A duty to act in good faith
o A timetable for negotiations
  – A set period of time, or
  – A period of time that can be extended upon written notice (usually from the employer).
o Basing the contractual negotiations upon a particular set of contract conditions:
  – A standard form of design and build contract (e.g. JCT DB 2016, NEC4, FIDIC Yellow Book 2017), or
  – Contract terms that were previously agreed upon between the parties.
o The price may be based on:
  – Prices previously agreed upon between the parties (adjusted to reflect the nature and conditions of the proposed project and adjusted for year and location), or
  – A maximum figure (either a fixed price, or a 'target cost') not to be exceeded.
o Procedure for progressively agreeing on parts of the proposed contract:
  – Priority of agreement (focusing on parts of the 'Contract Particulars' for JCT DB 2017, or the 'Contract Data' for FIDIC Yellow Book 2017 or NEC4, before negotiating over the contract conditions).
  – Recording agreements progressively (e.g. in a series of memoranda of understanding).
o That negotiations (and performing any works undertaken under the PCSA) do not oblige the employer to appoint the contractor to carry out the main project works. Consequently, if another contractor is appointed, the contractor will not be entitled to:
  – Loss of profit that would otherwise have been earned when carrying out the main project works, or
  – Any other loss of opportunity costs.

Contractors should take careful note of any provisions in a PCSA relating to either the main project's contract conditions or that may govern how the main project works may be carried out. This will assist the contractor to avoid:

o Any unsuitable contract conditions[50] (e.g. the works may be too complicated to use the JCT ICD 2016 contract, so the JCT DB 2016 contract may be more appropriate, or a target cost contract may be more appropriate than a fixed-price contract[51]).
o Being bound to or restricted by any unfavourable terms for the main project works that were imposed by the employer in the PCSA. For example, the employer may attempt to impose a programme or a budget for the main project works.

---

50 Part One of this book only considers standard form contracts (JCT DB 2016, FIDIC Yellow Book 2017, and NEC4) in their unamended forms. Part Two of this book considers amendments to these standard form contracts (refer to Chapter 11).
51 NEC4 Option C or D, instead of NEC4 Option A or B (or JCT DB 2017 or FIDIC Yellow Book 2017).

5.5   Tender Returns and Evaluation

Receipt of tenders (or tender return) and their evaluation should be carried out carefully and consistently by or on behalf of the employer. If the employer is represented by a chartered quantity surveying firm, they should have thorough procedures that comply with the RICS's guidance.[52]

5.5.1   Tender Receipt and Initial Checks

The employer may wish to attend the tender opening, although this is not always the case. The employer's consultants (normally, a quantity surveyor) should record when each tender is received. The employer's consultants should then proceed to evaluate all tenders received on or before the deadline and in a manner complying with the employer's tendering instructions. Some employers request hard copies of tender returns to be provided, even where the format for submission is by electronic means (such as via a portal or by email).

There may be two consultants receiving tenders on the employer's behalf. One may be the consultant who assembled the tender documents, and the other may be a senior member (such as a director or partner) present to witness and ensure that the tenders are received correctly. Many professional consultancies operate in this way for their own quality assurance reasons.

The consultants should firstly check to see if the returned tenders contain all the required documents. If documents are absent from a tender, then strictly this should be marked as non-compliant.

For private sector procurement, it is entirely the employer's choice whether or not he will allow non-compliant[53] tenders to be considered. The employer should provide clear instructions to his consultants if non-compliant tenders are to be considered.

5.5.2   Tender Evaluation

Once the initial compliance checks have been carried out, the tender evaluation process can begin. This stage of evaluation usually focuses on the pricing and financial aspects of the tenders as follows:

○ Arithmetic checks. Each tender return is arithmetically checked to identify:
  – The correct tendered price;
  – The constituent prices for the main works,[54] preliminaries, risk allocation, and overheads and profit;
  – Any disproportionately lower or higher prices, or unusual prices;
○ Any priced alternatives (refer to Section 3.6.3); and
○ Any exclusions or qualifications that affect the price.

---

52 RICS Guidance Note 'Tendering Strategies', 1st ed., (2014).
53 Tenders that are either received late and/or tenders that are incomplete or do not contain the required documents.
54 Possibly identifying Provisional Sums as a separate subtotal within the overall sum.

An initial cost benchmarking (or 'normalisation') exercise may take place once the consultants are satisfied that the tenders are arithmetically correct. This compares the submitted prices against each other and possibly against the employer's cost plan value. The comparison can be made against:

o The cost plan value,
o The highest submitted tender price, or
o The average (or other statistical method) tender price.

The initial cost benchmarking exercise can be useful to compare the initial prices with those after the correction of errors or any permitted price adjustments to remove (or reduce) exclusions or qualifications. The usual practice when pricing errors are discovered is for tenderers to either correct the error or accept it and stand by their original price. A second cost-benchmarking exercise can then be undertaken.

5.5.3   Evaluation of the Tenderer's Design

The evaluation of design may be a largely subjective process, particularly where the design includes a significant aesthetic element. In such cases, the preferred design may be simply a matter of the employer's choice, without any form of evaluation or analysis taking place.

A more objective and empirical evaluation of the design is possible where the Employer's Requirements (or Scope) set out clear functional requirements. This could be a performance specification or a requirement to design the works (or part of them) to particular dimensions – for example, in order to be accommodated by the existing site or fall within planning constraints.

Evaluation of the design may therefore include:

o Checking dimensions, specifications, and calculations to ensure the tenderer's design complies with the Employer's Requirements (or Scope).
o Assessing or validating materials, goods, equipment, and plant specified by the tenderer:
  – Including any alternative products; and
  – Scrutiny of manufacturer's certificates, quality assurance, and technical information.
o Considering any design alternatives offered by the tenderer.
o Compiling a list of design questions for the tenderer to answer:
  – Requesting further information for design alternatives; and
  – Seeking clarification of the source of materials or goods that may not have been expressly specified in the tenderer's design.
o Obtaining clarification of which specialist subcontractors or suppliers the tenderer proposes to engage.

Evaluating the tenderers' designs also provides the employer with an opportunity to consider how much of the contractor's (i.e. the successful tenderer's) design will have to be reviewed before the contract is entered into, and how much design needs to be reviewed after the contract is awarded (see Section 3.5.2):

o JCT DB 2016: Any Contractor's Design Documents[55] not included in (what is likely to become) the Contractor's Proposals will be subject to the Design Submission Procedure in Schedule 1 of the contract.

---

55 As defined in clause 1.1 of JCT DB 2016.

    ◦ FIDIC Yellow Book 2017: Any 'Contractor's Documents' that are not specified for review in either the Employer's Requirements or the contract Conditions.[56]

    ◦ NEC4: The contractor must submit all of his design to the project manager for acceptance. The employer (or client in the case of NEC4) along with the project manager should consider assigning a different 'period for reply'[57] to particular parts of the successful tenderer's (contractor's) design.

5.5.4 Evaluation of Other Aspects of the Tender

The pricing element of a tender usually attracts the most attention from the employer, but price may not always be of prime importance. A tenderer may have a good track record of similar projects and submit a commercially competitive bid, but other aspects may dissuade the employer from awarding the building contract to that tenderer. After any period of economic upheaval or change,[58] employers may look for the most financially stable contractor, so the strength of that tenderer's balance sheet may prove to be pivotal in their selection.

    A tenderer's outline programme may also be attractive to the employer. If a particular programme is seemingly too short or too long a duration, it may indicate that the tenderer either hasn't fully considered the likely programme or may not understand the project's logistics. It is therefore important that the tenderer provides a programme that realistically reflects the likely sequencing of the project (including any sectional completion or key dates), as it will indicate that the tenderer has a good understanding of the project.

    The tenderer's health and safety policies and record may also be a factor in contractor selection. It is unlikely that the employer will give comprehensive health and safety documentation and an exemplary health and safety record more weight than a competitive and compliant price, but a poor health and safety record may dissuade the employer from appointing that tenderer.

    A tenderer's method statement may be important to the selection process, particularly if it provides a good explanation of how an important or complex element of the project will be constructed.

5.5.5 Tender Interviews

Once the employer (and his team) have completed their initial checks, evaluated the received tenders (in terms of price, design, and other aspect), and obtained any clarifications arising from those procedures (that can be practically addressed in correspondence), then the employer can proceed to the interview stage. If there is little disparity between the returned tenders then it may be necessary to interview all tenderers. If only one or two tenders fall within the employer's range of acceptance criteria, then only those tenderers may be invited for interview.

    The interview should be consistent in terms of format and line of enquiry for all attendees, and the employer should clearly inform each attending tenderer of this beforehand.

---

56 Clause 5.2.2 of FIDIC Yellow Book 2017.
57 NEC4 clause 13.3 and Contract Data Part One.
58 Such as the 2008 recession or the Covid-19 crisis.

In doing so, the employer should specify which members of the tenderer's bid and proposed construction teams should attend, along with providing any outstanding information from the tender evaluation and query stage. This should help the employer understand the tenderer's proposed design, price, method, and programme in more detail before deciding upon the successful tenderer. For example, the employer can meet the tenderer's intended designers, site team, and perhaps any specialist subcontractors or suppliers who are likely to have a significant involvement in the project.

It is likely that the employer and the interviewed tenderers may reach further agreements – such as in terms of design alternatives (including different specifications or methods of construction), pricing adjustments, or amendments to the intended programme. Many (if not all) of these subsequent agreements may need to be incorporated into the building contract, so the following procedures should be followed:

o Accurately record the principle(s) of the agreement in writing, and obtain written confirmation to the extent and effect from the other party.
o Discuss how the agreement may be captured and incorporated into the building contract. Some examples could be:
  – For an agreed-upon change in specification or the employer agreeing to an alternative design offered by the contractor, the Employer's Requirements (or Scope) may need to be amended. Similarly, options in the Contractor's Proposal(s) (or contractor's design) that will not now apply should be deleted. It is important that the contractor's design satisfies the Employer's Requirements/Scope and that there are no discrepancies, divergencies, inconsistences, or inadequacies between them (refer to Section 3.3).
  – Any adjustments to the overall tendered price should be clearly changed in the pricing document, and appropriate cross references to the Employer's Requirements/Scope and Contractor's Proposal(s)/design added.
  – The intended programme should be amended accordingly, and corresponding entries in the Contract Particulars (JCT), Contract Agreement (FIDIC Yellow Book 2017), or Contract Data Parts One and Two (NEC4) made in draft.
o Make any other changes to the Articles of Agreement (JCT DB 2016), Contract Agreement (FIDIC Yellow Book 2017), or the chosen means of executing the NEC4 contract that reflect any subsequent agreements.

5.6 Contractor Selection and Appointment

The conclusion of the tender process, evaluation, and any further adjustments after the interview stage may not immediately result in the contractor's appointment. However, it may signal that the tender appraisal services provided by the employer's consultants are significantly completed. The consultants' purpose is to advise the employer so he has all the information he requires to make an informed choice between the tenderers who have made it through the tender process. Consultants generally should not recommend which contractor to appoint but should provide the employer with the necessary pros and cons of each contractor. A consultant who negligently makes a recommendation may be liable to the employer.

Once the employer has the necessary information to choose which contractor to appoint, then he has the following tasks:

o Inform the successful tenderer,
o Inform the unsuccessful tenderers, and
o Conclude any negotiations with the successful tenderer and formalise their contractual relationship by executing the building contract.

5.6.1 Inform the Successful Tenderer

The employer may just simply notify the successful tenderer that he will be appointed as contractor. It is advisable for the employer's legal advisors to either draft or have an input into drafting this notification, as if done incorrectly, it may inadvertently bring about the existence of a contract before all the relevant matters are concluded and captured in the intended building contract. For this reason, any post-tender, but pre-contract correspondence between the employer and the contractor are likely to state that any discussions are 'subject to contract'.

5.6.2 Inform the Unsuccessful Tenderers

Informing the unsuccessful tenderers must be carried out appropriately. In cases where the tendered 'price' is the overriding criteria for selecting the contractor, the employer will often inform the unsuccessful tenderers by setting out the tendered bids in increasing or decreasing order. Each tendered bid is set out anonymously, so each tenderer can identify their own price, but not be informed of their competitors or their prices.

For reasons of commercial sensitivity, hard copies of tenders are often returned, and any bids submitted by electronic means can be deleted from the relevant portal or server. The JCT Practice Note 'Tendering 2017' advises that tender documents that will not be in the public domain should not be used for any other purpose, either by the employer or its consultants. This is not always the case as competitively obtained cost information can be a valuable commodity, so any tenderers should specify that their documents are to not to be copied, returned, or deleted and rendered inaccessible and unusable.

Unsuccessful tenderers may also want to learn how they scored against the non-commercial criteria. In the interests of promoting improved competition, the employer should provide the tenderer's score and range of scores (anonymously) if the tenderer requests them.

5.6.3 Conclude Negotiations and Enter into the Building Contract

The majority of matters to be included and incorporated into the ultimate building contract may be agreed upon at the end of the tender period, but there may be some remaining matters to be concluded before both parties are satisfied that the building contract captures the relevant agreements.

The employer may require the contractor to start work before the building contract is concluded. There may be a number of reasons for this, such as the following:

o The employer may benefit from a grant if work commences before a certain date.

o The project's completion date may be immovable, and the tender process may have taken longer than envisaged, so the project needs to commence as early as possible.
o The project's funding criteria may only be activated once the contractor commences work. In such cases, the employer may be waiting for funding to pay consultants' fees so will want the first payment under the funding agreement as soon as possible.
o Certain items of work may require a long 'lead in' period, or manufacturing slots will have to be secured early so as not to delay the project's progress.

In such cases, the employer usually provides the contractor with a 'letter of intent' that will give some security to the contractor before the contract is signed. A letter of intent usually acts as a temporary contract that bridges the gap between selecting the contractor and concluding the building contract. Ideally, a letter of intent will state:

o The parties' positions in terms of what is and what is not agreed to in the negotiations leading to concluding the building contract.
o The letter of intent may stipulate that the contractor is only to undertake certain items of work (e.g. procuring design work or work early in the programme).
o That the employer will pay for any work carried out by the contractor:
    – On the basis of the rates and prices that will be included in the building contract (to the extent agreed between the parties);
    – Valued on the basis agreed in the building contract (e.g. the contract's valuation rules);
    – On a 'quantum meruit' basis for work where no rates and prices are agreed (or the basis of calculating them is not agreed);
    – Up to a financial limit set out in the letter of intent.
o The payment terms (if not already agreed as part of negotiations).
o The programme of work (if not already agreed).
o Any consequences of failure to complete either specific works or the entire works (in the event that a building contract cannot be agreed) by a particular date.
o The required insurances.
o That the letter of intent will be superseded by the ultimate building contract, and any sums paid by the employer (against the letter of intent) will be deemed to have been payments made under the terms of the building contract.
o Set out the procedures for serving notices, terminating the letter of intent and dispute resolution.

The parties should make every effort to conclude and execute the building contract as soon as possible.

6

Subcontracting

6.1 General

Subcontracting is an ever-present and essential part of contracting. Unless the organisation acting as the main contractor is carrying out all parts of the work itself (for example, including ground engineering, structural frame, and building services), then the majority of the physical work will be directly undertaken by subcontractors. As building technology advances, construction work becomes more complicated and increasingly specialised, so it will be rare to find a main contractor that also undertakes specialist works such a piling, ground stabilisation, building services, or IT installations. He or she will need to engage specialists. The work undertaken by subcontractors of all disciplines will effectively be delegated from the main contractor to the subcontractors. Contrary to some published views,[1] in the author's experience, the use of subcontractors has not caused a reduction in skills or innovation. Whilst it may be the case that particular skills and innovations are not driven by main contractors, they are certainly advanced by specialist subcontractors, often offering contemporary products and construction methods.

The effective delegation of work is not the only matter to be 'passed down' from the main contractor to the subcontractors. Particular obligations, risks, and liabilities owed by the main contractor to the employer under the 'main contract' will be passed down from the main contractor to the subcontractors. As all subcontracted work will be carried out under the umbrella of the main contract (for which the main contractor remains responsible to the employer), the main contractor will often manage risks such as liability for defects, or late completion, by transferring them (in whole or in part) to the subcontractors.

This transfer of risk is not only inherent in most standard forms of subcontract, but is also a characteristic of the widespread use of collateral warranties where the subcontractor makes a formal promise, outside the terms of the subcontract, directly to a beneficiary. The beneficiary is usually the employer, a funder, a purchaser, a tenant, an occupier, or operator, and the subcontractor promises to that beneficiary that he has correctly performed the obligations set out in the subcontract.

However, this is not to suggest that the role of main contractor is either an easy one or one where risks can simply be offloaded to others. The main contractor carries a great responsibility to both the employer and his supply chain (subcontractors and suppliers).

---

1 RICS draft guidance note – Subcontracting, 1st ed., (March 2020).

Design and Build Contracts, First Edition. Guy Higginbottom.
© 2024 John Wiley & Sons Ltd. Published 2024 by John Wiley & Sons Ltd.

He must properly carry out the works for the employer, and in doing so, ensure that all subcontractors' work is integrated, coordinated, and programmed correctly. For design and build contracts, the main contractor will have similar responsibilities in terms of the project's overall design. He must ensure that designing subcontractors are provided with sufficient information from which to prepare their design. He must also obtain these subcontractors' designs in sufficient time to for them to be integrated with one another and into the overall design. The main contractor and his design team (particularly the 'lead designer' – refer to Section 4.2.6) will be responsible for coordinating all the subcontractors' designs, before the work can be effectively constructed.

For these reasons, main contractors may feel they are caught between 'a rock and a hard place' – between demanding employers on one side and (often) equally demanding subcontractors on the other. Main contractors are often very adept at managing this 'squeeze', not least because they are used to being the conduit between the employer and the subcontractors carrying out the work.

Most main contractors enjoy very positive relations with their supply chain. Despite hearing reports of payment terms exceeding 90 days for some (possibly now defunct) main contractors, in the author's experience, a great many main contractors maintain subcontract or supply-chain 'charters' and operate far more favourable payment terms (many not exceeding 45 days). A strong, reciprocal relationship between a main contractor and the supply chain is key to successfully delivering a project, particularly where all parties are familiar with the nature of the work and with working with each other.

6.1.1 The General Nature of Subcontracts

A subcontract is the contract between the main contractor (i.e. the organisation directly engaged by the employer) and an organisation (the subcontractor) who, on the main contractor's behalf, accepts responsibility for carrying out part (or sometimes all) of the main contractor's work under the building contract. A subcontract does not create a direct contractual link between the employer and the subcontractor, as there is 'privity of contract' between the subcontractor and the main contractor. In simple terms, the relevant parts of the main contract conditions, along with the relevant parts of the main contract works, are 'stepped down' or formally delegated from the main contractor to the subcontractor.

Subcontractors under design and build (and traditional) main contracts should be contrasted from 'trade contractors' or 'works contractors' in management contracting, as they are engaged directly by the employer, and the contractor is engaged purely to manage those works or trade contractors.

Every subcontract should provide the following for each of the parties:

o For the main contractor:
   – The subcontractor complies with the terms of the main contract, both in terms of satisfying the required specification for the works and other administrative requirements (such as giving notices or entitlement to variations).
   – An ability to control the subcontract works to the same (or greater) extent than the employer (or the employer's representative) controls the main contractor's works.
   – The subcontractor performs its works in sufficient time and in a sequence compatible with the main contract works.

- The main contractor can rely upon the subcontractor to carry out a specific part of the main contract (notably for specialist works) including its design.
- The main contractor has recourse against the subcontractor if he breaches the subcontract's terms.
o For the subcontractor:
  - The main contractor properly coordinates the subcontract work with the work of others (including interfaces with the main contractor's other subcontractors and any contractors such as those carrying out statutory work or work on behalf of the employer).
  - The main contractor facilitates the subcontract work by providing adequate site accommodation, storage and loading areas, and other 'attendances' (such as attendant labour, disposal facilities, lighting, and power).
  - The subcontractor has recourse against the main contractor if his works are adversely affected by either the main contractor himself, another subcontractor, or others working on the site.

The three major forms of subcontract considered in this chapter are all principally intended to align with the corresponding main contracts, and they deal with the 'stepping down' concept in different ways. As will be seen, the JCT and FIDIC versions of these subcontracts provide an express obligation for the subcontractor to perform his subcontract works in accordance with the main contract, whereas the NEC4 subcontract relies upon the relevant parts of the Scope (as applying to the main contract) being included within the Subcontract Scope. Given the nature of NEC4 contracts, this is unsurprising, as they rely more on the employer providing comprehensive information to the main contractor (and then from the main contractor to the subcontractor) than on a 'catch all' clause obliging the subcontractor to comply with the main contract.[2]

The three corresponding subcontracts considered in this chapter are shown in Table 6.1.

Table 6.1 Subcontracts for JCT DB 2016, FIDIC Yellow Book 2017, and NEC4.

Main contract	Corresponding subcontract documents
o JCT DB 2016	o Design and Build Subcontract Agreement 2016 (DBSub/A 2016) incorporating o Design and Build Subcontract Conditions 2016 (DBSub/C 2016)
o FIDIC Yellow Book 2017	o Conditions of Subcontract for Plant Design-Build (1st ed., 2019)
o NEC3	o NEC4 Engineering and Construction Subcontract (June 2017)

The subcontract for use with JCT DB 2016 consists of two documents: the 'Agreement' (DBSub/A 2016) and the 'Conditions' (DBSub/C 2016). The Agreement contains the subcontract's 'specific information' (refer to Section 2.2.2) and the Conditions are the subcontract's 'conditions of contract' (see Section 2.2.3).

Clause 2.1.1 of the JCT subcontract conditions (DBSub/C) obliges the subcontractor to carry out the subcontract works in accordance with the Subcontract Documents, which

---

2 The JCT Intermediate Building Contract with Contractor's Design 2016 (JCT ICD 2016) is also considered in relation to 'Named Subcontractors' (refer to Section 6.3).

include the relevant parts of the main contract.[3] The FIDIC conditions of subcontract take a much simpler route. Subcontract clause 2.2 obliges the subcontractor to comply with the main contract.

Most subcontractors are referred to as 'domestic' subcontractors, as they are selected and engaged by the main contractor (see Section 6.2). The terms 'domestic subcontract' and 'domestic subcontractor' are still widely used, since subcontractors were engaged under the old 'DOM/1' and 'DOM/2' subcontracts that were used for the JCT Standard Form of Building Contract 1980 edition ('JCT 1980'), and its design and build counterpart (JCT CD81). Domestic subcontractors are simply engaged by the main contractor, and the employer leaves the main contractor (and the subcontractor) to get on with the work. They should be distinguished from subcontractors for whom the employer retains some involvement in the administration and possibly payment of those subcontractors (as for nominated or named subcontractors).

Nominated or named subcontractors (see Section 6.3 below) are ordinarily selected by or on behalf of the employer. They may be carrying out a particular part of the works where their early involvement is required (e.g. such as a lead-in period or some advanced design), or where the employer wants that particular subcontractor to construct a particular part of the works. The use of nominated subcontractors was popular under the traditional versions of the JCT standard form contracts (the 1963–1998 editions), but the practice was largely abandoned since the 2005 versions. For the design and build versions of the JCT contracts from 1998[4] onwards, subcontractors could be specified[5] as 'named' subcontractors.

6.1.2 Types of Subcontract for Each Main Contract

Different main contractors adopt different subcontracts. Some use their own forms of sub-contract, whereas others adopt the subcontracts that are intended for use with a particular main contract. Adopting a 'one size fits all' approach, irrespective of which main contract is being operated, can cause difficulties, as the subcontract may contain procedures that are either incompatible with the main contract, or it may not contain any directly compatible procedures at all.

Some main contractors use their own, bespoke forms of subcontract. These can vary significantly in terms of their suitability, complexity, and the way that risks are allocated between the main contractor and the subcontractor. For example, some bespoke subcontracts may be specifically drafted and intended for use with a particular main contract, whereas others may be little more than a one-page 'purchase' order, or a subcontract that, to a significant degree, is intended to operate independently of the main contract. These shorter and less-sophisticated subcontracts often contain very simple provisions for 'stepping down' the relevant parts of the main contract conditions and works to the subcontractor. They may often say that the subcontract is 'back to back' with the main contract.

---

3 Clause 1.1 of the Subcontract Conditions (DBSub/C) defines the 'Subcontract Documents' as the documents in Article 1 of the Subcontract Agreement (DBSub/A). This includes the 'Main Contract Information Schedule' which, at point 1.5, includes the 'Main Contract'.

4 JCT WCD98 – Supplemental Provision S4, JCT DB 2005 and JCT DB 2011– paragraph 2 of Supplemental Provision 2.

5 In the Employer's Requirements.

This approach should be avoided, as not only does it create uncertainty for the subcontractor (who may not have been made fully aware of all the relevant main contract provisions and requirements), but it also is unlikely to offer protection to the contractor. For example, if a subcontract is simply expressed as being 'back to back' with the main contract, the main contractor will have to ensure that any administrative duties and procedures under the main contract can be effectively carried out when applied to the subcontract. This may not be straightforward, as under JCT DB 2016, FIDIC Yellow Book 2017, and NEC4 the employer's representative is responsible for contract administration. The contractor will have either have to designate a particular person (either individually or by class or discipline) to carry out the equivalent tasks under the subcontract or include provisions in the subcontract that the main contractor also acts as the equivalent of the employer's representative under the subcontract. This may create conflicts of interest where the main contractor has to carry out 'certifying' duties under the subcontract that require him to act fairly, professionally, and possibly hold the balance between the main contractor and the subcontractor (see Section 4.2.8).

Conversely, if a subcontract that corresponds with the main contract is used, the subcontractor will have greater certainty regarding how the relevant obligations and works from the main contract affect the subcontract works, and the contractor will often benefit from the subcontractors being obliged to assist the main contractor in certain matters (e.g. providing notices for extensions of time) that affect him under the main contract. For example, the NEC4 subcontract provides that the subcontractor must notify the main contractor of a compensation event within seven weeks,[6] which allows the main contractor a further week to consider the subcontractor's notification before being obliged to notify the project manager of the same.[7] A bespoke subcontract may not be suitable for works where the subcontract is required to design part of the works. For example, it may not set out how or what the subcontractor must design, and it may not provide for any professional indemnity insurance obligations for the subcontractor.

Whilst none of the three main forms of design and build contract considered by this book are identical (NEC4, in particular, operates in a fundamentally different way[8] to JCT DB 2016 and FIDIC Yellow Book 2017), there are common principles that will apply to the subcontracts intended for use with each of these main contracts. They will all seek to 'step down' the contractor's obligations under the main contract to the subcontractors. The corresponding subcontracts all do this, and they include appropriate cross references to the relevant provisions in their corresponding main contract. However, if a main contractor is using his own, bespoke form of subcontract that is usually intended for lump-sum main contracts (such as JCT or FIDIC), he will have to amend its terms if the main contract is an NEC4 contract – particularly if one of the target cost options (Main Option Clauses C or D) is used.

---

6 NEC4 Engineering and Construction Subcontract, clause 61.3.
7 NEC4, clause 61.3 requires the contractor to notify the project manager of a compensation event within eight weeks after becoming aware of it. Refer to Section 8.8.4.2.
8 Particularly if Main Option Clauses C (Target Cost with Activity Schedule), D (Target Contract with Bill of Quantities) or E (Cost Reimbursable Contract) are used.

The corresponding obligations and procedures that may be 'stepped down' from the main contract to the subcontract are often as follows:

o Scope and quality of works
  – The subcontractor will be obliged to perform the relevant part of the main contract works to the standards and quality set out in the main contract. Ideally, the subcontractor should be able to identify (i) his works from the works of others from the drawings and specifications produced under the main contract and (ii) the interfaces between them.
  – The subcontractor may have equivalent obligations to identify discrepancies, errors, or inadequacies in the scope of work to the main contractor.
  – Right of objection[9] if the subcontractor considers that his design will be compromised by an instruction of the contractor.
  – Obligation to rectify parts of the subcontract works that are deemed non-compliant or defective under the main contract, and complete such rectification work so as not to cause the main contractor to be in breach of the main contract.
o Standard of design
  – Ordinarily the same standard as for the main contractor (usually exercising reasonable skill and care).
o Payment – The subcontractor may have to (i) apply for interim payments under the subcontract earlier than the main contractor applies for payment under the main contract, but (ii) wait slightly longer for payments under the subcontract than the main contractor has to wait for payments under the main contract:
  – To allow the main contractor to use the subcontractor's application in the payment application under the main contract.
  – To assist with the main contractor's cash flow (giving the main contractor a period between receiving the employer's payments and having to pay the subcontractors).
o Programme – The subcontractor is often obliged to carry out and complete the subcontract works:
  – By a completion date or series of completion dates set out in the subcontract.
  – In accordance with the programme included or described in the subcontract (this may be a range of dates, or particular activities for the subcontractor's work on the main contractor's programme).
  – In accordance with the subcontractor's programme, or
  – To maintain the reasonable progress of the main contract works.
o Carrying out the subcontract works (normally set out in the parts of the employer's requirements that are relevant to the subcontract works instead of in the subcontract conditions)
  – Complying with the employer's health and safety and logistical requirements.
  – Observing the employer's site procedures (such as dates and times for working, noise levels, etc.).
  – Contributing to and complying with the main contractor's risk assessments and method statements.

---

9 Equivalent to the main contractor's right of objection to an instruction changing the kinds or standards of materials or goods (JCT DB 2016, clause 3.5.1).

o Administrative procedures – The subcontractor may be obliged to:
  - Issue notices in a format that complies with the format in the main contract, so the main contractor can incorporate the subcontractor's notice within his own.
  - Notify the main contractor in sufficient time to allow the main contractor to comply with any stipulations as to time (including 'time bar' clauses) when providing his own notifications to the employer or the employer's representative.
  - Arrange for collateral warranties and bonds.
o Equivalent entitlements
  - Variations (or changes or compensation events) for the same reasons as under the main contract.
  - Extensions of time, and loss and expense (or compensation events) under the same circumstances as the main contractor becomes entitled under the main contract.
  - Protection from insurance policies under the main contract.
  - Ability to be present during inspections of the works (on or off site) and measurement.[10]
  - May terminate the subcontract if the main contract is terminated.
  - Limitations of liability.

6.2 Domestic Subcontractors

The majority of subcontractors are treated as domestic subcontractors and are selected and engaged by the main contractor. In terms of selection, it will be entirely the main contractor's choice who is appointed to carry out the relevant parts of the main contract works. The term 'domestic' subcontractor is often used to describe a contractual relationship that is entirely between the main contractor and the subcontractor; it is one which has no link (direct or otherwise) to the employer. Under the terms of the main contract, the main contractor remains responsible to the employer for the performance of a domestic subcontractor as though the main contractor had carried out the subcontracted works himself.

6.2.1 Domestic Subcontractor Selection

The main contractor is likely to have a list of 'tried and trusted' subcontractors for most trade disciplines; and in most cases, the subcontractor ultimately carrying out the relevant subcontract works will be chosen from this list. The list usually contains subcontractors with whom the main contractor already has a good working relationship and those with whom the main contractor would like to develop one. This latter category will often be asked to bid for work, and if their price is competitive, the main contractor will usually engage them at some stage. The main contractor's list is likely to be augmented by adding new subcontractors in the following circumstances:

o Specialist requirements. Specialist works may be required that the main contractor has not carried out before, thus the need to engage a specialist subcontractor.
o To maintain commercial competition. Adding a new subcontractor can 'freshen up' a tender list by encouraging competition between the subcontractors.

---

10 JCT DBSub/C 2016, clause 5.4.

o Operational reasons. Owing to existing commitments (market conditions), a number of the main contractor's regular subcontractors may be either unavailable or unable to meet the project's programme dates.

o Replacement. A subcontractor may become uncompetitive, perform poorly, or become insolvent, so a replacement will have to be found.

The criteria for selecting a subcontractor for a particular project are likely to vary, but the domestic subcontractor's price is usually the deciding factor. This is the simplest way for the main contractor to be profitable when the main contract will be let on a fixed-price basis.

There are also ever-growing sources of information from which a main contractor can appraise and approve a subcontractor. Organisations such as Constructionline provide main contractors and subcontractors with vetting information for one another. Also, credit reference agencies can provide valuable and up-to-date financial information, so the parties can enter into a subcontract having the benefit of information that's been obtained independently.

6.2.2 Employer's Approval of Subcontractors

The JCT DB 2016, FIDIC Yellow Book 2017, and NEC4 contracts all contain provisions for the employer (or the employer's representative) to approve any domestic subcontractors proposed by the main contractor. In general, these clauses are intended to prevent the main contractor from subcontracting the whole of the works (so the project is effectively carried out by another contractor), but they may impose additional restrictions. Under clause 4.4 of the FIDIC Yellow Book 2017, the employer may restrict the main contractor from subcontracting more than a certain percentage of the works' value[11] or from subcontracting certain parts of the works at all.[12] The criteria for subcontracting under the forms of main contracts are summarised in Table 6.2.

Table 6.2 Subcontracting criteria under JCT DB 2016, FIDIC Yellow Book 2017, and NEC4.

Subcontractor criteria	JCT DB 2016	FIDIC Yellow Book 2017	NEC4
Employer's/Employer's representative's consent required to subcontract	3.3.1	4.4	26.2
Contractor to provide subcontractor's contact details, detailed particulars, and relevant experience	—	4.4	26.2
Contractor to provide further information if requested	—	4.4	26.2
Contractor encouraged to use a particular form of subcontract	3.4	—	26.3
Statement from the main contractor and subcontractor that they will act in a spirit of mutual trust and cooperation	—	—	26.3

---

11 FIDIC Yellow Book 2017 – Particular Conditions Part A – Contract Data, entry for clause 4.4(a).
12 FIDIC Yellow Book 2017 – Particular Conditions Part A – Contract Data, entry for clause 4.4(b).

Somewhat unsurprisingly, JCT DB 2016 and NEC4 encourage the use of their own, corresponding subcontracts.[13] This not only makes practical sense for both the main contractor and subcontractor (because the subcontract will be written to facilitate the main contract works), but also commercial sense for JCT and NEC, as they will want to promote their own suite of contracts. The FIDIC Yellow Book 2017 does not encourage the use of the FIDIC Conditions of Subcontract for Plant Design-Build (First Edition 2019), but that is probably explained by the subcontract being drafted two years later than the main contract.

Clause 3.4 of JCT DB 2016 provides that the JCT subcontract should be used 'Where considered appropriate', whereas NEC4 (clause 26.3) goes further, by requiring the main contractor to provide the project manager with copies of the subcontract documents,[14] unless:

o The main contractor intends to use the NEC4 subcontract; and
o The proposed subcontract contains no amendments other than those in the Z clauses (additional conditions of contract) in the NEC4 main contract.

Whilst NEC4 appears to police the use of onerous or amended subcontracts (as the main contractor will be obliged to disclose copies), the use of onerous or amended terms in themselves do not appear to be grounds for the project manager not to accept the subcontract documents, unless they either do not contain the 'mutual trust and co-operation' statement, or their use will somehow prevent the main contractor from providing the works. Given that main contractors amend their subcontracts in order to exert greater control over the subcontractor and the subcontract works, it seems doubtful that subcontract amendments of this nature would be a valid ground for non-acceptance.

6.2.3 Administration of a Domestic Subcontract

The main contractor is left to administer a domestic subcontract without any further involvement (or perhaps, perceived interference) from the employer. He will be responsible for agreeing to the subcontract sum or price, along with the programme and method statement for the subcontract works. The main contractor is usually entirely responsible for controlling the subcontract works, so the subcontractor can only act upon receipt of the main contractor's instruction or direction.[15] The employer ordinarily does not see any evidence of either the subcontract price or any payments made to the subcontractor, unless they constitute 'Defined Cost' within the target cost options (C or D) of the NEC4 main contract.

In the event of a dispute, this is ordinarily between the main and subcontractor only, and if the subcontractor commits a serious default, then the main contractor may terminate the subcontractor's employment without having to either notify or gain the employer's consent to do so. Similarly, the subcontractor may terminate his employment if the main

---

13 JCT DB 2016 encourages the use of the Design and Build Subcontract Agreement 2016 (DBSub/A 2016), and NEC4 encourages the use of the NEC4 Engineering and Construction Subcontract (June 2017).
14 Save for the pricing information.
15 JCT DBSub/C 2016 – refer to clauses 3.4 and 3.5. FIDIC Conditions of Subcontract 2019 clauses 2.3 and 3.1, NEC4 Subcontract, clause 27.3.

contractor commits a serious default. The main contractor may also terminate the subcontractor's employment, if his employment under the main contract has been terminated.[16]

The main contractor may have obligations to notify the employer (or a beneficiary under a collateral warranty – see Section 7.2.1) if he's considering terminating a subcontractor's employment, but these are normally contained in separate agreements (such as collateral warranties between the subcontractor, main contractor, and the employer or another beneficiary) and not in the subcontract itself.

6.3 'Nominated' or 'Named' Subcontractors

The process whereby the employer (or someone acting on his behalf – such as the employer's representative or a member of the employer's design team) selects a particular subcontractor is, in itself, not overly controversial. In many ways, there seems to be little difference to specifying a particular product or material for the works, only extending the specification to include the design, workmanship, and products provided by a particular subcontractor. Initially, this appears to be a straightforward concept, allowing the employer a greater opportunity to discuss and decide upon what may be a central feature or a substantial element of the works. The employer and the subcontractor may be able to agree on matters of price and programme, and develop the employer's requirements into a detailed specification for the relevant work. However, the following problems can arise when the contractor becomes involved:

o The relationship between the contractor and the employer-selected subcontractor may be non-existent or very poor, so the contractor is either reluctant to work with the subcontractor or refuses to work with him.
o The main contractor may feel that the subcontractor is the 'employer's man', who has no allegiance to the contractor, and he may perceive a conflict of interest will arise.
o Owing to the prior relationship between himself and the employer, the subcontractor may perceive that he is entitled to preferential treatment from the main contractor.
o The subcontractor's intended method may conflict with the method in which the contractor proposes to design and/or construct the works.
o The programme for the subcontractor's works may have been established prior to the employer engaging the (main) contractor, so it may need to be adapted or completely revised if it's incompatible with the main contractor's programme.

These problems can be compounded if the contractual framework for the main contractor engaging the subcontractor is on terms that make the main contractor responsible for the employer-selected subcontractor's performance, or if it provides little recourse for the main contractor in the event that the subcontractor commits a breach or default.

6.3.1 Nominated or Named Subcontractor Selection

Historically, a number of main contracts provided for 'nominated subcontractors' to be selected by the employer. For engineering projects, this concept was used in the 1987

---

16 JCT DBSub/C, clause 7.9; FIDIC Subcontract 2019, clause 15.1; NEC4 Subcontract, clause 91.9.

and 1992 versions of the FIDIC *Conditions of Contract for Works of Civil Engineering Construction* (Red Book), and continued with the 1999 and 2017 versions of the FIDIC Yellow Book. Similarly, the last version of the ICE *Conditions of Contract,* (7th ed., 1999) provided for nominated subcontractors, and this continues with its current incarnation – the *Infrastructure Conditions of Contract 2014*[17] *Edition.*

For construction projects, editions of the JCT contracts between 1963 and 1998 all provided for the employer to nominate subcontractors. When the JCT revised the format for the main contracts in 2005, the concept of nominated subcontractors (as was it understood up to the 1998 version) was almost abandoned. The only part of the old regime that remained after 2005 was a somewhat diluted form of nomination called 'Named subcontractors' found in the Supplemental Provisions[18] to JCT D&B 2005 and in a similar but slightly more comprehensive selection procedure in the JCT Intermediate Building Contract 2005.[19] The 'Named subcontractor' procedures for the current versions of the JCT DB 2016 and for JCT Intermediate Building Contract with Contractor's Design 2016 ('JCT ICD 2016') remain materially the same as their 2005 predecessors.

The procedure for the employer when selecting the nominated or named subcontractor varies. The FIDIC Yellow Book 2017 includes two, relatively simple means of selection:[20]

o The nominated subcontractor is named in the Employer's Requirements; and
o The engineer can instruct[21] that the work included as a Provisional Sum is carried out by a nominated subcontractor.

JCT DB 2016 adopts a similar procedure to FIDIC, whereby the employer identifies the 'Named Subcontractor' in the Employer's Requirements.

The JCT ICD 2016 contains a much more formal selection process which in relation to JCT DB 2016 seems somewhat peculiar, as the former is generally perceived to be a simpler contract. The employer informs the contractor that he requires certain work to be undertaken by a Named Subcontractor by stating this in the Contract Bills, Specification, or Work Schedules.[22] However, the employer must have not only selected the 'Named Subcontractor' but also should have issued the formal tender documents (the JCT ICSub/NAM Intermediate Named Subcontract Tender and Agreement 2016)[23] to him. The ICSub/NAM 2016 documents inform the selected subcontractor that he will have to complete the Intermediate Named Subcontractor Agreement, which incorporates the Named Subcontract Conditions (JCT ICSub/NAM/C 2016). The Named Subcontractor procedure under JCT ICD 2016 seems to be a throwback to the nomination procedure used

---

17 CECA and the ACE revised the ICE contracts, now ICE focuses on the NEC4 contracts. The current version of what used to be the ICE contract is the Infrastructure Conditions of Contract ('ICC'). ACE also publishes a Design and Construct Version (June 2018) of the ICC.
18 JCT Design and Build 2005 – Schedule 2 Supplemental Provisions, paragraph 2 ('Persons named as subcontractors in Employer's Requirements').
19 JCT Intermediate Building Contract 2005 – refer to clause 3.7 and Schedule 2 (Named Subcontractors).
20 FIDIC Yellow Book 2017, clause 4.5.
21 FIDIC Yellow Book 2017, clause 13.
22 JCT ICD 2016, clause 3.7.
23 JCT ICD 2016, Schedule 2 (Named Subcontractors), paragraph 5. The Intermediate Named Subcontract Tender and Agreement 2016 comprises the Invitation to Tender (ICSub/NAM/IT), the Tender (ICSub/NAM/T), and the Agreement (ICSub/NAM/A).

in JCT contracts up to their 1998 revision. The JCT SBC 2016 may include the Contractor's Designed Portion,[24] but the use of 'Named Specialists' does not apply to contractor-designed work.[25]

NEC4 contains no equivalent procedure for the employer to select nominated or named subcontractors.[26] If the employer wants to select a particular subcontractor to carry out part of the work, then it appears that all he must do is include a clear reference to that subcontractor in the Scope. Additionally, the employer may want to introduce some Z clauses to assist with specifying a subcontractor in a similar way to the nominated or named process under JCT DB 2016 or JCT ICD 2016. This may include the introduction of 'provisional sums' for the relevant subcontract work[27] and how their expenditure is dealt with, along with any additional 'nomination' provisions that may be necessary. Additionally, some amendment to NEC4 clause 26.1 may be required, as it appears that all subcontractors[28] fall within the contractor's responsibility. Owing to the way that NEC4 operates, a procedure similar to nomination or naming may be possible without extensive amendments, particularly if the nominated or named subcontract work was not included in the Scope but was introduced by a project manager's instruction. The contractor could be entitled to compensation events under:

o 60.1(1) for a change in the Scope;
o 60.1(6) in the event of a subcontractor's termination or insolvency, and the project manager fails to re-nominate within the period for reply.

If the contractor was absolved of responsibility or the client was to accept responsibility[29] for the specified subcontractor, then the contractor may be entitled to the following compensation events:

o 60.1(18) a breach of contract by the client; or
o 60.1(19) if the subcontractor's default (as neither the fault of the client nor the contractor) prevented the contractor from completing by the planned completion date on the accepted programme.

Further, as NEC4 envisages that the main contractor will use the NEC4 subcontract for all subcontracted work (refer to Section 6.2.2 above), the conditions for engaging employer-selected subcontractors should not be controversial.

6.3.2 Contractor's Objection to Nominated or Named Subcontractors

Both FIDIC and the JCT contracts allow the contractor to object to the nomination or naming of a particular subcontractor, on specific grounds.

---

24 Refer to Article 9, JCT SBC 2016.
25 JCT SBC 2016, Schedule 8: Supplemental Provisions, paragraph 9 (Named Specialists).
26 Refer to *Managing Reality*, Book One (2018), p. 35.
27 Particular for NEC4 Option A (Priced Contract with Activity Schedule) – refer also to *Managing Reality*, Book One, 3rd ed., (2018), pp. 26–27.
28 NEC4, clause 11.2(19) defines 'Subcontractor', so a new definition could be introduced – such as 'Nominated Subcontractor' to distinguish from a clause 11.2(19) Subcontractor.
29 There would need to be some amendment to clause 26.1.

The grounds for objection in the FIDIC Yellow Book (clause 4.5.1) are far more specific than are set out in either JCT DB 2016 or JCT ICD 2016. FIDIC allows the contractor to object on the following grounds:

o That a nominated subcontractor may not have sufficient resources, competence, or financial stability.[30]

o The subcontractor will not have the same obligations and liabilities to the main contractor (under the subcontract) as the main contractor has to the employer under the main contract.[31]

o The subcontractor will not indemnify[32] the main contractor against:
  – Negligence,
  – Misuse of goods, or
  – Subcontractor's failure to perform its obligations and liabilities in accordance or in connection with the main contract.

o Other matters (although it is thought that they will have to be closely related to the above grounds – for example, the nominated subcontractor may not be able to comply with the contractor's programme, or its proposed method of working may conflict with the main contractor's method).

It appears that once the works are underway, the main contractor may only object to a nominated subcontractor that was instructed by the engineer, as under clause 4.5.1 of the FIDIC Yellow Book 2017, the main contractor can only give his notice of objection (with detailed particulars) not later than 14 days after the engineer's instruction. It would therefore be a prudent step for main contractor to thoroughly check the Employer's Requirements for nominated subcontractors before executing the main contract. The main contractor can either reject a proposed nominated subcontract or require an indemnity from the employer (see clause 4.5.1 and below) in order to engage one.

The FIDIC *Conditions of Subcontract for Plant Design and Build*, (1st ed., 2019) ('the FIDIC subcontract') expressly provide that the subcontractor has the same obligations and liabilities as the contractor[33] and the indemnities[34] referred to above. This offers the main contractor additional protection from the ordinary position under the Supply of Goods and Services Act 1982 (SGA82), whereby the contractor warrants that he has supplied materials of satisfactory quality,[35] even if those materials are selected by the employer. The contractor may have additional obligations to those implied by the SGA82 if expressly provided in the Employer's Requirements. If both the employer and contractor know that materials can only be provided by a nominated subcontractor, and the conditions under which they are provided limit the subcontractor's liability to the potential detriment and risk of the main contractor, then the main contractor's warranty for satisfactory quality may be reduced or excluded. However, the subcontractor's express obligations and indemnity to

---

30 FIDIC Yellow Book 2017, clause 4.5.1(a).
31 FIDIC Yellow Book 2017, clause 4.5.1(c)(i).
32 FIDIC Yellow Book 2017, clauses 4.5.1(b) and 4.5.1(c)(ii).
33 FIDIC Subcontract clause 2.2 (Compliance with Main Contract), and clause 5.3, which applies clause 5.3 of the FIDIC Yellow Book 2017 (Contractor's Undertaking) to the subcontractor.
34 FIDIC Subcontract clause 2.2 (Compliance with Main Contract) subject to clause 17.4 (Limitation of Liability) and clause 7.3 (Indemnity for Misuse).
35 Supply of Goods and Services Act 1982, sections 4(1) and 4(2).

the main contractor under the FIDIC subcontract should obviate the main contractor's risk in this regard.

FIDIC Yellow Book 2017 also provides the employer with two straightforward options if the contractor objects to nomination. Either he can accept the objection and the main contractor will not be obliged to employ a nominated subcontractor, or he provides an indemnity to the main contractor against the consequences of the subject matter of the main contractor's objection.

Under the JCT DB 2016 contract, the main contractor may object to a named subcontractor for 'any reason' provided that it is reasonable.[36] It is likely that reasonable reasons or grounds for objection will be similar to those set out in clause 4.5.1 of the FIDIC Yellow Book 2017. The main contractor must immediately notify the employer, who can then instruct accordingly. JCT ICD 2016 is perhaps a little more onerous, as the only ground for objection is limited to which of the particulars in the Contract Documents impedes the named subcontract from being executed.[37]

For JCT DB 2016 and JCT ICD 2016, the employer[38] has broadly the same options when issuing instructions following the main contractor's objection to a named subcontractor, and these are summarised in Table 6.3.

Table 6.3 Instructions for 'Named' subcontractors under JCT DB 2016 and JCT ICD 2016.

Nature of instruction	JCT DB 2016 (Schedule 2 para)	JCT ICD 2016 (Schedule 2 para)
Remove the grounds for objection	1.1.2.1	—
Change the particulars in the Contract Documents to remove the impediment to executing ICSub/NAM/A	—	2.1
Direct that the relevant work be carried out by either the main contractor or a domestic subcontractor	1.1.2.2	—
Omit the relevant work	1.1.2.3	2.2
Omit the relevant work and substitute a provisional sum	—	2.3

6.3.3 Risks Associated with Nominated or Named Subcontractors

Nominated or named subcontractors present risks for both the main contractor and the employer. The historic risks for the employer[39] were that firstly, the employer took on some of the responsibility for nominating the subcontractor in good time;[40] secondly, he could be liable to the contractor for some defaults by the nominated subcontractors;[41] and finally,

---

36 JCT DB 2016, Schedule 2 Supplemental Provisions, Part 1, paragraph 1.1.2.
37 JCT ICD 2016, Schedule 2 Named Subcontractors, paragraph 2.
38 The Architect/Contract Administrator for JCT ICD 2016.
39 Under the JCT 'nominated subcontractor' procedure up to the 1998 versions.
40 *Percy Bilton v Greater London Council*, 1 WLR 794, HL, [1982].
41 Under JCT 1998 Standard Form of Building Contract (Private with Quantities), a delay by a Nominated Subcontractor or Nominated Supplier was a Relevant Event (under clause 25.4.7), and if the architect/contract administrator failed to issue a clause 35.9.2 instruction within a reasonable time, the contractor may be entitled to damages.

without a collateral contract or warranty with the nominated subcontractor, the employer may have no direct remedy against him. These risks could arise when replacing a nominated subcontractor if the original subcontractor became insolvent, the main contractor had rightly determined his employment under the subcontract, or he had repudiated the subcontract.

Owing to the relatively simpler procedure under clause 4.5.1 of the FIDIC Yellow Book 2017, the only feasible delays in nomination could occur following an engineer's instruction to expend a provisional sum under clause 13.4, and either (i) the engineer rejecting the contractor's clause 4.5.1 objection (presumably, on the ground that it was not 'deemed reasonable'), or (ii) any delay in the employer agreeing to indemnify the contractor against the matters in clauses 4.5.1 (a) to (c). Such delays could entitle the contractor to an extension of time under clause 8.5(e).

JCT DB 2016 and JCT ICD 2016 deal with named subcontractor defaults in similar ways. For JCT DB 2016, the employer may instruct a change under paragraphs 1.1.2.1 to 1.1.2.3 of Schedule 2 (see Table 6.3 and Section 6.2 above), which may entitle the contractor to the following:

o An adjustment of the Contract Sum (see clause 5.2), including where the Named Subcontractor's employment has been correctly terminated[42] by the Contractor, and the Contractor elects to complete the balance of the Named Subcontractor's work himself,[43] although the Contractor is obliged to offset any monies recovered from the Named Subcontractor in this regard[44] from the costs of the Change to the Employer;
o An extension of time (under clauses 2.23–2.6); and
o Loss and Expense (under clauses 4.19–4.23).

In the event that the named subcontractor is employed under the JCT DBSub 2016 subcontract, the contractor may determine the named subcontractor's employment for any of 'specified' default or defaults in clause 7.4.1, provided that he complies with the procedure and timetable for issuing notices in clause 7.4.2 and those notices are issued to the named subcontractor in accordance with clause 1.7.4.[45] The contractor may not issue any of the clause 7.4.2 notices under JCT DBSub 2016 without the employer's consent[46] and must send a copy of each clause 7.4.2 notice to the employer.[47]

The above applies in the event of the named subcontractor repudiating the subcontract, whereupon the contractor must notify and consult with the employer before terminating the named subcontractor's employment. It should be noted that there is no express provision in clause 7 (termination) to deal with a repudiatory breach[48] (i.e. it is not one of the

---

42 Unless the termination is caused by the contractor (JCT DB 2016 Schedule 2 Supplemental Provisions, Part 1 Named Subcontractors, paragraph 1.4.2), then completing the balance of the Named Subcontractor's work is not treated as a Change.
43 JCT DB 2016 Schedule 2 Supplemental Provisions, Part 1 Named Subcontractors, paragraph 1.4.2.
44 JCT DB 2016 Schedule 2 Supplemental Provisions, Part 1 Named Subcontractors, paragraph 1.4.3.
45 JCT DBSub/C clause 1.7.4 provides such notices are delivered by Hand, by 'Recorded Signed for', or 'Special Delivery Post'.
46 JCT DB 2016 Schedule 2 Supplemental Provisions, Part 1 Named Subcontractors, paragraph 1.3.2.
47 JCT DB 2016 Schedule 2 Supplemental Provisions, Part 1 Named Subcontractors, paragraph 1.3.3.
48 Envisaged by JCT DB 2016 Schedule 2 Supplemental Provisions, Part 1 Named Subcontractors, paragraph 1.3.

'specified' default or defaults in clause 7.4.1). However, if the main contractor is thinking of terminating the subcontractor's employment, it would be advisable for the contractor to comply with the same procedure, timetable, and means of serving notices to the named subcontractor as he would otherwise have to do in the event of a 'specified' default (or defaults) under clause 7.4.1.

The procedure under JCT ICD 2016 for dealing with named subcontractor default, termination, and repudiation is a little lengthier than for JCT DB 2016, but nevertheless, it is in a similar vein. The main contractor can terminate the named subcontractor's employment if he commits a 'specified default', becomes insolvent or commits an act of corruption,[49] and provided he obtains the architect/contract administrator's consent. Additionally (and presumably to allow the architect/contract administrator to try and exert some influence over the named subcontractor in order to reduce or stop his default and avoid termination), the contractor should notify the architect/contract administrator of the events likely to entitle him to terminate the named subcontractor's employment as soon as he becomes reasonably aware of them. The main contractor should comply with termination notice provisions in clause 7 of the ICSub/NAM/C conditions (which are the same as for JCT DB 2016). If the subcontract works are uncompleted, the architect/contract administrator then instructs the main contractor to either engage another named subcontractor, carry out the relevant work himself, or omit the uncompleted balance of the named subcontractor's work. With the exception of instructing a new named subcontractor, the architect/contract administrator's instructions in this regard are Variations,[50] which may entitle the main contractor to an extension of time and loss and expense, along with adjusting the Contract Sum to reflect the cost of completing the relevant work. An instruction to employ a new named subcontractor is similar, save that it does not entitle the main contractor to loss and expense, nor will the main contractor be entitled to payment for the new subcontractor rectifying defects present in the previous named subcontractor's work.[51] As with JCT DB 2016, the main contractor is expected to take reasonable steps to recover any costs of the default from the previous named subcontractor, and to account to the employer for any amounts so recovered.[52] If the named subcontractor's employment is terminated for reasons other than in clauses 7.4, 7.5, or 7.6 of the ICSub/NAM/C conditions, or without the architect/contract administrator's consent, then the main contractor will be liable to repay the employer for the additional costs.[53]

6.3.4 Administration of a Nominated or Named Subcontract

Administering a nominated or named subcontract often requires the employer's involvement to some degree. Historically under JCT contracts, the architect/contract administrator would show amounts in respect of each amount due for nominated subcontractors' work separately on an Interim (payment) Certificate to the contractor.[54] For named

---

49 JCT ICD 2016 Schedule 2 paragraphs 6 and 8, referring to clause 7.4 (specified defaults), 7.5 (insolvency of Named Subcontractor) and 7.6 (Corruption) of the ICSub/NAM/C conditions 2016.
50 JCT ICD 2016, clause 5.1.
51 JCT ICD 2016 Schedule 2, paragraph 8.1.
52 JCT ICD 2016 Schedule 2, paragraph 10.2, but is not required to invoke the dispute resolution procedures in clause 8 of ICSub/NAM/C.
53 JCT ICD 2016 Schedule 2, paragraphs 10.1 and 10.2.4.
54 JCT SBC 1998, clause 35.13.1.

subcontractors under the JCT DB 2016 and JCT ICD contracts, this practice has been abandoned, so the employer (or his representative) has no visibility of payments made by the main contractor to the named subcontractor. The only exceptions would be in the event of the main contractor having to account to the employer for amounts recovered from the named subcontractor following termination.[55]

Clause 4.5.3 of the FIDIC Yellow Book 2017 contains a different procedure. Its focus seems to be more towards the employer reducing the likelihood of payment disputes between the main contractor and the nominated subcontractor, by (the engineer) requiring the main contractor to provide the following:

o Evidence that the nominated subcontractor has received payment of the amounts included in the Interim Payment Certificate[56] for his work, or
o An explanation of why the main contractor is entitled to withhold or deduct such amounts from those otherwise due to the nominated subcontractor, and that he has notified the nominated subcontractor of this entitlement.

If the main contractor cannot provide the above information, then the employer may pay the nominated subcontractor directly, less any amounts ('applicable deductions') for which the contractor provides supporting evidence. For each subsequent Interim Payment Certificate (and presumably, for as long as the employer deems that direct payments to the nominated subcontractor are appropriate, given the main contractor's failure to provide the aforementioned information), the engineer notifies the main contractor of amounts paid directly by the employer, and shows this as a deduction in the relevant certificate.

In terms of programme, clause 8.3(c) of FIDIC Yellow Book obliges the main contractor to show details of the nominated subcontractor's work on the programme (refer to Section 3.5.3 above). In practical terms, most main contractors are likely to produce a programme if JCT DB 2016 or JCT ICD 2016 are the main contract,[57] and this should show the named subcontractors' work if it is significant. As explained in Section 6.3.1 above, nominated subcontractors are not intended to be employed under NEC4. However, if the client did include provision for nominated or named subcontractors in the Scope and the Z clauses, the contractor's programme under NEC4 would have to show the nominated or named subcontractor's work.

Aside from the effects of terminating a nominated or named subcontractor under FIDIC Yellow Book 2017 or JCT DB 2016 (or JCT ICD 2016), the contractor evidencing payment to the nominated subcontractor or the employer paying the nominated subcontractor directly (both under FIDIC), the main contractor will administer the nominated or named subcontract in the same way as for a domestic subcontract.

6.3.5 Decline of Nominated or Named Subcontractors

The two main reasons for the decline in nominated or named subcontractors are (i) that the employer often has to take on additional risks, and (ii) he may need to participate in a

---

55 JCT DB 2016, Schedule 2 Supplemental Provisions Part 1, paragraph 1.4.3, and JCT ICD 2016 Schedule 2 Named Subcontractors, paragraph 10.2.2.
56 FIDIC Yellow Book 2017, clause 14.6.
57 The only reference to 'programme' is in the 'Cost savings and value improvements' Supplemental Provisions at Schedule 2, paragraph 7 of JCT DB 2016, and Schedule 5, paragraph 3 of JCT ICD 2016.

formal tendering procedure for the relevant subcontracted works that would otherwise be carried out by the main contractor.

The risks carried by the employer for nominating or naming a subcontractor seem unnecessary, when considering that he can specify that an element of work is undertaken by a particular subcontractor in the Employer's Requirements (JCT DB 2016[58] or FIDIC Yellow Book 2017) or in the Scope (NEC4). In the event that the main contractor does not use that specified subcontractor to carry out the relevant work, then under each of those contracts, he may face sanctions from the employer of varying severity. The employer may consider that failing to comply with the Employer's Requirements (or Scope) constitutes a defect, and may instruct the main contractor to replace the allegedly defective work[59] with work carried out by the specified subcontractor. Alternatively, the employer may seek to reduce (abate) the amount payable to the main contractor.[60] Failure to comply with the employer's instructions can lead to termination under JCT DB 2016, FIDIC Yellow Book 2017, and NEC4.[61]

By either simply 'specifying' a subcontractor or not approving a subcontractor (see Section 6.2.2 above), the employer may be able to exert a similar degree of control over the works as he would if he opted for a nominated or named subcontractor, whilst avoiding the risks associated with the following:

o Indemnifying the contractor.[62]
o Additional costs and delays for a nominated or named subcontractor's default[63]:
  - Extra-over costs of completing the previous named subcontractor's works,
  - Main contractor's loss and expense, and
  - Loss of liquidated (or delay) damages.

The employer can still select a particular subcontractor without having to go through the formal 12 tendering procedures required by contracts such as JCT ICD 2016. Unless the employer is familiar with the named subcontract tender procedures, it is likely that his advisors will have to complete the relevant documents on the employer's behalf, and he may incur the costs of doing so. These costs would be avoided if the relevant subcontractor was simply specified in the Employer's Requirements or Scope, as it would be the main contractor's responsibility to obtain prices and conclude a 'domestic' subcontract with him.

6.4 Early Subcontractor Involvement

Section 5.4 in Chapter 5 explored early contractor involvement from the perspectives of the employer and the contractor, highlighting the advantages and disadvantages. In general terms, the advantages and disadvantages for the employer and contractor can be respectively applied to the contractor and the subcontractor, albeit with some differences.

---

58 Or Contract Bills, Specification, or Work Schedules in JCT ICD 2016.
59 JCT DB 2016, clause 3.13.1; FIDIC Yellow Book 2017, clauses 7.5 and 7.6; and NEC4, clause 44.
60 JCT DB 2016, clause 2.35.2 and NEC4, clause 45.
61 JCT DB 2016, clause 8.4.1.3; FIDIC Yellow Book 2017, clause 11.4(d); and NEC4, clause 91.2.
62 FIDIC Yellow Book 2017, clause 4.5.1.
63 JCT DB 2016 and JCT ICD 2016. In the event that the employer indemnifies the contractor under clause 4.5.1 of FIDIC Yellow Book 2017, he may face similar costs.

When bidding for design and build projects, it is comparatively rare for a main contractor to assemble his bid purely from his own estimating and technical resources, without going to the subcontract market for pricing and technical input (e.g. during RIBA Stage 2 – Concept Design to Stage 4 – Detailed Design). Relative to an employer (see Section 5.4.5), the contractor may be exposed to a lesser risk of reduced competition as:

o It is generally accepted that main contractors will seek early (or tendering) involvement from a greater number of subcontractors, compared to the number of main contractors with whom an employer may seek early involvement.
o In terms of the commercial effect of reduced competition:
   – Many subcontractors clearly set out a period during which their quoted price can be accepted without adjustment, so the main contractor has a clear timeframe during which to conclude negotiations over the price; and
   – The main contractor may already have comparable (or 'benchmark') pricing information for similar subcontract works, so unless the relevant work is particularly unique or state of the art, then the subcontractor's price can easily be checked.

The early involvement between the main contractor and subcontractor will be more analogous to early involvement between an employer and main contractor, when the market conditions dictate that only a small number of subcontractors can carry out the relevant work. These conditions may arise in the following circumstances:

o The relevant subcontract work is unique. This may be for a variety of reasons:
   – A subcontractor may have sole intellectual property rights, or an exclusive licence to carry out his particular work; or
   – Few other subcontractors have the expertise to carry out the works.
o The subcontractor may be selected by the employer – as a 'nominated' or 'named' subcontractor (see Section 6.3 above).
o A subcontractor may be the only one available to meet the main contractor's programme or price; or
o Other subcontractors are unavailable because their resources are fully engaged for the relevant period.

7

Collateral Warranties, Third-Party Rights, Bonds, and Guarantees

7.1 General

Collateral warranties, third-party rights, and bonds are legal agreements in which either a party to a contract, or a third party (who is not a party to the contract) can obtain financial security or greater protection of their rights in respect to the building or structure that is the subject of the building contract between the employer and the contractor.

The requirement to provide a collateral warranty or bond, or to identify a party to whom third-party rights are given, will usually be a written condition of the particular contract. Under that contract (be it the main contract, a subcontract, or a contract for design), the party providing the works or its design will usually be obliged to provide the warranty or bond in the form appended to that contract to the employer or to a beneficiary.

7.1.1 Collateral Warranties – General Points

In contracts which are between two parties, where one of those parties either manufactures or constructs something and the other party is the initial purchaser, problems can arise when whatever is manufactured or constructed is passed on to a third party. By 'passed on', this means the benefit of the original contract (i.e. the building or structure itself) is either sold or leased to a third party. This is less common in engineering projects as the employer usually retains the works to operate them, so he will retain his rights against the contractor for the appropriate limitation period. However, the ability of third parties such as purchasers, tenants, or funders to take action against contractors in construction projects is commonplace. This is the problem caused by 'privity of contract', where a third party (who is not a party to the contract), cannot sue one of the parties to that contract in the event that the building or structure is defective because of their default.

The preceding paragraph considers a third party who is higher up the contractual chain; for example, someone to whom the employer sells or leases the building or from whom he obtains funding for the project. However, the same principle applies down the contractual chain, from the main contractor to the subcontractor. If the main contractor completes the works but then becomes insolvent, the employer ordinarily has no right of recourse against any of the subcontractors or designers in the event that the default causing the employer to suffer loss was the fault of one of those parties.

Design and Build Contracts, First Edition. Guy Higginbottom.
© 2024 John Wiley & Sons Ltd. Published 2024 by John Wiley & Sons Ltd.

The 'privity of contract' hurdle can be overcome by a collateral warranty. In this case, 'collateral' means 'additional, but subordinate' or 'secondary',[1] and warranty means 'a written guarantee[1]' or a promise. The 'additional', 'subordinate', or 'secondary' element refers to the collateral warranty making reference to the principal contract, such as the main contract (between the employer and main contractor), a subcontract (between the main contractor and a subcontractor), or a contract for design (between the main contractor or a subcontractor, and a designer). The 'guarantee' or 'promise' is made to the third party by a party to the principal contract, the promise being that it has properly performed its obligations under the principal contract. This allows the third party (the beneficiary) to sue, not under the principal contract, but under the promise (i.e. the warranty) given by the party making that promise (the warrantor or promisor).

7.1.2 Third-Party Rights – General Points

The 'privity of contract' rule, and the problems emanating from it, was considered by the Law Commission in 1996 when it published its report entitled 'Privity of Contract: Contracts for the Benefit of Third Parties'.[2] Its introduction contained the following passage from Steyn LJ's judgement in *Darlington Borough Council v Wiltshier Northern Ltd*,[3] which summarised the criticism that the law (as it then was) faced regarding third parties:

> 'The case for recognising a contract for the benefit of a third party is simple and straight-forward. The autonomy of the will of the parties should be respected. The law of contract should give effect to the reasonable expectations of contracting parties. Principle certainly requires that a burden should not be imposed on a third party without his consent. But there is no doctrinal, logical, or policy reason why the law should deny effectiveness to a contract for the benefit of a third party where that is the expressed intention of the parties. Moreover, often the parties, and particularly third parties, organise their affairs on the faith of the contract. They rely on the contract. It is therefore unjust to deny effectiveness to such a contract. I will not struggle with the point further since nobody seriously asserts the contrary.'

Parliament enacted the Contract (Rights of Third Parties) Act 1999 (CRoPTA), which filled the void described by Steyn LJ. CRoPTA provided that the parties to a contract could, when agreeing to that contract, include provisions to allow third parties to gain rights under that contract – in the same way as if the third party was a beneficiary under a collateral warranty. It does this in quite a straightforward way, so the third party must be identified in the contract, either by name, or 'as a member of a class answering a particular description'.[4] The third party does not have to be in existence when the principal contract is entered into, so in the example where the employer will sell or transfer the building to a purchaser or landlord, the purchaser or landlord can be set up after the employer and the contractor agree to the contract. Commonly used 'classes' of third parties are financial organisations (providing funds for the project), purchasers, operators, landlords, and tenants.

---

1 *Oxford Dictionary of English* (2010).
2 EWLC 242, (1996).
3 1 WLR 68, [1995], at paragraph 78.
4 Contract (Rights of Third Parties) Act 1999, ss 1(1) and 1(3).

7.1.3   Bonds and Guarantees – General Points

A bond is a contract of guarantee, which in some ways operates in a similar way to an insurance policy, although there are clear differences. Unlike collateral warranties or third-party rights, a bond doesn't provide a third party with rights of action under or in connection with a contract. Instead, it provides that party with financial protection up to an agreed limit, either in the event that a party to a contract is in default or under another express term of the bond. The financial security is usually limited (often to 10% of the contract value) and provided by the bondsman or surety. The party seeking the bond's financial security is usually the employer, who pays the contractor to take out the bond. In a bond where the surety has guaranteed the main contractor's performance to the employer, he will be liable to the employer for losses up to the agreed limit that were caused by the contractor's default. Similarly, a parent company guarantee (PCG) is taken out where one of the parties to the contract is a subsidiary of another, larger company. The larger (or parent) company may have all the working capital (cash) and assets, so the smaller company (the party to the contract) may appear unlikely to fulfil its liabilities to the other party under the contract. The other contracting party may seek that the PCG covers the performance or meets the liabilities of the smaller company in the event of the smaller company's default. There are several different types of bonds commonly used in construction and engineering projects:

o Bond to provide security for performance
  – Performance bond
  – On-demand bond
  – PCG
o Security for payment
  – Advance payment bond
  – Retention bond
  – Bond for payment of off-site goods and materials
  – PCG

7.2   Collateral Warranties

Collateral warranties are used to create a contractual link between one party to a contract (who is carrying out the works or its design) and a third party. The collateral warranty is normally used to create a contractual 'bridge' between two parties who are otherwise contractually separate by a single tier, such as between:

o An operator, funder, purchaser, or tenant, with the main contractor (and his designer);
o An employer and a subcontractor (and his designer); or
o A main contractor and a sub-subcontractor.

However, if a lower-tier party is providing a significant or substantial part of the works, he also may have to provide a warranty to a third party in a higher contractual tier, such as:

o A subcontractor providing a warranty to an operator, funder, purchaser, or tenant; or
o A sub-subcontractor providing a warranty to the employer.

7.2.1 Collateral Warranties Under the Main and Subcontracts

The three design and build contracts deal with collateral warranties in very different ways.

JCT DB 2016 not only includes clear obligations (in clause 7.4 – Rights Particulars[5]) for the contractor to provide a collateral warranty, but also provides two, standard forms of warranty: one for purchasers or tenants (CWa/P&T), and the other for funders (CWa/F).

Whilst both JCT warranties contain similar core obligations, there are distinct obligations that apply to the warranty for a purchaser or tenant and for a funder. A warranty for a purchaser or tenant is likely to focus on rectifying defects in the building once it's completed and to cover any ancillary costs incurred by the purchase or tenant whilst the rectification works are carried out. A funder is likely to be more concerned with completing the building to protect his investment. In the event that the employer becomes insolvent or has committed a breach that entitles the contractor to terminate his employment under the building contract, most warranties for funders allow the funder an opportunity to intervene; and even take over ('step in'), in place of the employer.

The rights and obligations under the JCT collateral warranties are as follows:

o Core obligations
   – The contractor's, subcontractor's, or designer's promises (to the purchaser, tenant, or funder) that has complied with the building contract with the employer and has only used and will only use permitted materials[6];
   – The relevant subcontractors have also provided warranties to the purchaser, tenant, or funder;
   – The contractor's liability is limited by a net contribution clause[7] and may be further limited to a specific figure. Further, the contractor has no liability to the purchaser, tenant, or funder for a delay in completing the works[8];
   – The purchaser, tenant, or funder has no authority to instruct the contractor under the building contract[9];
   – Any design documents prepared by the contractor may be used by the purchaser, tenant, or funder[10];
   – The contractor must maintain professional indemnity insurance[11];
   – If the warranty is executed under hand or by deed, the limitation periods are 6 and 12 years, respectively, so no action can be brought against the contractor after these periods expire,[12] and no action can be brought by a third party[13];
   – The warranty may be assigned[14];

---

5 And its corresponding entry in the Contract Particulars. Clause 7C relates to the warranty for purchasers and tenants, and clause 7D relates to the funder's warranty.
6 CWa/P&T clause 1.1 and clause 2, CWa/F clause 1 and clause 2.
7 CWa/P&T clause 1.3 – with the limitation of liability capped to 'Maximum liability' in the Warranty Particulars, CWa/F clause 1.1.
8 CWa/P&T clause 9, CWa/F clause 13 (save where the funder 'steps in' to replace the employer under clauses 5 or 6.4).
9 CWa/P&T and CWa/F clause 3.
10 CWa/P&T clause 4, CWa/F clause 8.
11 CWa/P&T clause 5, CWa/F clause 9.
12 CWa/P&T clause 8, CWa/F clause 12.
13 CWa/P&T clause 10, CWa/F clause 14.
14 CWa/P&T clause 6, CWa/F clause 10.

– Notices to and from the contractor and the purchaser, tenant, or funder must be sent by 'Recorded Signed For' or 'Special Delivery' post, or delivered by hand[15]; and

– The law of England applies, and the English Courts have jurisdiction.[16]

o The warranty for the purchaser or tenant (CWa/P&T) provides that the contractor is only liable for the 'reasonable costs' or remedial work and any other losses incurred by the purchaser or tenant up to the maximum (if stated[7]).

o The funder warranty (CWa/F) contains provisions for the funder to 'step in' to replace the employer, if (i) the funder cancels the funding agreement with the employer, or (ii) the contractor is entitled to terminate his employment under the building contract. The warranty provides that:

– If the funder terminates the funding agreement and steps in to replace the employer under the building contract, the contractor shall accept instructions from the funder (or from the funder's nominee) in this regard, and the employer cannot claim that accepting such instructions is breach of the building contract by the contractor; or

– If the contractor is in the process of terminating his employment under the building contract[17] because of an employer default, he must copy the required termination notices (required by the building contractor to effect termination) to the funder, and shall not treat his employment as terminated (or the contract as repudiated) for 14 days from the funder's receipt of the termination notice. This is to allow the funder an opportunity help resolve the employer's default so the building contract continues without the funder stepping in; and

– If the funder does step in and replace the employer, he (or his representative) becomes liable to the contractor for paying sums that were due and payable at the time the funder notified the contractor that he was 'stepping in'.

JCT subcontracts also provide for collateral warranties. Clause 2.26 of the subcontract (DBSub/C) contains similar provisions to clause 7.4 of JCT DB 2016, whereby the subcontractor is obliged to execute the collateral warranties as listed in the 'Numbered Documents'.[18] If the entry in the Numbered Documents is blank, the applicable warranty could be bespoke, although JCT publishes the following 2016 editions of collateral warranties for subcontractors:

o JCT Subcontractor Collateral Warranty for a Funder (SCWa/F)

o JCT Subcontractor Collateral Warranty for a Purchase or Tenant (SCWa/P&T)

o JCT Subcontractor Collateral Warranty for Employer (SCWa/E)

Perhaps owing to its more 'international' nature where it could be used in a number of different legal jurisdictions, the FIDIC Yellow Book 2017 does not contain any reference to the use of collateral warranties. FIDIC does permit modifications to be included in the 'Special Provisions',[19] but encourages that they comply with the 'FIDIC Golden Principles'

---

15 CWa/P&T clause 7, CWa/F clause 11.

16 CWa/P&T clause 11, CWa/F clause 15.

17 Or accepting that the employer has repudiated the building contract.

18 Refer to the JCT Subcontract Agreement (DBSub/A), section 18 of the Subcontract Particulars.

19 FIDIC Yellow Book 2017, 'Special Provisions' (defined in clause 1.1.78), which is incorporated as 'Part B of the Particular Conditions'.

(set out in the 'Particular Conditions Part B – Special Provisions'). FIDIC Yellow Book 2017 – General Conditions could be modified to provide for collateral warranties as follows:

o Introducing an enabling clause, obliging the contractor, his designer, and a subcontractor to provide the relevant collateral warranty (e.g. in the form[s] appended to or included with the Special Provisions) within a prescribed period after either the Commencement Date[20] or after receiving a notice from the employer or engineer; and

o Providing the relevant forms of collateral warranty within the Special Provisions. If the parties agree to use collateral warranties containing similar wording to the various JCT warranties (CWa/P&T, CWA/F and SCWa/E, SCWa/F and SCWa/P&T), then this should satisfy the FIDIC Golden Principles.

Similar modifications can be included in the corresponding FIDIC conditions of subcontract and the FIDIC Client/Consultant Model Services Agreement, 5th ed., (2017).

NEC4 is similar to FIDIC Yellow Book 2017 as it doesn't provide any forms of collateral warranty itself. Given NEC4's clear intention to provide an aligned suite of contracts to suit all contractual relationships involved with a project, this seems unusual, and may be something the authors of NEC4 will address in the future (see Section 12.5 in Chapter 12). However, NEC4 does provide an enabling clause in Option X8, whereby the main contractor is obliged to give an 'undertaking to Others' which can include providing a collateral warranty if set out in Contract Data Part One. The NEC4 subcontract and the NEC4 PSC also contain a similar Option X8, so the main contractor's, subcontractor's, and consultant's (or designer's) obligations are summarised below:

o NEC4 Contract Data (main contract) Option X8
  – Contractor provides a warranty (i.e. the undertaking) to others.
  – Subcontractor provides a warranty to others and/to the client.
o NEC4 Subcontract Data Option X8
  – Subcontractor provides a warranty to others and/or to the client.
o NEC4 Professional Services Contract (PSC) Option X8
  – Consultant provides a warranty to others (who may include the client).

As could be the case with warranties for use with the FIDIC Yellow Book 2017 (and its corresponding subcontracts and consultant model service agreement), similar wording to the JCT collateral warranties could be adopted for collateral warranties provided under NEC4 Option X8 as undertakings to the others (and to the client).

7.2.2   Collateral Warranties for Designers

Designers engaged by the design and build contractor and subcontractors are often obliged to provide collateral warranties to third parties in a higher tier in the contractual chain. This usually follows a similar pattern to that described in Section 7.2 but between the following parties:

o An operator, funder, purchaser, or tenant, with the main contractor's designer;
o The employer and the main contractor's designer (if not novated – see Section 12.2);

---

20 FIDIC Yellow Book 2017, 'Commencement Date' is defined in clause 1.1.6.

o The employer and a subcontractor's designer; and

o The main contractor with a subcontractor's designer or a sub-subcontractor's designer.

The conditions of engagement for consultants (discussed in Chapter 4) deal with collateral warranties as follows:

o RIBA Standard PSC 2020
  – Clause 4.4 obliges the designer to provide a collateral warranty within 14 days of the client's request. The RIBA PSC[21] envisages that the designer will execute collateral warranties in favour of the main contractor, employer, purchaser, tenant, or funder, either in the CIC (Construction Industry Council) form,[22] or in a bespoke form appended to the conditions of engagement.
o FIDIC Client / Consultant Model Services Agreement, 5th ed., (2017)
  – In its briefing note from 2019,[23] FIDIC discouraged the use of collateral warranties between consulting engineers and project financiers.
o Association of Consulting Engineers (ACE) Professional Services Agreement 2017
  – Clause 14 of the ACE PSA provides that the designer shall provide collateral warranties to the third parties and in the forms either appended or referred to in the schedule.
o NEC4 PSC (June 2017 with Jan 2019 amendments)
  – NEC4 PSC includes Option X8 whereby the designer is to provide an undertaking to others, which can include collateral warranties as described in the PSC Contract Data.

The rights and obligations under designers' warranties are very similar in terms to those in the various collateral warranties provided by a main contractor or a subcontractor. Warranties provided by the designer to the third parties in a higher tier (e.g. the employer, funder, or main contractor), often allow them to 'step in' to replace the contractor (or subcontractor), in the event that the designer intends to terminate his appointment or accept it as repudiated. Contractors or subcontractors who engage designers may have similar obligations in their warranties with third parties in a higher contractual tier – not to terminate their designers' appointment without informing the third parties.

7.3 Third-Party Rights

Under the Contract (Rights of Third Parties) Act 1999 ('CRoTPA'), the parties to a contract can agree to bestow rights under that contract to a third party.[24] As explained in Section 7.1.2, it is neither essential that the third party (or parties) be identified by name as they can be identified by class[25] (such as purchaser, tenant, or funder), nor does any third party have to be in existence at the date when the contracting parties agree to bestow such rights.

---

21 In 'Item M' – Supplementary Rights.
22 CIC Collateral Warranties – Employer (2018), Purchaser (2018), or Funder (2018).
23 Collateral Warranties, FIDIC Briefing Note (June 2019).
24 CRoTPA s1(1).
25 CRoTPA s1(3).

JCT DB 2016 provides for third-party rights[26] to be conferred on those parties identified in the 'Rights Particulars',[27] with further details provided in Parts 1 and 2 of Schedule 5 as follows:

o Part 1: Third-Party Rights for Purchasers and Tenants
o Part 2: Third-Party Rights for a Funder

Importantly, Schedule 5 of JCT DB 2016 provides a purchaser or tenant or a funder with equivalent rights as they would have under their respective collateral warranties[28] (refer to Section 7.2.1), which are additional to rights that a third party would have if they were purely bestowed rights under CRoTPA. The corresponding subcontract (DBSub/A and DBSub/C 2016) includes equivalent third-party rights to purchasers, tenants, and funders,[29] but with the addition of employer rights.[30]

Therefore, it appears that under JCT DB 2016 and its subcontract, a third party will have equivalent third-party rights under clause 7.4.1 (and Schedule 5) as it would otherwise have under the respective collateral warranty, but without the need to execute the warranties themselves. This alleviates the difficulties experienced by both the employer and contractor in drafting collateral warranties and ensuring they're correctly signed and returned. Considering the relative ease of bestowing equivalent rights on third parties, it is surprising that the use of collateral warranties (including the JCT warranties) remains so widespread.

FIDIC Yellow Book 2017 does not contain any provision for third-party rights. This may be because the CRoTPA is UK legislation, and FIDIC contracts are intended to be used internationally. However, FIDIC does contemplate 'Modifications' to the General Conditions,[31] to account for 'local requirements', such as a particular law of the country identified in the Contract Data.[32] A clearly worded 'Special Provision' dealing with third-party rights could be easily incorporated into the FIDIC Yellow Book 2017.

NEC4 contains third-party rights in 'Option Y(UK)3: The Contract (Rights of Third Parties) Act 1999', which is found in NEC4's main contract, subcontract, and designer's contract.[33] In all cases, the third party is identified in Contract Data Part One as the 'beneficiary' along with the 'terms' from which he may enforce.[34] It therefore appears that the client (when completing Contract Data Part One) may select any particular terms that a third party may enforce. At its simplest, the client is likely to select clause 20 (Providing the Works) and 21 (the Contractor's Design) to be enforced by a third party.

---

26 JCT DB 2016, clause 7.4.1.
27 JCT DB 2016, 'Rights Particulars' are as described in the Contract Particulars' entry for clause 7.4.
28 JCT warranty for a Purchaser or Tenant is the CWa/P&T, and for a funder is CWa/F.
29 JCT DBSub/C 2016, clause 2.26, ad Schedule 6; Part 1 (P&T Rights) and Part 2 (Funder Rights).
30 JCT DBSub/C, Schedule 6: Part 3 (Employer Rights), which are equivalent to the SCWa/E collateral warranty.
31 Clauses 1 to 21 of FIDIC Yellow Book 2017.
32 FIDIC Yellow Book 2017, clause 1.4 (Law and Language).
33 NEC4 PSC.
34 NEC4 Y(UK)3, clause Y3.1.

7.4   Bonds and Guarantees

The three design and build contracts provide for various types of bonds and guarantees as summarised in Table 7.1.

Table 7.1   Bonds and guarantees under JCT DB 2016, FIDIC Yellow Book 2017, and NEC4.

Description of bond or guarantee	JCT DB 2016	FIDIC Yellow Book 2017	NEC4
(a) Advance payment bond	Schedule 6, Part 1	Annex E	–
(b) Off-site materials/goods bond	Schedule 6, Part 2	–	–
(c) Retention bond	Schedule 6, Part 3	Annex F	–
(d) Performance bond	–	Annex C, Annex D	–
(e) Employer's payment guarantee	–	Annex G	–
(f) Parent company guarantee	–	Annex A	–
(g) Tender security	–	Annex B	–
Enabling clauses	(a) 4.6 (b) 4.15 (c) 4.17	(a) 14.2.1 (c) SP[a)]14.9 (d) and (f) 4.2 (e) SP14 (g) N/A	(a) Option X14 (d) Option X13 (f) Option X4

a)   FIDIC Yellow Book 2017 – 'SP' denotes the clause is a suggested modification of the General Conditions and is found in the Special Provisions.

The relevant enabling clauses for each contract are cross referenced to the type of bond or guarantee described in rows (a)–(g) of Table 7.1.

7.4.1   Security for Performance

Employers and contractors often seek additional protection from the others' breaches of contract by entering into separate agreements with third parties such as the parent companies of either of the parties or a surety (or bondsman).

Performance bonds or guarantees are commonly used for projects that are let under JCT DB 2016, FIDIC Yellow Book 2017, and NEC4 contracts. From Table 7.1, it is clear that the FIDIC Yellow Book provides the most comprehensive suite of bonds or guarantees within its pages, but that should not give a false impression regarding the use of bonds or guarantees for JCT DB 2016 or NEC4 projects. For many construction projects let under JCT DB 2016, and particularly where the employer is a private developer, they will often amend the contract to oblige the contractor to provide performance bonds and PCGs that are not drafted by JCT. NEC4 adopts a similar approach insofar as it does not provide a sample NEC4 PCG, performance bond, or advance payment bond. However, it does provide for the contractor to execute such documents in Options X4, X13, and X14, respectively. The contractor could also be obliged to provide additional forms of bond or guarantee as an 'undertaking to Others' in Option X8 (in the same way as providing a collateral warranty – refer to Section 7.2.1). The guarantee (in Option X4) and bonds (in Options X13 and X14) must all be accepted by the project manager. This could cause the project manager difficulty because if he is providing financial advice himself, he may need to be regulated by the Financial Conduct Authority (FCA). Therefore, it is advisable that the project manager and the client agree that a suitably qualified third party (such as a FCA regulated consultant, solicitor, or insurance broker) accept the PCG, performance bond, or advanced payment bond.

In the case of NEC4, copies of all applicable bonds or guarantees should be appended to the Scope, whereas for JCT DB 2017, any bonds or guarantees in addition to those in Schedule 6 will have to be incorporated into the contract by reference or be appended to the contract documents.

There are two main types of bonds to give additional security for performance; and in the majority of cases, it is the employer seeking the security for the contractor's performance. These types of bonds are:

○ Performance bonds
○ On-demand bonds

A performance bond provides the beneficiary (typically, the employer) with additional security in the form of a payment from the surety (or bondsman), in the event that the contractor commits a default under the terms of the building contract. These types of bond are also referred to as 'conditional' bonds. In this example, the surety will be liable to the beneficiary for losses arising from the contractor's default, up to the amount set out in the performance bond, which is often up to 10% of the contract price. Depending upon the terms of the performance bond, the surety's liability may cease once practical completion (or taking over) has taken place, or may reduce by 50%, until the rectification or defects period has expired. Again, depending upon the terms of the bond, the surety may be liable to the beneficiary for the contractor breaching any terms of the building contract, or only for prescribed breaches (such as failure to complete on time). In order to make a claim under the bond (often referred to as 'calling the bond'), the beneficiary may have to notify the surety within a specific period after the contractor's default, and also in a specific manner (such as by delivering the notice by hand or sending it by recorded post).

On-demand bonds are much rarer, as they are far more onerous for the surety. Subject to its wording, an on-demand bond may be payable by the surety to the beneficiary even if the contractor has not committed a default; however, the beneficiary will normally have to provide the surety with particular documentation required by the bond.

Most bonds for construction or engineering projects in the UK are performance (or conditional) bonds. Surprisingly, in the author's experience, sureties don't always insist upon the terms of their own bonds, so they will often underwrite the terms of a bond drafted by the employer's legal team.

FIDIC Yellow Book contains two differing forms of performance security. The entries in the Contract Data (against clause 4.2) set out in what amounts, and in what currencies, the contractor must provide the performance security to the employer. If the Contract Data contains no entries against clause 4.2, then clause 4.2 does not apply,[35] and the contractor will not be obliged to provide any performance security. Annex C of the FIDIC Conditions contains an example 'Demand Guarantee'. The employer (as beneficiary) must submit a demand in writing to the surety, along with a written statement indicating what breach of the building contract the contractor has committed. This latter requirement seems to put the Annex C example halfway between a performance bond and an on-demand bond, as on one hand, the beneficiary must provide some evidence of a breach by the contractor, but on the other, this may only have to be an 'indication' and not firm evidence of a breach. The Annex C example requires the parties to agree upon the maximum amount payable by the surety to the beneficiary ('the Guaranteed Amount'), along with a provision to increase or decrease this amount, in proportion with any increase or decrease of the Accepted Contract Amount. Annex C expresses the Guaranteed Amount in figures and words and not as a percentage of the Accepted Contract Amount. This is not wholly consistent with the Contract Data entry performance security (clause 4.2), which clearly expresses the amount of performance security as a percentage of the Accepted Contract Amount. The parties should therefore check that the 'Guarantee Amount' expressed in the Annex C Demand Guarantee precisely equates to the relevant percentage of the Accepted Contract Amount that is set out in the Contract Data. Any adjustments to the Guaranteed Amount have to be evidenced by the engineer.[36] Claims under the Annex C example must include authenticated copies of the beneficiary's signatures and can be made up to 70 days after the Defects Notification Period, whereafter, the Annex C guarantee shall expire.

Annex D to the FIDIC Yellow Book 2017 contains a 'Surety Bond' example. This has all the characteristics of a 'performance bond', as the surety's liability to the beneficiary must be firstly triggered by the contractor's default[37] or any of the circumstances set out in clause 15.2.1.[38] It does not appear that the beneficiary has to comply with any notification requirements (such as the notice being in writing) save that any claim must be made before the surety bond expires. Similar to the Annex C example, the surety's liability (the 'Bond Amount' in the Annex D example) is expressed in figures and words, so again, the parties should check that the 'Bond Amount' is the same as the percentage of the 'Accepted Contract Amount' as set out in the Contract Data. The parties can agree upon the amount (expressed as a percentage) by which the Bond Amount reduces upon Taking Over. Most sureties would expect a 50% reduction once works are complete and the contractor's major remaining obligation is rectifying defects. The Annex D Surety Bond becomes effective upon the Commencement Date,[39] and the surety's liability expires six months after the Defects Notification Period, so this provides a slightly shorter duration than the Annex C example.

---

35 FIDIC Yellow Book 2017, clause 4.2 (Performance Security).
36 FIDIC Yellow Book 2017, clause 4.2.1.
37 Expressed in the Annex D example as '…Default by the Principal' (who is the contractor).
38 The circumstances entitling the employer to issue its clause 15.2.1 Notice – set out in clause 15.2.1 (a) to (h).
39 There is no stated commencement date for the Annex C example.

A PCG is slightly different from a bond, insofar as it is (usually) the contractor's parent company that guarantees the contractor's performance and agrees to indemnify the employer against any losses arising from the contractor's default. Unless the contractor's liability to the employer is limited under the building contract,[40] it is unusual to find any similar limitations of liability in a PCG, in the same way that a surety's liability is limited to a maximum amount under a performance bond. The principal protection a PCG offers to the employer is in the event of the contractor's insolvency; then the parent company can meet any liabilities that the employer incurs. The PCG example in Annex A of the FIDIC Yellow Book 2017 is fairly typical in these respects.

NEC4 refers to the PCG as the 'Ultimate Holding Company Guarantee' but it has the same meaning as FIDIC Yellow Book 2017's Annex A. The form of PCG is provided in the Scope unless the contractor proposes an alternative (which must be accepted by the project manager). The project manager may reject the PCG if the parent (or ultimate holding) company's financial position is too weak to bear the PCG. As stated in Section 7.4.1, it is doubtful whether most construction professionals acting as project manager have sufficient expertise to adequately assess the financial position of a company, so the project manager and client should agree that acceptance of the PCG should be carried out by an appropriately qualified third party.

7.4.2 Security for Payment

Both JCT DB 2016 and FIDIC Yellow Book 2017 include measures which provide additional security for payment (principally for the employer), as both contracts provide bonds in respect of advance payment[41] and retention.[42] JCT DB 2016 also provides a bond for payment of materials or goods off site,[43] and FIDIC Yellow Book 2017 contains an employer's payment guarantee.[44]

The advance payment bonds for both JCT DB 2016 and FIDIC Yellow Book 2017 operate in similar ways. Where the employer makes an advance payment to the contractor (e.g. as a deposit for expensive long lead-in items or to assist with a costly site establishment), the contractor will repay the advance payment to the employer through deductions from interim payments. In both cases, the contractor must provide the advance payment bond to the employer before the advance payment is made. Under JCT DB 2016, the employer makes the advance payment to the contractor on the date set out in the Contract Particulars,[45] whereas the FIDIC Yellow Book 2017 provides that the employer makes the advance payment after receiving the Advance Payment Certificate from the engineer.[46] The engineer is only permitted to issue the Advance Payment Certificate after the employer has received

---

40 FIDIC Yellow Book 2017, clause 1.15 (Limitation of Liability).

41 JCT DB 2016 – Schedule 6, Part 1 and FIDIC Yellow Book 2017 – Annex E.

42 JCT DB 2016 – Schedule 6, Part 3 and FIDIC Yellow Book 2017 – Annex F.

43 JCT DB 2016 – Schedule 6, Part 2.

44 FIDIC Yellow Book 2017 – Annex G.

45 JCT DB 2016, clause 4.6. The advance payment may not always be issued in advance of the contractor commencing work; it may be issued in advance of starting a particular element of the work (e.g. midway through the contract period).

46 FIDIC Yellow Book 2017, clauses 14.2 and 14.2.2. The 'Advance Payment Certificate' is defined in clause 1.1.2.

both the Performance Security and the Advance Payment Guarantee[47] from the contractor and the ngineer has received the contractor's advance payment application.[48] Both the JCT and FIDIC contracts provide for the employer and contractor to agree the amount of each repayment instalment, albeit JCT DB 2016 enables the parties to agree the dates when repayments are made.[49] The FIDIC Yellow Book 2017 contains default procedures[50] in the event that the advance payment entries aren't completed in the Contract Data, which is something that is missing from JCT DB 2016. FIDIC's default procedure applies when the total of all interim payments exceeds 10% of the Accepted Contract Amount; then 25% of each interim payment is deducted to repay the employer until the advance payment is repaid. For advance payments under JCT DB 2016 and FIDIC Yellow Book 2017, the surety will pay to the employer the amount of the advance payment that remains unpaid, until his liability under the bond expires.[51] Both bonds stipulate how the employer must make a claim for the repayment. The JCT DB 2016 bond includes a 'Schedule' which sets out the form of the employer's claim, and the bond in FIDIC Yellow Book 2017 requires the employer to provide a written statement,[52] with two signatures that must be authenticated.

NEC4 Option X14 provides for the option of an advanced payment bond, but as stated above, it will have to be drafted by the employer or the surety,[53] as no standard form of bond is issued by NEC4. The bond must be in a form that is acceptable to the project manager (see Section 7.4.1). Once the bond is provided, the client makes the advance payment to the contractor at the next 'assessment date'.

FIDIC Yellow Book 2017 also contains an option[54] whereby a guarantor will pay the contractor any amount that the employer has failed to pay. This 'reverse' payment guarantee may seem unusual in the context of UK projects, but considering that FIDIC is intended for international use and the employers may include emerging economies, aid agencies, or organisations arranging emergency work, they may be less financially secure. The 'Payment Guarantee' requires the guarantor (such as a bank, fund, or government), to pay the contractor any amounts up to the total stated in the guarantee. In order for the guarantor to pay, the contractor must submit a written demand, containing authenticated signatures, stating that the employer has failed to pay within 14 days of the period set out in the contract the amount that the employer has failed to pay, along with the evidence[55] of the contractor's entitlement to be paid the sum claimed under the guarantee.

NEC4 provides for an advance payment bond under Option X14, but as stated above, it does not set out the form of such a bond.

---

47 FIDIC Yellow Book 2017, clause 14.2.2(a).
48 FIDIC Yellow Book 2017, clause 14.2.2(b).
49 JCT DB 2016, Contract Particulars entry for clause 4.6.
50 FIDIC Yellow Book 2017, clause 14.2.3.
51 JCT DB 2016 – when the advanced payment is repaid either by the contractor or by the surety, or a date agreed by the parties. FIDIC Yellow Book 2017 – when the advance payment is repaid by the contractor, or a date not later than 70 days after the Time for Completion.
52 FIDIC Yellow Book 2017.
53 NEC4 Option X14 refers to 'bank' or 'insurer' instead of the more general 'surety'.
54 Annex G – Example Form of Payment Guarantee.
55 The Annex G example provides for the parties to agree on what evidence is required (e.g. an Interim Payment Certificate – or in the UK, a contractor's payment notice complying with s110B[2] of the Housing Grants Construction and Regeneration Act 1996, as amended).

7.4.3 Retention Bonds

JCT DB 2016 and FIDIC Yellow Book 2017 both contain provisions for retention bonds, but each deals with the issue of retention in a different way. JCT DB 2016 contains the option[56] for the employer to obtain a bond instead of deducting retention from payments due to the contractor. The retention bond provides the employer with security for the retention deductions he would otherwise have been entitled to retain in the event of the contractor's default under the building contract.

If the JCT DB 2016 retention bond option is selected, the contractor must firstly provide the retention bond in the form set out in Schedule 6, Part 3 in accordance with clause 4.17.2, in order to disapply the retention deduction clauses 4.14 and 4.18. If the contractor provides the retention bond after one or more interim valuations, then the retention deducted under those interim payments becomes due to the contractor on the next interim valuation after he provides the bond. Once the contractor obtains the retention bond, the surety promises to pay to the employer the amount of retention that would otherwise have been deducted. In order to make a claim (or call) on the bond, the employer has to submit a written demand to the surety that meets the detailed requirements[57] of clause 4 of the retention bond, with authenticated signatures. The surety's liability firstly reduces by 50% upon practical completion, and then ceases on issue of the Notice of Completion of Making Good Defects or on satisfaction of the bond. If the employer makes its demand without the documents required by clause 4 of the retention bond, then the bond will terminate.

The retention bond under FIDIC Yellow Book 2017 provides less of a benefit to the contractor than under its JCT DB 2016 equivalent. If the parties choose the FIDIC retention bond, they are encouraged to incorporate clause 14.9 from the Special Provisions.[58] This provides that once the amount of retention retained by the employer reaches 60% of the limit of Retention Money,[59] the employer releases 50% of the limit of retention money to the contractor. This may leave the employer only holding 10% of the limit of retention money, which may be insufficient if the contractor fails to rectify defects. The retention bond protects the employer by the surety guaranteeing to pay the employer the amount of retention (50% of limit of retention money) that the employer has paid to the contractor. The employer's demand must contain authenticated signatures, and state which obligations the contractor failed to perform. The surety's liability will end 70 days after the Defects Notification Period (as the employer cannot make a demand after this date), or after paying the retention to the employer.

---

56 JCT DB 2016, Contract Particulars entry for clause 4.17.
57 Schedule 6, Part 3: Retention Bond. Clause 4 requires the employer to provide the demand to the surety's office, and include details of (i) the amount of retention that would otherwise be deducted, (ii) the amount demanded, (iii) the costs incurred by the employer, (iv) any insurance premiums paid by the employer, (v) any liquidated damages to be paid by the contractor, (vi) any of the employer's expenses or loss or damage arising from him terminating the contractor's employment, (vii) any other costs that the employer is entitled to deduct, and (viii) a statement that the employer has claimed the amount in (ii) above from the contractor, but the contractor has not paid within 14 days.
58 Special Provisions are contained in 'Particular Conditions B' of FIDIC Yellow Book 2017.
59 FIDIC Yellow Book 2017 Contract Data entry for clause 14.3(iii).

7.4.4 Off-Site Materials/Goods Bond

Clause 4.15 of JCT DB 2016 allows for the value of materials or goods (Listed Items) to be included in an interim or final payment to the contractor[60] when those materials or goods have been neither included in the cumulative value of stages completed (payment Alternative A – clause 4.12.1.1) nor within the work properly executed (payment Alternative B – clause 4.13.1.1). The Contract Particulars' entries for clause 4.15 provides two categories of Listed Items – those that are uniquely identified[61] and those that are not[62] – and includes their respective values. The descriptions of the listed items falling into each of category are listed and annexed to the Employer's Requirements.[63] Examples of uniquely identified items may be large or complicated items of plant such as a generator, air-conditioning units, or a building management system. Listed items that may not be uniquely identified could include materials for finishes, such as marble.

Under the terms of the bond in Schedule 6, Part 2, the surety is liable to the employer for the values of each category of Listed Items (as set out in the Contract Particulars) in the event that the employer includes their values in a payment, but the contractor fails to deliver them to or adjacent to the works. In order to make the claim, the employer must issue a 'Notice of Demand' in the form set out in the 'Schedule to the Bond', which must include the employer's authenticated signatures. The surety must make payment to the employer within five business days of receiving the employer's Notice of Demand, irrespective of whether the employer has agreed to waive any terms of the contract, has amended the terms of the contract, or has granted an extension of time to the contractor.[64] The surety's liability under the bond shall cease on the earlier of either (i) the date when the listed items are delivered by the contractor, or (ii) a longstop date agreed on by the parties.

7.4.5 Other Bonds

FIDIC Yellow Book 2017 also contains a form of 'tender security' at Annex B. The tender security is a guarantee or a 'bid bond' that is intended to compensate the employer in the event that the successful tenderer either (i) withdraws his offer without prior agreement, (ii) fails to correct any errors in his offer, (iii) doesn't sign the Contract Agreement within 35 days after receiving the Letter of Acceptance,[65] or (iv) fails to provide the Performance Security.[66]

Tender security or bid bonds are comparatively rare in the UK but are often used for international projects. One of the reasons for their rarity in UK projects could be the increase in projects being awarded under frameworks, the conditions of which prevent the circumstances described in (i)–(iv) (in the previous paragraph) arising. Many contractors expend

---

60 Included within the Gross Valuation under clause 4.12.3 (Alternative A – Stage Payments) or clause 4.13.3 (monthly Interim Payments).
61 JCT DB 2016, refer to the Contract Particulars entry for clause 4.15.4.
62 JCT DB 2016, refer to the Contract Particulars entry for clause 4.15.5.
63 JCT DB 2016, 'Listed Items' definition in clause 1.1.
64 JCT DB 2016 Schedule 6, Part 2, clause 7 of the bond.
65 FIDIC Yellow Book 2017, clause 1.6 (Contract Agreement).
66 FIDIC Yellow Book 2017, clause 4.2.

significant money and resources in securing a place on frameworks, so it's unlikely they will commit a default under the framework's terms.

However, tender security may be attractive in the private developer market. Insisting upon a bond that effectively guarantees that the tendering contractor will comply with the tender conditions may provide developers with a shorter and smoother period between concluding the tender process and entering into the building contract. Obtaining a bid bond may alleviate any last minute increases to the tender sum, or late attempts by one party to change the risk allocation in the building contract.

The 'tender security' in Annex B of FIDIC Yellow Book 2017 has similar requirements to the other bonds or guarantees to those described in Sections 7.4.1–7.4.4. The employer (also referred to as the beneficiary) must submit a written demand for the sum set out in the bond, and this must contain his authenticated signature. The demand must also contain a statement of which of the circumstances in (i)–(iv) above, where the tenderer is in default, and be submitted not later than 35 days after the Letter of Tender's[67] period of validity expires.

---

67 FIDIC Yellow Book 2007, 'Letter of Tender' defined in clause 1.1.51. The period of validity is the period for which the tenderer confirms (in the Letter of Tender) that the offer remains valid.

8

Construction

8.1 General

This chapter examines how the three design and build contracts provide for the activities generally required for RIBA Stage 5 (Manufacturing and Construction). It will assume that the preceding stages, such as RIBA Stage 4 (Technical Design), are sufficiently completed to allow the works to proceed without any significant interruptions or adverse effects on progress. There may be some activities that overlap with the subsequent RIBA Stage 6 (Handover), such as completion in sections or partial possession (occurring whilst the remaining RIBA Stage 5 works continue).

8.1.1 Pre-start Obligations

The pre-start obligations for the contractor and employer are considered as steps leading up to the building contract being formally agreed upon and the contractor commencing the physical work. They generally fall into the following categories:

o Contract agreement and performance security obligations
o Financial obligations
o Design obligations

8.1.2 Contract Agreement and Performance Security Obligations

There aren't any obligations to 'agree' to in the contract under JCT DB 2016 or NEC4 before formally executing the contract. For both, the contract documentation is prepared and the specific information[1] is completed (refer to Section 2.2.2), although unlike JCT DB 2016 (where the parties sign the 'Attestation' that is printed in the contract), NEC4 leaves the method of execution up to the parties (refer to Section 2.2.2).

---

1 JCT DB 2016 – The Recitals, Articles, Contract Particulars along with the Employer's Requirements, Contractor's Proposals, and Contract Sum Analysis; NEC4 – Contract Data Parts One and Two, the Scope, the Programme, and the activity schedule (Option A or C), bill of quantities (Option B or D), and Schedule of Cost Components.

Design and Build Contracts, First Edition. Guy Higginbottom.
© 2024 John Wiley & Sons Ltd. Published 2024 by John Wiley & Sons Ltd.

Clause 1.6 of the FIDIC Yellow Book 2017 requires the parties to sign the 'Contract Agreement' within 35 days after the contractor has received the 'Letter of Acceptance'[2] (the employer's acceptance of the contractor's tender). The date of receiving this letter is important. It triggers the 28-day period[3] within which the contractor is obliged to provide the 'Performance Security'[4] to the employer, as well as triggering other obligations (see Section 8.2.2). The provision of Performance Security (under clause 4.2) is very important as it also triggers the contractor's entitlement to receive an Interim Payment Certificate (or IPC) from the engineer (see clause 14.6[a] and Section 8.7.2 in this chapter). Clearly, FIDIC intends the performance security to be in place at least one week before the contract is executed. If the employer does not issue the Letter of Acceptance, FIDIC requires the contractor to provide the performance security within 28 days of signing the Contract Agreement[2].

8.1.3 Financial Obligations

The financial obligations will depend upon what the parties have agreed to. Under JCT DB 2016, the parties may have agreed that in return for the contractor providing the appropriate bonds,[5] the employer is to make an advance payment, will pay the contractor for off-site materials or goods, or will not deduct retention. All of these agreements are set out in the relevant entries in the Contract Particulars (see Chapter 7[6]). The advance payment bond and the off-site materials/goods bond may not always be required either before or at the start of the contract. For example, the advance payment may be required to secure a manufacturing slot or to pay for a particularly expensive piece of equipment that may not be required for some time after the project has commenced. JCT only envisages that, ideally, the contractor provides the retention bond before the works start[7] (i.e. before the Date of Possession) but allows it to be provided later, albeit the contractor will lose some of the bond's benefit, as the employer is entitled to deduct retention until the retention bond is provided.[8]

The mechanism for providing the advance payment under the FIDIC Yellow Book 2017 can only commence once the engineer has provided the Advance Payment Certificate (see Section 7.4.2), which he can only issue once the contractor has provided the Performance Security (if it applies – see Section 7.4.1 and above) and the Advance Payment Guarantee (see Section 7.4.2). If the employer is obliged to provide the Payment Guarantee[9] to the contractor, this must be provided within 28 days after the parties have entered into the Contract Agreement.[10]

---

2 FIDIC Yellow Book 2017, clause 1.6. 'Letter of Acceptance' is defined in clause 1.1.50.
3 FIDIC Yellow Book 2017, clause 4.2.1.
4 FIDIC Yellow Book 2017, 'Performance Security' as set out in the Contract Data against sentry 4.2. This may include the parent company guarantee and performance bonds (at Annexures A, C, and D).
5 JCT DB 2016, Schedule 6: Part 1 (Advance Payment Bond), Part 2 (Off-site Materials/Goods Bond), or Part 3 (Retention Bond).
6 JCT DB 2016: Advance Payment Bond (see Section 7.4.2), Off-site Goods/Materials Bond (see Section 7.4.4), and Retention Bond (see Section 7.4.3).
7 JCT DB 2016, clause 4.17.2.
8 JCT DB 2016, clause 4.17.3.
9 FIDIC Yellow Book 2017, Annex G 'Example form of Payment Guarantee by Employer'.
10 FIDIC Yellow Book 2017, Particular Conditions Part B – Special Provisions, clause 14.

8.1.4 Design Obligations

The contractor may be obliged to submit his design for the employer (and/or his representative) to review. Section 3.5.2 provides details of the review procedures for JCT DB 2016, FIDIC Yellow Book 2017, and NEC4. For JCT and FIDIC, the employer only reviews the design documents that are either not contained in the Contractor's Proposals (JCT DB 2016[11]) or are stated to be for review in the Employer's Requirements (FIDIC Yellow Book 2017[12]). If the design for early work activities (e.g. piling or foundation design) falls into these categories, then the contractor may be prevented from commencing critical work until the relevant design has been approved. Therefore, the contractor should allow at least 14 days (JCT DB 2016) or 21 days (FIDIC Yellow Book 2017) for the design to be approved. Under FIDIC, the 21-day period may be concurrent with the 35-day period for the parties to sign the Contract Agreement; but for JCT, the 14-day period may, in practical terms, be the first activity (e.g. foundation design approval) on the contractor's programme.

Under NEC4, it appears that all the contractor's design must be submitted to the project manager for approval (refer to Sections 3.5.2 and 3.3.3). The design can be submitted in its entirety (clause 21.2), although the more realistic provision in clause 21.3 allows the contractor to progressively submit design for approval, as long as each part can be fully assessed.

8.2 Commencement

The date upon which the contractor commences work (and usually, commences work on site), is very important as it generally signifies the date from which the contractor's major obligations begin. The most obvious will be the remaining time in which the contractor has to complete the design and construction of the works. Other obligations, such as the contractor's obligation to insure the works, may also begin on this date. The three design and build contracts deal with commencement in slightly different ways.

8.2.1 JCT DB 2016

The parties to a JCT DB 2016 contract are expected to have agreed the commencement date (called the 'Date of Possession[13]') and included it in the Contract Particulars, in the entry against clause 2.3. Clause 2.3 itself provides that on the Date of Possession, the employer shall give possession of the site to the contractor, who shall then commence the physical works. If the work is to be completed in sections,[14] the entries for the 'Dates of Possession of Sections' (again, against clause 2.3) in the Contract Particulars must be completed. The description of the sections is set out in the entry for the 'Fifth Recital' in the Contract Particulars.

---

11 JCT DB 2016, Schedule 1 (Design Submission Procedure).
12 FIDIC Yellow Book 2017, clause 5.2.2.
13 JCT DB 2016, 'Date of Possession' is defined in clause 1.1 (Definitions).
14 JCT DB 2016, Fifth Recital and corresponding entry in the Contract Particulars.

The Date of Possession is also relevant to the following:

o Clause 2.4 (Deferment of Possession). The employer may delay giving possession of the site to the contractor if the Contract Particulars state that clause 2.4 applies.[15]

o Clause 4.7.2 (Interim Payment – Interim Valuation Dates). If the first 'Interim Valuation Date' is not set out in the Contract Particulars, then by default, the first Interim Valuation Date is one month after Date of Possession.

o Clause 4.17.2 (Retention Bond). Clause 4.17.2 encourages the contractor to provide the retention bond on or before the Date of Possession.

o Clause 6.7.2 (Insurance). Whilst the Date of Possession is not specifically mentioned, the period of insurance in clause 6.7.2 (applying to either the employer or the contractor in Insurance Options A, B, or C[16]) appears to begin on the Date of Possession.

o Clause 8.9.2 (Termination). The start of the contractor's entitlement to terminate employment if the employer suspends performance for longer than the period stated in the Contract Particulars[17] (entry for clause 8.9.2).

8.2.2 FIDIC Yellow Book 2017

Under clause 8.1 (Commencement of the Works) of the FIDIC Yellow Book 2017, the engineer must notify the contractor when he may commence the works (the Commencement Date[18]). The engineer's notice shall be issued not less than 14 days prior to the Commencement Date which, unless the Contract Data states otherwise, shall be within 42 days after the contractor receives the Letter of Acceptance. Clause 2.1 provides that the employer shall grant the contractor access and possession of the site within the period stated in the Contract Data.[19] This period is measured from the time the contractor receives the Letter of Acceptance. As the Contract Data contains no entry against clause 8.1, it appears that the period stated in the Contract Data (the clause 2.1 entry), should expire on the Commencement Date. The parties should check to see that these dates align before signing the Contract Agreement. FIDIC does contemplate[20] the contractor having early access to the site (e.g. to carry out site investigations), and suggests that if early access is required, this should be set out in the Employer's Requirements. However, given the commencement complexities[21] imposed by the FIDIC Yellow Book 2017, it may be simpler if any necessary investigative work is dealt with as either:

o A separate (or enabling) contract, such as under a PCSA (see Section 5.4.5 above), either by a third party or by the contractor, so any investigations are completed prior to the Commencement Date; or

---

15 JCT DB 2016, The Contract Particulars entry against clause 2.4, the employer may defer for a maximum period of six weeks.
16 JCT DB 2016: Schedule 3, Insurance Option A (New Buildings – All Risks Insurance of the Works by the Contractor), Insurance Option B (New Buildings – All Risks Insurance of the Works by the Employer), and Insurance Option C (Joint Names Insurance by the Employer of Existing Structures in or Extensions to them).
17 JCT DB 2016, Contract Particulars entry for clause 8.9.2 (if no period is stated, the period of suspension is two months).
18 FIDIC Yellow Book 2017, 'Commencement Date' is defined in clause 1.1.6.
19 FIDIC Yellow Book 2017, Contract Data entry for clause 2.1.
20 FIDIC Yellow Book 2017, Special Provisions – Clause 2.1 (Right of Access to the Site).
21 Compared to JCT DB 2016 and NEC4.

o Included in the Employer's Requirements as the Works' initial activities (i.e. after the Commencement Date but preceding the majority of the Works).

These steps should avoid the difficulties created by giving early obligations to the contractor before the Contract Agreement is signed.

The contractor is also obliged to appoint the 'Contractor's Representative' before the Commencement Date unless he is already named in the Contract (clause 4.3). Curiously, the Contract Data does not contain any entry against clause 4.3 (Contractor's Representative), so the contractor should ensure that he either appoints his representative in good time, or that he is named in the contract (perhaps, by a manuscript addition under clause 4.3 in the Contract Data). The consequences of failing to appoint the Contractor's Representative are onerous, as the Employer is not obliged to make any payment to the contractor, unless he has done so (see Section 8.7.2).

A summary of the FIDIC Yellow Book 2017s obligations taking place prior to the Commencement Date is shown in Figure 8.1.

Whilst FIDIC Yellow Book 2017 provides for the works to be completed in sections[22] (see Section 8.3.2 below), it does not appear to provide for separate commencement dates for each section. On this basis, it seems that the contractor may commence work on any section on or after the Commencement Date, and will not breach his obligations to complete the works if he completes each section on or before the relevant sections' Time for Completion has expired.

8.2.3 NEC4

NEC4 provides two distinct dates on which the contractor is able to commence elements of the work. The first is the 'starting date', and it seems clear that the latter will be the

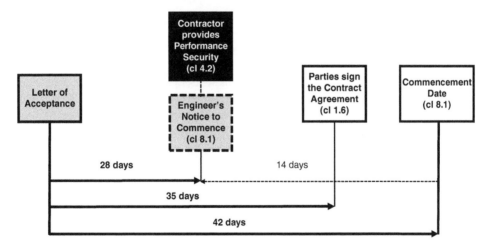

Figure 8.1 Pre-commencement stages under the FIDIC Yellow Book 2017.

---

22 FIDIC Yellow Book 2017, 'Sections' defined in clause 1.1.76, with further definition of 'Sections' in the Contract Data Part A, namely (i) a description of each Section, (ii) the Section's value (as a percentage of the Accepted Contract Amount), (iii) the Section's 'Time for Completion', and (iv) the Delay Damages for each Section.

first 'access date'. The distinction between the two is that the contractor may commence any other work after the starting date, whereas he is not permitted to start work on the site until the first of the access dates. Such 'other work' may include off-site tasks such as preparing the design (the 'Part Two Scope'[23]) for the project manager's approval (under clause 21.2) and procuring long lead-in items, with the following obligations commencing after the starting date:

o Early warning risk register issued and first early warning meeting.[24]
o Activities carried out at the intervals stated in the Contract Data:
  – Revising the programme,[25]
  – Providing forecasts of defined costs[26] (under Options C, D, E, and F).
o Providing insurance,[27] including professional indemnity under Option X15 (the contractor's design[28]).

Reading the NEC4 contract as a whole indicates that the starting date is intended to occur before the first access date. This allows the contractor to carry out the obligations mentioned above, and by virtue of clause 50.1 (Assessing the Amount Due), the contractor is entitled to be paid for carrying out certain obligations that occur after the starting date but before the first access date when the contractor may commence work on site.

Contract Data Part One (Section '3 Time') provides for several access dates for particular parts of the site. This provides an access date for each part of the works in the following circumstances:

o When the contractor will progressively be given access to parts of the works but must complete the whole of the works on or before the completion date; or
o Where Option X5 (Sectional Completion) is selected, access dates can be allocated to each of the sections, as only the sections' completion dates are provided in Contract Data Part One (against the entries for Option X5).

8.3   Completion in Sections (Subdividing the Whole Works)

Not every engineering or construction project benefits from a single completion date, and employers may want greater control over the progress of work and the sequence in which it is carried out. This may be for many reasons. The employer may have a particular funding arrangement, where funds are provided after particular stages of the project have begun or are completed. It may be logical to complete a project in a particular order. For example, a project consisting of several accommodation blocks may be completed sequentially, so the employer can earn revenue from progressively letting or selling each block as soon as it is completed instead of having to wait for them all to be finished.

---

23 Refer to Section 3.3.3.
24 NEC4, clause 15.2.
25 NEC4, clause 32.2.
26 NEC4, clause 20.4 for Option C (Target Contract with Activity Schedule), Option D (Target Contract with Bill of Quantities), Option E (Cost Reimbursable Contract), and Option F (Management Contract).
27 NEC4, clauses 83 and 84 (insurance by the contractor), clause 86 (insurance by the client).
28 NEC4, Option X15, clauses X15.5 and X15.6.

The project may also benefit from the contractor completing particular activities by a specific date. This may include a statutory body approving part of the works, or achieving a particular level of performance or function. These specific dates are often referred to as 'Milestones' (or 'Key Dates' under NEC4).

It is essential that *all* of the works are contained within the sections, and no part of the works remains unallocated. The format of these descriptions will be dictated by how intricately the works (as a whole) are subdivided into sections. If the subdivision is relatively straightforward, a simple written description should suffice. For example, a construction project consisting of two office blocks and surrounding external areas may be subdivided into sections as follows:

o Section 1 – Office block 1
o Section 2 – Office block 2
o Section 3 – External works and landscaping

For more complicated subdivision, a written description may not provide the required clarity, so annotated drawings should be used that clearly show which parts of the works are contained within each section. This allows for more intricate subdivision. Two examples are as follows:

o For a rail project, certain infrastructure diversions (such as the re-routing of roads and pipelines) could comprise one section, with the other section being the construction of the railway line itself.
o For a building refurbishment project, specific rooms could form part of one section, common areas (such as stairways and corridors) could form another section, and any external areas form the final section.

If delay damages are applicable to the sections (in the same way as they would be to a project with a single completion date), then it is essential that the sections are properly described.[29]

8.3.1   JCT DB 2016

JCT DB 2016 allows the works to be divided into sections. In order to operate correctly, the contract requires the following information to be provided (Table 8.1).

JCT DB 2016 effectively treats each section of the works as a 'mini-project', so the same provisions relating to the matters in Table 8.1 (which would otherwise apply to a project where the works are not subdivided) apply to each section. By allowing the employer to defer possession, apply different rates of liquidated damages, take partial possession, and include separate rectification periods for each section, the employer has considerably more flexibility and control over the sequence in which the contractor must complete the works. The relevant clauses in the Conditions (i.e. those dealing with the date of possession, dates for completion, liquidated damages, etc.) are drafted to also account for the works being completed by sections.

---

29 *Taylor Woodrow Holdings Ltd v Barnes & Elliott Ltd*, EWHC 3319 (TCC), [2004].

Table 8.1 JCT DB 2016 entries for completion by sections.

Information required for each section	Relevant contract entry
o Statement that works are divided into sections	Agreement: Fifth Recital[a]
o A clear description of each section	Contract Particulars: Fifth Recital
o The date of possession	Contract Particulars: clause 2.3
o The dates for completion of sections	Contract Particulars: clause 1.1
o Maximum periods of deferment of possession	Contract Particulars: clause 2.4
o Rates of liquidated damages	Contract Particulars: clause 2.29.2
o 'Section Sum' – the value of each section (for apportioning liquidated damages where the employer has taken partial possession of a section)	Contract Particulars: clause 2.34
o The rectification period for each section	Contract Particulars: clause 2.35

a) JCT DB 2016 – If the works are not to be divided into sections, the Fifth Recital is to be deleted.

JCT DB 2016 contains no provision for the contractor to complete separate milestones (e.g. in the same manner that NEC4 provides for Key Dates – see Section 8.3.3), so any individual tasks will have to be included as sections within the Contract Particulars.

8.3.2 FIDIC Yellow Book 2017

FIDIC Yellow Book 2017 also divides the works into sections,[30] although it does not provide an immediately obvious indication that the whole of the works will be subdivided in the same way that JCT DB 2016 does with the Fifth Recital (see Section 8.3.1 above). The parties will have to go looking, as the only entry is found in 'Particular Conditions Part A – Contract Data', and that is buried right at the end[31] of the contract documents.

The 'Contract Data' entry is in the form of a neat table, which sets out the following criteria for the sections:

o The description of the section.
o The value of work for that section[32] expressed as a percentage of the Accepted Contract Amount for the purposes of calculating the retention to be released on its completion.
o The Time for Completion[33] of the section.
o The Delay Damages[34] that apply to that section.

The table is certainly too small to allow sufficiently comprehensive descriptions of the sections (the entries in the column for 'Subclause 1.1.76'), so it is likely that it will have to be reproduced in an expanded form to allow detailed descriptions of the work forming each section to be given.

---

30 FIDIC Yellow Book 2017, clause 1.1.76 defines 'Section' and refers to the Contract Data.
31 On the last page before 'Particular Conditions Part B – Special Provisions'.
32 FIDIC Yellow Book 2017, clause 14.9 (Release of Retention Money).
33 FIDIC Yellow Book 2017, clause 1.186 defines 'Time for Completion'.
34 FIDIC Yellow Book 2017, clause 8.8 (Delay Damages).

Similar to JCT DB 2016, the relevant clauses in FIDIC Yellow Book 2017 that deal with commencement, completion, retention, delay damages, and so on are all written to also allow for the works to be completed in sections. It appears that clause 10.2 (Taking Over Parts) can be operated in an equivalent way as 'partial possession' under JCT DB 2016, as the Taking Over Certificate reduces the contractor's liability for the 'taken over' part, and clause 10.2 also provides for the apportionment of Delay Damages.

Unlike JCT DB 2016, sectional completion under the FIDIC Yellow Book 2017 does not appear to provide for the following:

o A separate 'Commencement Date' for each section (see Section 8.2.2 above);
o Separate periods for 'deferment' of possession for each section[35]; or
o Defects Notification Periods that differ for each section.[36]

FIDIC Yellow Book 2017 does not provide for milestones within the General Conditions, but does include an optional or 'Additional Subclause – Milestones' (4.24) within the Special Provisions part of the Guidance section. This provides additional entries in the Contract Data, new definitions of 'Milestone' and 'Milestone Certificate' (in clause 1.1) and the operative provisions for the contractor achieving the milestones. The Contract Data entries for 'Milestones' provide a description of the milestone, the period of time (expressed in days) in which the contractor must complete the milestone, and the delay damages (expressed as a percentage of the final Contract Price to be levied for each day of overrun).

As well as being a specific part of the works, a milestone can be an item of 'Plant.'[37] This makes good practical sense, as the project may benefit from having defined milestone dates for matters such commissioning, obtaining permanent power, testing an important component of the works, or provision of important documents. The milestone provisions operate in a similar way to the 'sectional completion' in that each one is effectively a mini-project. So, for example, each programme (clause 8.3) must show the milestone, and the 'Advance Warning' (clause 8.4) and 'Extension of Time for Completion' (clause 8.5) provisions also apply to the milestones. FIDIC seems to envisage that the milestone may need to accommodate fairly substantial work, and this is reflected in the periods for the contractor to give notice to the engineer for completing the milestone (14 days prior to when he thinks the milestone will be complete), and for the engineer to issue the 'Milestone Certificate' (within 28 days after the contractor's notice).

8.3.3 NEC4

NEC4 provides for Section Completion in secondary option X5. The first part (1 – General) of Contract Data Part One will set out which secondary options apply, so this will alert the parties to the need to complete the works in sections.

---

35 FIDIC Yellow Book 2017, clause 8.9 (Employer's Suspension) allows the contractor to be instructed to suspend 'part' of the works, but the suspension seemingly applies once the works are underway.
36 FIDIC Yellow Book 2017, clause 1.127 defines 'Defects Notification Period' as one year (from the Date of Completion as set out in the Taking Over Certificate), unless the Contract Data states otherwise.
37 FIDIC Yellow Book 2017, clause 1.1.66 provides the definition of 'Plant' as essentially any equipment or machinery that is intended for the Permanent Works.

Similar to FIDIC Yellow Book 2017, the following relevant entries are found in Contract Data Part One:

o For Option X5 (Sectional Completion):
   – A description of each section, and
   – The completion date for each section.
o If Option X7 (Delay Damages) is selected:
   – A description of each section, and
   – The amount of daily delay damages for each section. In the event of the client taking over part of a section, it appears that clause X7.3 allows for delay damages to be proportionately reduced.

As with the table in FIDIC Yellow Book 2017, it seems that the section descriptions (for Options X5 and X7) will have to be written separately, if a sufficiently comprehensive or accurate description is to be relied upon. Whilst not contained in the entry for Options X5, NEC4 provides for each section to have its own start date and defects correction period[38] as follows:

o Each section can be given its own access date in the entry for '3 Time' in Contract Data Party One, by giving an equivalent description to the 'part of the site' as is provided against Option X5, along with the specific access date.
o The entry for '4 Quality Management' provides for a different 'defect correction period' for parts of the work. Again, the entry should be described consistently with the description of the section in Option X5.

If each section is to have its own access date, this will have to be written into Part 3 (Time) of Contract Data Part One,[39] as the entry for Option X5 only provides the sections' completion date. The NEC4 equivalent of JCT DB 2016's 'deferred possession' is likely to be a project manager's instruction 'not to start work'.[40]

Unlike JCT DB 2016 and FIDIC Yellow Book 2017 (unless the Supplemental Provisions' Milestone option is adopted), NEC4 provides for completing milestones, such as discrete elements of work or achieving degrees or states of completion. These milestones are referred to as 'Key Dates' in clause 30.3. Key dates, and the conditions for meeting them, are identified in Contract Data Part One (against '2 The Contractor's Main Responsibilities'), and must also be set out in the programme and any revision to it (clause 31.2). Clause 25.3 provides that the project manager decides if the contractor has satisfied the particular 'Condition' for meeting a key date by the date stated in Contract Data Part One. If, in the project manager's opinion, the contractor has failed to satisfy the key date and the client incurs (or is likely to incur) the additional costs of either carrying out the work himself or paying others to carry it out, then the contractor pays these additional costs to

---

38 Similar to JCT DB 2016's providing an individual 'date of possession' or 'rectification period' for each section.
39 NEC4, Contract Data Part One, Section '3 Time' – the 'section' should be identified as the 'part of the Site' and a separate access date ascribed to each.
40 NEC4, clause 34.1.

the client. Clause 25.3 expressly restricts the client's losses to the two classes of 'additional cost' set out in the clause, so he will be unable to recover any other losses or damages he may incur as a result of the contractor's failure to hit the key date.

8.3.4 Summary of Sectional Completion Provisions

Table 8.2 summarises the three contracts' sectional completion provisions.

Table 8.2 Summary of sectional completion provisions.

| Description | Entries in the particular contract[a] | | |
	JCT DB 2016 Contract Particulars	FIDIC Yellow Book 2017 Contract Data	NEC4 Contract Data Part One
Statement that the works are to be subdivided into sections	Fifth Recital	No	1 General
Description of each section	Fifth Recital	Table in Contract Data	Option X5
Dates for possession of each section	Clause 2.3	No	3 Time
Dates for completing each section	Clause 1.1	Table in Contract Data	Option X5
Deferment of possession/starting each section	Clause 2.4	*Clause 8.9*	*Clause 34.1*
Rates of liquidated damages for each section	Clause 2.29.2	Contract Data	Option X7
Apportionment of liquidated damages for partial possession of a section	Clause 2.34	Clause 10.2	Clause X7.3
Different rectification periods for each section	Clause 2.35	No	4 Quality Management
Milestone or key dates (See Table 8.3)	No	Yes, in Supplemental Provisions	Key dates 2 – The Contractor's Main Responsibilities *Clause 30.3*

a) Entries in the contract shown in Table 8.1 are (i) JCT DB 2016 – 'Contract Particulars', (ii) FIDIC Yellow Book 2017 – 'Contract Data', (iii) NEC4 – 'Contract Data Part One', or (iv), if written in *italics*, in that part of the contract.

8.3.5 Summary of Milestone or Key Date Provisions

Table 8.3 summarises the milestone or key date provisions in the Special Provisions of FIDIC Yellow Book 2017 and in NEC4 respectively. Nothing is shown for JCT DB 2016 as does not contain any similar provisions.

Table 8.3 Summary of milestone/key date provisions.

Description	FIDIC Yellow Book 2017 Supplemental Provisions (subclause 4.24)	NEC4
Definition	Insert in clause 1.1	Clause 11.2(11)
Description of milestone/key date (including the criteria for achieving it)	Contract Data	Contract Data Part One
Programme to show milestone/key date	Yes	Yes, clause 31.2
Contractor to notify that milestone/key date is complete	Yes, within 14 days prior to completing the milestone	No
Event signalling the milestone/key date has been met	Milestone Certificate	Project manager's decision (clause 25.3)
Time limit for the engineer to notify completion	Yes, Milestone Certificate issued within 28 days after the contractor's notice	No
Employer's/Client's remedy for contractor's failure to achieve the milestone/key date	Delay damages (as % of final contract price per day)	Additional costs of either (i) Client completing himself, or (ii) employing others to complete
Milestone or key date deemed completed by default	Yes, if the engineer fails to issue Milestone Certificate within 28 days of the contractor's notice	No
Extension of time for milestone or key date	Yes, see clause 8.5	Yes, see clause 60.1

8.4 Progress

In general, to progress means to develop something with the aim of reaching an advanced or completed state. In the context of construction and engineering projects, it is often expressed as a rate of progress (particularly where carrying out the physical works is linear or repetitive in nature[41]), or a manner of progress.

8.4.1 Rate of Progress

Fairly rudimentary construction economics refers to an 'S' curve, where the value of work completed is plotted on the Y-axis against the passage of time on the X-axis. Following the shape of the resulting flattened 'S', it can often be observed that progress expressed as

---

41 For example, progress of a construction project such as a hotel refurbishment could be expressed in terms of 'rooms completed per week', or the progress of an engineering project such as a road or railway could be expressed as 'km per week'.

'value/time' is slower (flatter) towards the project's start and finish but usually increases rapidly (i.e. showing a steeper gradient) to a peak, somewhere in the middle. Owing to the varied nature of construction and engineering projects, it is not always possible to express or measure an overall rate of progress as many different activities will proceed concurrently, but at different rates, which may not be comparable.

The rate of progress can be expressed in subjective terms, often in relation to the project's completion date or in relation to intervening periods during which work has been completed. Comparing progress to a programme can be a useful tool, but it is rare that the contractor is bound to complete the work strictly in accordance with the programme. His principal obligation to complete the works is to complete by the completion date, so any progress he makes in order to do that is usually considered satisfactory and will not be a breach of the contract. If the contractor is at fault for the works progressing too slowly, he risks failing to complete by the completion date and may face delay damages, so the incentive to maintain sufficient progress is clear.

Working strictly in accordance with a programme exposes the parties to additional risk. If the contractor is falling behind on a particular task but is not jeopardising the completion date, then without additional controls,[42] it is unlikely that the employer can seek compensation. Conversely, if the employer fails to provide something to the contractor (such as sufficient or unimpeded access, or items of plant and materials), and prevents him from maintaining progress in accordance with the programme, the employer will be in breach. If the contractor's progress is interrupted by the employer but the completion date is still achieved, the employer may be exposed to a disruption[43] claim from the contractor. An employer who seeks to control the contractor's progress can impose individual targets for parts of the works. He can do this by dividing the works into sections, so that the contractor has to complete the sections in a particular order or by particular dates. He can also impose milestones (or key dates[44]), where the contractor must complete a particular task – such as a part of the works being ready for testing or receiving equipment.

8.4.2 Manner of Progress

The manner of progress describes the way in which the contractor proceeds with the work. The employer is likely to impose general obligations which oblige the contractor to carry out the work competently and properly and to apply a consistent effort:

o Competently and properly – carrying out the work in accordance with the contract (e.g. with employer's requirements, drawings, and specification), any applicable laws (such as the Building Regulations); and, often, a further obligation of working in accordance with 'good practice'[45] is added or implied.

---

42 Such as working to Key Dates (NEC4).

43 Disruption claims are usually made by the contractor to recover the additional expenditure (caused by the employer's breach) of carrying out the work when the completion date is achieved.

44 NEC4, 'Key Date' is defined in clause 11.2(11).

45 'Good practice' is a subjective term which can mean 'The accepted method of building, with the standard and quality being satisfactory to all requirements and regulations' (*Oxford Dictionary of Construction, Surveying, and Civil Engineering*, 2020). Importantly, it should be added that 'good practice' will vary over time, with some previously accepted methods becoming obsolete or superseded.

o The 'frequency' obligation intends that the contractor's efforts should be consistent and uninterrupted. The employer will not want to see periods of inactivity on site.

Further details of the manner of progress should be provided in the contractor's proposals, in documents such as method statements, the programme, and any technical information. As a general rule, the contractor's method of work is his own, and this usually means he is entitled to adopt the most economical method of working for the particular contract, provided that the chosen method complies with the contract, any restrictions imposed by it (e.g. working hours), and health and safety legislation.

8.4.3   JCT DB 2016

As described in Section 3.5.3, of the three design and build contracts considered by this book, JCT DB 2016 contains the least detailed requirements in terms of monitoring the progress of works to completion. The contractor's principal obligation as to time is that he must complete the works, or a section before the relevant completion date; otherwise he will be liable for delay damages, if he is not entitled to have a later completion date to be fixed. The manner of progress is set out in clauses 2.1.1, 2.3, and 2.25.6.1 – applying to *all* the works, either per section or as a whole. The relevant parts of these clauses are as follows:

o Clause 2.1.1 – '…carry out and complete the works in a proper and workmanlike manner…'
o Clause 2.3 – '…regularly and diligently proceed with and complete the [Works] on or before the relevant Completion Date.'
o Clause 2.25.6.1 – 'constantly use his best endeavours to prevent delay…'

Clause 2.1.1 essentially repeats the implied obligation[46] that services will be carried out with 'reasonable care and skill'. Clause 2.3, however, is much more wide-ranging, as the contractor must proceed both 'regularly' and 'diligently', and failure in each respect may put him in breach. The contractor's obligations were described as follows:

> Taken together, the obligation upon the contractor is essentially to proceed continuously, industriously and efficiently with appropriate physical resources so as to progress the works steadily towards completion substantially in accordance with the contractual requirements as to time, sequence and quality of works…[47]

It appears that an obligation to proceed regularly and diligently would prohibit the contractor from working in a disjointed, 'stop-start' or haphazard manner. The obligation as to quality is expressly set out in clause 2.1, as the contractor must also carry out the works in accordance with the Contract Documents.[48]

Clause 2.25.6.1 is also important as it imposes a further obligation to prevent delay, irrespective of whose fault that delay may be. Seemingly, the 'best endeavours' obligation

---

46 Supply of Goods and Services Act 1982, s13.
47 *West Faulkner Associates v London Borough of Newham*, 71 BLR 1, (1994).
48 JCT DB 2016, 'Contract Documents' is defined in clause 1.1 as the Agreement, the Conditions, the Employer's Requirements, the Contactor's Proposals, the Contract Sum Analysis, and any applicable BIM Protocol.

applies to preventing 'likely' delays,[49] as well as preventing an existing delay from becoming worse. Employers should not assume that 'best endeavours' means that the contractor should take every imaginable step, irrespective of cost.[50] However, it seems that the contractor should neither remain passive nor make a minimal or token effort. He should take prudent steps such as advising the employer of delays, along with rescheduling the works and the programme if appropriate. However, it seems that the extent of using 'best endeavours' is measured by reference to someone acting in their own interests[51] and not fulfilling a contractual obligation. In other words, the contractor's best endeavours are focused on helping himself avoid delays. The scope of the contract's variation clause provides a useful guide to whether the contractor has used constant 'best endeavours' as he will be entitled to additional payment if the 'endeavours' fall within the scope of the variation clause.[52] Under a design and build contract, a 'best endeavours' obligation may place more risk with the contractor than under a traditional contract,[53] as variation clauses under the former may be less detailed and more restrictive than under the latter, reflecting the contractor's additional responsibility to carry out the design.

8.4.4 FIDIC Yellow Book 2017

The contractor's obligations as to progress are set out in clauses 7.1 (Manner of Execution), clause 8.1 (Commencement of the Works), clause 8.3 (Programme), and clause 8.7 (Rate of Progress). Clause 7.1 sets out that the contractor must undertake the following as he carries out the Works:

o For Plant[54] and Materials[55]:
  – Manufacture (or 'produce' in the case of materials)
  – Supply
  – Test
o For Plant only:
  – Commission
  – Repair
o Carry out all other operations:
  – In any manner set out in the contract[56];
  – In a 'proper workmanlike and careful manner', in accordance with accepted good practice[57];
  – Using 'properly equipped facilities'; and
  – Unless otherwise specified in the contract, using non-hazardous materials.[58]

---

49 When read alongside JCT DB 2016, clause 2.24.1.
50 *Keating on Construction Contracts*, 10th ed., at 20–152.
51 Noble, A., (1997), 'Contractors' and Sub-Contractors' Constant Best Endeavours to Prevent Delay', *Construction Law Journal* 13: 293.
52 JCT DB 2016, clause 5.1.2.
53 A contract in which the employer is responsible for the design.
54 FIDIC Yellow Book 2017, 'Plant' is defined in clause 1.1.66.
55 FIDIC Yellow Book 2017, 'Materials' is defined in clause 1.1.53.
56 FIDIC Yellow Book 2017, clause 7.1(a).
57 FIDIC Yellow Book 2017, clause 7.1(b).
58 FIDIC Yellow Book 2017, clause 7.1(c).

Clauses 8.1 and 8.3 provide express obligations for the contractor as follows:

o Clause 8.1 – the contractor proceeds 'with due expedition and without delay'; and
o Clause 8.3 – the contractor proceeds 'in accordance with the Programme'.

The phrase 'due expedition' was used in the ICE Conditions of Contract,[59] and this wording survives[60] in the current incarnation of this contract – the Infrastructure Conditions of Contract (ICC) (2014 Edition). Both the ICE and ICC contracts contained a similar requirement to clause 8.3 of the FIDIC Yellow Book 2017 for the contractor to provide a programme, and the commentators at the time[61] felt that the contractor's programme could be used to measure his progress and determine whether or not he was proceeding with 'due expedition'.

Clause 8.3 requires the contractor to proceed in accordance with the programme (see Section 3.5.3), and this appears to provide the contractor's principal obligation as to progress for the following reasons:

o The programme (under clause 8.3) must show the time for completion[62] and, in detail, the sequence in which the contractor intends to carry out the works.[63]
o The contractor's clause 8.1 obligation to proceed with 'due expedition' is likely to hinge on how closely actual progress compares with the programme.
o When assessing the contractor's progress under clause 8.7 (Rate of Progress), the engineer may consider one or both of the following:
  – If insufficient actual progress is being made to complete within the time for completion; and/or
  – If the contractor's actual progress has, or is likely to, fall behind – relative to the programme.

The wording of clause 8.7 is important as, seemingly, assessing the sufficiency of the contractor's progress must be done objectively and not subjectively by relying upon the discretion, opinion, or even the reasonable opinion of the engineer. Clause 8.7 seems to have been drafted on the basis that, unless slower progress is being achieved than is shown on the programme because of a cause that entitles it to an extension of time,[64] then the engineer may instruct the contractor to provide a revised programme, setting out his proposed methods for recovering lost progress. Clause 8.7 confirms that the contractor's proposed 'methods' are not limited to changing the method of construction,[65] but also can include increasing the contractor's working hours and the physical resources on site. Clause 8.7 is also very clear that the contractor's proposed methods of recovering progress are entirely at his own risk and cost, so if the employer incurs additional costs because of their implementation, then the contractor is liable to the employer for those additional costs.

---

59 The final version was the 7th edition in 1999. Clause 41(2) required the contractor to proceed with 'due expedition and without delay'.
60 The Infrastructure Conditions of Contract (2014 ed.), clause 10.1.
61 Keating (7th ed.) for the ICE, and Keating (10th ed.) for the ICC commentaries.
62 FIDIC Yellow Book 2017, clause 8.3(a), and under clause 8.2, the contractor is bound to complete the Works or a Section (if applicable) within the relevant Time for Completion.
63 FIDIC Yellow Book 2017, clause 8.3(c) describes the sequence of work, and clause 8.3(g) requires the programme to be logic-linked and show any critical path(s).
64 FIDIC Yellow Book 2017, clause 8.5 (Extension of Time for Completion).
65 For example, changing from in situ to pre-cast or pre-fabricated construction.

8.4.5 NEC4

Unlike JCT DB 2016 and FIDIC Yellow Book 2017, NEC4 does not expressly set out the manner or rate of progress that the contractor must achieve. Instead, clause 30.1 appears to imply that the contractor must achieve sufficient progress in order to complete the works (including key dates) on or before the relevant, contractual completion dates set out in Contract Data Part One.[66] This rate of progress should also be sufficient to meet key dates (clause 30.3) and complete any section of the works if Option X5 (Sectional Completion) is selected.

The employer can monitor progress through the contractor submitting programmes for the project manager to accept (under clause 31.2), but an insufficient rate of progress is not one of the grounds for rejection. For each revised programme, the contractor is obliged to show the actual progress for each programmed activity,[67] and as he must show any measures to overcome delays, his programme ought to identify steps to increase the rate of progress, provided that the contractor's insufficient progress is the cause of delay.

8.4.6 Reporting Progress

The three design and build contracts contain varying provisions for the contractor reporting progress. In Section 3.5.3, the requirements for the contractor providing a programme with the contractor's proposals were briefly set out. JCT DB 2016 was described as having virtually no requirements for the contractor providing a programme, whereas FIDIC Yellow Book 2017 and NEC4 contain much more detailed provisions. However, in the context of the contractor providing a programme to the employer (or his representative) with his tender, that programme will not show any progress as the contractor would not have commenced work.

Once the works are underway, FIDIC Yellow Book 2017 and NEC4 require the contractor to submit a revised programme to the employer's representative that shows the actual progress of the works[68] (see Section 8.6.4).

Clause 8.3(k) of FIDIC Yellow Book 2017 also requires the contractor to produce a report to accompany the revised programme, which, in relation to progress, should identify any changes against the previous programme (e.g. arising from variations or changes in the works' sequence and timing) and describe any measures the contractor intends to carry out to overcome any[69] delays. This would appear to be the 'monthly report' required by clause 4.20 (Progress Report), although oddly, clause 8.3(k) does not include a cross reference to clause 4.20. However at clause 4.20(h), the monthly report must compare planned and actual progress, giving details of matters which may adversely affect progress

---

66 NEC4 – Contract Data Part One, Section 2 (The Contractor's main responsibilities) 'key date', and Section 3 ('Time') the 'completion date for the whole of the works'. If Option X5 (Sectional Completion) is selected, the 'completion date' is for each 'section'.
67 NEC4, clause 32.1. This is usually shown as a 'drop line', recording progress ahead or behind the baseline programme dates, or by comparing the baseline programme with activities' current progress.
68 FIDIC Yellow Book 2017, paragraph 1 of clause 8.3 provides a general requirement to show progress, then more specifically at clause 8.3(j). NEC4, clause 32.1 contains a similar requirement.
69 FIDIC Yellow Book 2017, clause 8.3(k)(v) – 'any' delays will include overcoming delays caused by the employer, the contractor or by a neutral event – such as adverse climatic conditions – clause 8.5(c).

within the time for completion, or in accordance with the programme. Essentially, the monthly report seems intended to provide a summary of the project's status, as it should contain:

o Progress of the works in general, including:
  – Progress of the physical works being undertaken[70] (compared against the programme) both on and off site, including photographs or video[71] and the Contractor's Records (under clause 6.10)[72];
  – Progress of design[73];
  – Progress of manufacturing, including deliveries, and inspections.[74]
o Testing and compliance documentation.[75]
o Health and safety matters.[76]
o List of variations.[77]
o Administrative matters, such as notices given by either party.[78]

8.4.7 Advance or Early Warning

The three design and build contracts deal with prospective events in different ways. JCT DB 2016 places the responsibility of notifying events that are 'likely' to delay the works' progress firmly with the contractor[79] (refer to Section 8.10), as part of the process for seeking an extension of time.[80] FIDIC and NEC4 contain more detailed provisions for the employer, contractor, and the supervising officer to warn each other of matters that could affect the project.

Clause 8.4 (Advance Warning) in FIDIC Yellow Book 2017 obliges the 'Parties' (i.e. the employer and the contractor), along with the engineer to advise the parties of any potential matter that may adversely affect the following:

o Work being carried out by the 'Contractor's Personnel'[81];
o The execution of the Works (this includes a Section, and seemingly relates to Rate of Progress and the Time for Completion)[82];
o The performance of the completed Works; and
o The Contract Price (the adverse effect being an increase).

---

70 FIDIC Yellow Book 2017, clause 4.20(h).
71 FIDIC Yellow Book 2017, clause 4.20(b).
72 FIDIC Yellow Book 2017, clause 4.20(d).
73 FIDIC Yellow Book 2017, clause 4.20(a).
74 FIDIC Yellow Book 2017, clause 4.20(c).
75 FIDIC Yellow Book 2017, clause 4.20(e).
76 FIDIC Yellow Book 2017, clause 4.20(g).
77 FIDIC Yellow Book 2017, clause 4.20(f).
78 FIDIC Yellow Book 2017, clause 4.20(f), although clause 14.3 confirms that the monthly report itself is not considered as a Notice (as defined in clause 1.1.56).
79 JCT DB 2016, clause 2.24 (Notice by Contractor of Delay to Progress).
80 JCT DB 2016, clause 2.5 – the employer expresses an extension of time by fixing a new (later) Completion Date, as opposed to giving a longer period (e.g. another six weeks).
81 FIDIC Yellow Book 2017, clause 1.1.6, definition of 'Contractor's Personnel' which includes the contractor's staff and his subcontractors.
82 FIDIC Yellow Book 2017, clauses 8.2 (Time for Completion) and 8.7 (Rate of Progress).

The description of 'work of the Contractor's Personnel' is curious, as it does not refer to the 'Works', the 'Permanent Works', or the 'Temporary Works' (which are all defined terms in clause 1.1). If this description is intended to cover matters of quality, then if strictly interpreted, it may only cover 'workmanship' or 'design',[83] and not 'Contractor's Equipment', 'Employer's Equipment', 'Employer's Personnel',[84] 'Employer Supplied Materials' (after all, the obligation under clause 8.4 is a mutual one), 'Goods', 'Materials', or 'Plant'. The performance of the completed works may not cover matters where (for example) aesthetic work may not be in accordance with the contract, but ultimately, the Works' performance is unaffected. The engineer may ask the contractor to provide a proposal to reduce or avoid the potential matter's likely effects under clause 13.3.2 (Variation by Request for Proposal – see Section 8.8.3).

NEC4's 'Early Warning' procedure (in clause 15) expands FIDIC's 'Advance Warning' by including a formal risk management procedure that requires an 'Early Warning Register' and regular 'Early Warning Meetings'. Similar to FIDIC's 'Advance Warning' procedure, the project manager (not the client) and the contractor have the mutual obligation of providing early warning notices (EWNs) to each of prospective matters which could adversely affect the project. The four criteria in clause 15.1 are essentially the same as in clause 8.4 of FIDIC Yellow Book 2017, with the following exceptions:

o There appears to be no requirement to issue a EWN if either the quality of the Works or its design are likely to be affected, as long as the Works achieve their desired performance; and

o An EWN should be issued for any likely adverse effect on a key date[85] (as well as on completion).

No later than one week after the starting date (see Section 8.2.3), the project manager issues the 'Early Warning Register' to the contractor. This is prepared by the project manager and should contain all early warning matters, unless that matter was previously notified as a Compensation Event (see Section 8.8.4). Contract Data Part One also sets out which matters are to be included in the Early Warning Register, along with the frequency[86] of subsequent early warning meetings. The first early warning meeting is held not later than two weeks after the starting date – presumably to allow the project manager and contractor (and any subcontractor) one week to review the Early Warning Register. If either the contractor or the project manager feels it is appropriate to hold additional meetings, they can instruct the other accordingly. The meetings' purpose is to arrive at a collaborative solution[87] in accordance with the contract, to reduce or avoid the potential adverse matter or its effects.

---

83 FIDIC Yellow Book 2017, clause 1.1.80 defines 'Sub-Contractor' as including a subcontractor or designer.
84 FIDIC Yellow Book 2017, clause 1.1.32 defines 'Employer's Personnel' as including the engineer and, if appointed, the Engineer's Representative.
85 See Sections 8.3.3 and 8.3.4.
86 NEC4, clause 15.4 provides that following an early warning meeting, the project manager amends the Early Warning Register and issues it to the contractor within one week – which implies that the frequency of subsequent early warning meetings is at least weekly.
87 NEC4, echoing clause 10.2 (spirit of mutual trust and cooperation).

EWNs are important if a potential matter does occur and becomes a Compensation Event. If, during the procedure for assessing Compensation Events (see Section 8.8.4), the project manager notifies the contractor that in his opinion, the contractor should have issued an EWN[88] (thereby, forewarning the project manager, and enabling them along with any sub-contractors to try and avoid or alleviate the potential matter's effects) but did not, then under clause 63.7, the project manager assesses the Compensation Event as if the contractor had issued an EWN. The extreme consequences for the contractor could be that he is not entitled to any change in the prices or the completion date, if an experienced contractor would have issued an EWN and was able to avoid the effects of the potential matter. A less extreme example would be that there could be a reduction in what would otherwise be the project manager's assessment of a Compensation Event to reflect the contractor's failure to mitigate the matter's effects, so the prices and the completion date may not be adjusted by as much as the contractor would like.

8.5 Acceleration

Construction and engineering contracts include more provisions managing the effects of the decrease in the rate of progress (or managing and avoiding delays) than they do for the rate of progress increasing. In this context, an increase in the rate of progress is referred to as 'acceleration', with the aim of completing the project earlier than its completion date. Achieving an earlier completion date in this section refers to the employer instructing the contractor to accelerate to either finish earlier or to overcome a delay for which the employer is culpable. This section does not consider acceleration in order to overcome a delay caused by the contractor, as this is deemed to be 'mitigation' (see Section 8.10).

The general nature of contracts is to penalise or punish a failure to perform, rather than reward performance which exceeds the contract's requirements. This is understandable from the employer's perspective in two aspects.

Firstly, the employer may suffer significant inconvenience and financial losses if he cannot use the works from the agreed completion date. In such cases, the contract usually provides for the employer to claim delay damages, and an extension of time mechanism to preserve the employer's entitlement to delay damages, if he (and not the contractor) is responsible for part of that delay.

Secondly, the employer is normally content to pay a certain price if the contractor satisfactorily completes the works by a definite date. In some instances, the employer may not want the contractor to complete the works before the agreed completion date, as the employer usually becomes responsible for the works at this stage, so depending upon the particular contract's terms, he may become liable for the costs of insurance, maintenance, operation, and security and may have to repay any deducted retention to the contractor earlier than he planned.

The JCT DB 2016, FIDIC Yellow Book 2017, and NEC4 contracts deal with acceleration in slightly different ways and in varying degrees of detail, but in each case, the contractor provides the employer with a quotation, setting out the costs of accelerating along with

---

88 NEC4, clause 61.5.

the earlier completion date that may be achieved. For example, none of the three contracts provide any procedure for how the contractor arrives at his price for acceleration (as they otherwise do for valuing variations), so there is no framework that the parties can rely upon if they cannot reach agreement.

8.5.1 JCT DB 2016

With the exception of its Management Contract, JCT contracts did not provide for acceleration as a means of maintaining the completion date (or finishing earlier) until the 2011 editions of its contracts (which included JCT DB 2011[89]). Both JCT DB 2016 and JCT SBC/Q 2016 include provisions for acceleration within the Supplemental Provisions of those contracts.[90] However, there are no acceleration provisions found in the 2016 design and build versions of the JCT Intermediate or Minor Works contracts.

Under JCT DB 2016, the option to accelerate progress can only be exercised by the employer. This may be due partly to the contractor's obligation to continuously use his 'best endeavours'[91] to avoid delays (see Section 8.4.1). The employer firstly invites the contractor to confirm either:

o If it is practicable to achieve an earlier completion date; and if so,
o The time that can be saved and the contractor's costs for accelerating (the 'Acceleration Quotation'). The contractor is entitled to the reasonable costs of providing an Acceleration Quotation, irrespective of whether the employer accepts it and instructs him to accelerate.

Unless the parties agree to longer or shorter periods, the contractor has 21 days from the employer's invitation in which to provide the Acceleration Quotation,[92] and the employer must have at least seven days to accept it. The employer must provide a 'Confirmed Acceptance' of the contractor's Acceleration Quotation in order to implement the proposed acceleration measures.[93] This must confirm the amount by which the contract sum is adjusted,[94] the adjustment to the completion date[95] (or completion dates of any relevant section) and any other conditions set out in the employer's invitation.[96] The Acceleration Quotation may not be used for any other purpose if a Confirmed Acceptance is not issued.

8.5.2 FIDIC Yellow Book 2017

FIDIC Yellow Book 2017 adopts a similar but expanded procedure to JCT DB 2017. Under clause 13.3.2,[97] the engineer may request the contractor to propose to accelerate

---

89 JCT DB 2011, Schedule 2, Supplemental Provisions, Part 2, paragraph 6 (Acceleration Quotation).
90 JCT DB 2016, Schedule 2, Supplemental Provisions, Part 2, paragraph 4 (Acceleration Quotation), and JCT SBC/Q 2016, Supplemental Provisions, Part 2, paragraph 2 (Acceleration Quotation).
91 JCT DB 2016, clause 2.25.6.1.
92 The employer may ask for amended proposals during the 21-day (or extended) period for the contractor to provide the Acceleration Quotation (JCT DB 2016, Schedule 2, Supplemental Provisions, Part 2, paragraph 4.1.3).
93 JCT DB 2016, Schedule 2 Supplemental Provisions, Part 2, paragraph 4.3.
94 JCT DB 2016, Schedule 2 Supplemental Provisions, Part 2, paragraph 4.3.1.
95 JCT DB 2016, Schedule 2 Supplemental Provisions, Part 2, paragraph 4.3.2.
96 JCT DB 2016, Schedule 2 Supplemental Provisions, Part 2, paragraph 4.3.3.
97 FIDIC Yellow Book 2017, clause 13.3.2 (Variation by Request for Proposal).

completion,[98] or under the 'Value Engineering' clause 13.2(a),[99] the contractor may make such a proposal of his own volition. The engineer's power under clause 13.3.2 appears to extend to requesting that the contractor provide a clause 13.2(a) accelerated completion proposal. Similar to JCT DB 2017, FIDIC also allows the contractor to set out its reasons why acceleration cannot be achieved.[100] The purpose of FIDIC's Value Engineering clause (13.2) seems to be to provide the employer with additional value[101] for the project. As well as accelerating completion, this may include simply reducing project cost, improving the project's efficiency, or providing another benefit to the employer. The circumstances which may give rise to an acceleration proposal will depend upon the nature of the project. For example, a hydroelectric dam project may be overrunning because of events caused by the employer, and he may face the risk of considerable costs for the project overrunning. The contractor's costs of accelerating may be significantly less than the employer would otherwise pay if the project was completed late.

The contractor's clause 13.2(a) acceleration proposals shall (irrespective of whether the engineer requested it) include the following information required by clause 13.3.1 (Variation by Instruction):

o Any addition or reduction to the resources employed, or any different resources or proposed methods of working.[102]
o A programme showing changes to the time for completion[103] and to its clause 8.3 programme.[104] The requirement to show changes to the time for completion is important, as it suggests that the employer may reduce the time for completion if the contractor's acceleration proposal is accepted. If the prospect of acceleration is likely, then the employer may wish to amend the definition of 'time for completion' to allow it to be reduced in these circumstances.
o The adjustment to the contract price including appropriate substantiation.[105]

If the engineer accepts (by giving consent to) the contractor's acceleration proposal, he must instruct a Variation (under clause 13.3). If the contractor made the clause 13.2 acceleration proposal of his own volition, he will not be entitled to the costs of its preparation.

8.5.3 NEC4

The purpose of the acceleration clause 36 in NEC4 could not be clearer – it is to complete the works before the completion date. Whilst the contractor and the project manager may

---

98 FIDIC Yellow Book 2017, clause 13.2(a).
99 FIDIC Yellow Book 2017, clause 13.2 (Value Engineering).
100 FIDIC Yellow Book 2017, clause 13.3.2(b) – similar to JCT DB 2016, Schedule 2 Supplemental Provisions, Part 2, paragraph 4.1.1.2.
101 Value can simply be expressed as (function/cost).
102 FIDIC Yellow Book 2017, clause 13.3.1(a).
103 FIDIC Yellow Book 2017, clause 1.1.86 defines 'Time for Completion' as including any extensions thereto under clause 8.5.
104 FIDIC Yellow Book 2017, clause 13.3.1(b).
105 FIDIC Yellow Book 2017, clause 13.3.1(c).

discuss accelerating the works, the acceleration quotation process must commence by the project manager instructing the contractor to provide a quotation (which includes details of any key dates to be changed). The contractor provides the clause 36 quotation within three weeks thereafter, and it must include[106]:

o A revised programme showing the earlier completion date and any key dates that are to be changed and
o The changes to the prices.

The project manager replies to the clause 36 quotation within three weeks, and if it is accepted, he must change the completion date (and any key dates) and the prices accordingly, along with accepting the revised programme.

8.5.4 Summary of Acceleration Provisions

Table 8.4 summarises the three contracts' acceleration provisions.

Table 8.4 Summary of acceleration procedures.

Description	JCT DB 2016	FIDIC Yellow Book 2017	NEC4
Acceleration provisions (clause or section)	Schedule 2 Supplementary Provisions, Part 2, para 4	Clause 13.2(a) Clause 13.3.2	Clause 36
Acceleration process initiated by employer	✓	✓	✓
Contractor may initiate acceleration process	–	✓	–
Contractor provides a quotation	✓	✓	✓
Period for contractor's quotation	21 days	–	3 weeks
Quotation includes a programme		✓	✓
Quote includes earlier completion date	✓	✓	✓[a]
Quote includes costs of acceleration	✓	✓	✓
Quote acceptance period	7 days	As soon as practicable	3 weeks
Means of acceptance	Confirmed Acceptance	Clause 13 Instruction	Acceptance
Prohibition of using rejected quotation	✓	–	–

a) NEC4 – Also includes changes to any key dates.

---

106 NEC4, clause 36.2. The quotation must also provide 'details of the assessment' but it is not clear what this means.

8.6   Programme

Section 3.5.3 gave a brief description of how JCT DB 2016, FIDIC Yellow Book 2017, and NEC4 deal with the contractor's programme. In relation to the programme requirements of the FIDIC Yellow Book 2107 and NEC4, it is surprising how little reference JCT DB 2016 makes to the contractor's programme. However, despite the relative absence of contractor's programming obligations, it is good practice (and in practice, extremely likely) that a programme that contains similar detail to that which would be provided under a FIDIC or NEC4 contract be provided for a JCT DB 2016 contract. It is almost universally accepted that the proper operation of JCT DB 2016's extension of time provisions almost necessarily depends upon programming techniques (such as Critical Path Analysis[107]) being adopted.

8.6.1   Generating the Programme

It is rare for any modern engineering or construction project to be carried out without some sort of programming or planning shown in Gantt chart form. In the last 20 or so years, the use of sophisticated planning software generating Gantt charts has become commonplace. Software such as Elecosoft UK Ltd.'s Powerproject and Microsoft Project are commonly used to programme (or schedule) many construction and engineering projects. Oracle Corporation UK Ltd.'s Primavera is often the standard software for large engineering and infrastructure projects, although this is more of a database for project controls, which has a Gantt chart function (albeit a very good one) for programming and scheduling.

Modern programming (also referred to as scheduling) software greatly assists construction professionals by providing a single-platform tool to generate, monitor, and analyse the sequence and timing of a project's activities. Common functions to generate a project's sequence and timing are:

o Generating activity durations:
  – A duration estimated by the software user.
  – A calculated duration, using the works' quantities and rates of production (or output).
o Linking activities in a logical manner:
  – Successive activities linked 'finish to start', or the successor activity begins at some point after the preceding activity has commenced.
  – Activities starting concurrently linked 'start to start'.
  – Activities finishing together may be linked 'finish to finish'.
o Showing and calculating float:
  – Free float
  – Total float
  – Terminal float
o Showing constraints on activities (e.g. an activity may not be able to start or finish until a certain date).

---

107 See Section 8.10 (Extensions of time).

o Showing and calculating the Critical Path.[108]
o Showing milestones (usually shown as having zero duration – i.e. occurring instanta-
  neously).
o Calculating the completion date.
o Establishing the original or 'baseline' programme against which changes in the project's
  duration and sequence can be measured.

The planner should thoroughly check through the project's drawings and specifications
and liaise closely with the project team to ensure that the programme depicts a sequence
of work that is practicable and achievable. The planner along with the project team should
identify any inconsistencies, discrepancies, ambiguities, and instances of poor coordination between drawings and specifications of different disciplines (for example, between
the architectural, civil engineering, structural engineering, or building service engineering
information).

The sequence and timing of works carried out by specialist subcontractors should be carefully incorporated into the programme. It is common for many of the critical activities[109]
(such as ground engineering, foundations, structural steelwork or concrete, and building
services) to be undertaken by specialist contractors, so the sequencing and timing of their
works may dictate the programme of all other works. Many specialist contractors will have
equivalent planning (or scheduling) resources to those of a main contractor, so they will
often be able to provide a comprehensive programme for their works, often using the same
planning software used by the main contractor. If the specialist contractor does not have
appropriate planning resources, then the main contractor's planner (along with members
of its site team) should consult with the specialist, so a practical, workable, and achievable
programme for his works can be produced and subsequently incorporated into the main
contractor's programme.

8.6.2   Programme Requirements for Each Design and Build Contract

Section 3.5.3 highlighted the extent to which JCT DB 2016, FIDIC Yellow Book 2017,
and NEC4 require the contractor to provide a programme for the works. The differences
are stark. At one end of the scale, JCT DB 2016 contains virtually no references to a

---

108 Definitions of 'Critical Path':

> (1) 'The longest sequence of activities through a project network from start to finish, the sum of whose
> durations determines the overall project duration. There may be more than one critical path depending
> upon workflow logic. A delay to progress of any activity on the critical path will, without acceleration
> or re-sequencing, cause the overall project duration to be extended, and is therefore referred to as a
> "critical delay"'. Society of Construction Law. (2017). *Delay and Disruption Protocol*, 2nd ed.
> (2) 'The longest sequence of logically linked activities from the start of a schedule, or a part of a
> schedule, to its finish that determine the earliest possible achievement of a key date, sectional
> completion date or the completion date'. Chartered Institute of Building. (2018). *Guide to Good
> Practice in the Management of Time in Major Projects*, 2nd ed.

109 Activities that form part of the Critical Path.

programme,[110] whilst at the other, FIDIC Yellow Book 2017 and NEC4 both contain extensive requirements for what the contractor's programme must show. Despite FIDIC Yellow Book 2017 containing more prescriptive and comprehensive programme requirements, it does not use the programme to the same extent as NEC4. Under NEC4, the programme becomes central to calculating extensions of time (i.e. delays to the completion date or to a key date[111] – see Sections 8.7, 8.9.3.2, and 8.10.5).

FIDIC Yellow Book 2017 and NEC4 contain the requirements for the contractor's programme shown in Table 8.5.

Whilst JCT DB 2016's conditions contain none of the requirements for the contractor's programme that are found in either FIDIC Yellow Book 2017 or NEC4, it is rare for the contractor to escape having to provide any programme information at all. The Employer's Requirements will often oblige the contractor to provide a programme that shows many of the requirements summarised in Table 8.5 (see Sections 3.4.1,[112] 3.5.1, and 3.5.3[113]).

8.6.3 Revising the Programme and Employer's Acceptance

Both FIDIC Yellow Book 2017 and NEC4 have the aim of trying to ensure that the contractor keeps the programme up-to-date and accurate. They try and achieve this by the contractor revising the programme. NEC4's *Managing Reality*[114] gives a pragmatic explanation of why a contractor may want to revise his programme. It considers that the contractor's initial programme (or 'first programme' in NEC4[115]) may only be an extension of the contractor's preliminary or tender programme. Once the contractor commences work (and particularly when the design is developed further and becomes more coordinated) he may find that the original sequence and timing of operations needs to be amended to ensure the works can be constructed logically and efficiently. Many less experienced consultants, particularly those unfamiliar with the concepts of the NEC contract, find the concept of a changing programme to be difficult, as they often treat the programme as the contractor's binding commitment to undertake the works in precisely the sequence and timing shown thereon. This is rarely the case, as in general, the contractor is obliged to complete the works by a particular date (see Section 8.4) or by dates (see Section 8.3) if the works are subdivided into sections, and not in accordance with his current programme and certainly not in accordance with the programme he initially provided.

FIDIC Yellow Book 2017 obliges the contractor to submit a revised programme within 14 days after the engineer's request. NEC4 goes one step further, so in addition to the contractor providing a revised programme within the 'period for reply'[116] following the project manager's request, the contractor must also provide a revised programme in any event, at the interval specified in the Contract Data. Table 8.6 summarises the procedures for revising the programme under FIDIC Yellow Book 2017 and NEC4.

---

110 JCT DB 2016 only contains a fleeting reference to the contractor's programme in Schedule 2 (Supplemental Provisions) at paragraph 7.1 (Cost savings and value improvements).
111 NEC4 – clause 63.5 (Assessing compensation events).
112 Table 3.5 in Chapter 3.
113 Table 3.7 in Chapter 3.
114 NEC4, *Managing Reality*, Book Three – *Managing the Contract*, 3rd ed., (2017), at 2.6.2.
115 NEC4, clause 31.1.
116 NEC4, clause 13.3 (Communications).

Table 8.5 Requirements of the contractor's programme (FIDIC and NEC4).

The Contractor's programme must show:	FIDIC Yellow Book 2017	NEC4 (clause 31.2)
The particular planning software to be used[a)]	Yes – clause 8.3	No
The specific format of the programme	No[b)]	Yes
Showing the following information:		
o Logic links and the Critical Path	Yes – clause 8.3(g)	No
o Earliest and latest start and finish dates for activities	Yes – clause 8.3(g)	No
o Commencement Date	Yes – clause 8.3(a)	Yes
o Time for Completion (including for each Section), or Date for Completion	Yes – clause 8.3(a)	Yes
o Planned Completion[c)]	No	Yes
o Key Dates (including the contractor's planned dates for meeting the Condition for a Key Date)	No	Yes
o Dates for possession of, or access to the Site (or each part of the Site)	Yes – clause 8.3(a)	Yes
o The sequence and timing of the works, including;	Yes – clause 8.3(c)	Yes[d)]
– Each design stage	Yes	Yes
– Information to be provided by others	No	Yes
– Preparing and submitting the contractor's design docs	Yes	Yes
– Procurement	Yes	Yes
– Manufacture and inspections	Yes	Yes
– Delivery	Yes, also see clause 8.3(i)	Yes
– Construction	Yes	Yes
– Nominated sub-contractor elements of work	Yes	-
– Testing[e)], further inspections and commissioning	Yes, also see clauses 8.3(e), and 9.1	Yes
– Work carried out by, or on behalf of the Employer	No	Yes
– Time contingencies (including float)	No	Yes
– Health and safety obligations	No	Yes
o Periods for reviewing documents or submissions	Yes – clause 8.3(d)	Yes

Table 8.5 (Continued)

The Contractor's programme must show:	FIDIC Yellow Book 2017	NEC4 (clause 31.2)
o Any other requirements set out in the Employer's Requirements or Scope	No	Yes

a) FIDIC Yellow Book 2017. Clause 8.3 provides that the planning software to be used is either set out in the Employer's Requirements, or if not, as accepted by the Engineer.
b) FIDIC Yellow Book 2017 only specifies the software to be used, not the format of the programme itself. However, it can be argued that by clause 8.3 providing an obligation for the contractor to use a particular form of "programming software", it should be implied that invariably, a Gantt chart format will be used. However, see clause 8.3(g) which specifies that the programme must show activities to the level of detail set out in the Employer's Requirements – including logic links, critical path and earliest and latest start and finish dates.
c) NEC4. Whilst here is no definition of "planned Completion", it means the date when the contractor is aiming to complete the work, and this may occur before the Completion Date, it be the same as the Completion Date, or if the progress of work is delayed by events which are not Compensation Events, it may occur after the Completion Date.
d) NEC4 – Clause 31.2 provides that the contractor must show all "operations" that comprise carrying out the Works. This clause is wide enough to cover all the specifically mentioned stages or operations set out in clause 8.3 of FIDIC Yellow Book 2017.
e) FIDIC Yellow Book 2017, testing is required to be shown on the programme, and a "detailed test programme" must be provided by the contractor for "Tests on Completion" (see clause 9.1)

Table 8.6 Requirements of revising the contractor's programme (FIDIC and NEC4).

Requirement of revising the contractor's programme	FIDIC Yellow Book 2017	NEC4 (clause 32)
Revised programmes should be submitted for acceptance:		
o Period after the Engineer/Project Manager notifies or instructs the contractor to do so:	14 days[a]	Yes – clause 32.2[b]
o Revised programmes submitted in any event: – Period for the contractor to submit:	No –	Yes – clause 32.2[c] Prescribed interval
Revised programmes should show or include:		
o Remedial works (correcting defects)	Yes – clause 8.3(f)	Yes – clause 32.1
o Actual progress including delays	Yes – clause 8.3(l)	Yes – clause 32.1
o Effect of actual progress on remaining activities	No	Yes – clause 32.1, 62.2
o A supplemental report	Yes – clause 8.3(k)	No[d]
o Contractor's proposed changes to the programme	No	Yes – clause 32.1
The programme is used to calculate extensions of time (to either the Date for Completion, including Sections, and Key Dates)	No	Yes – clause 63.5

a) FIDIC Yellow Book 2017, clause 8.3 (last paragraph)
b) NEC4. The "period for reply" is set out in Contract Data Part One (Data Provided by the Client)
c) NEC4. The prescribed interval are set out in Contract Data Part One (Data Provided by the Client) and are expressed as *"intervals no longer than"*, which permits the contractor to provide revised programmes before the expiry of the relevant interval.
d) NEC4 – clause 31.2 stipulates that the contractor must state (i) the major resources and equipment he needs, and (ii) how he intends to carry out the work, but as clause 31.2 states that this information must be shown on the programme, it implies that the contractor's statements in this regard are sufficiently brief so as to fit on the programme, and not in a supplemental report.

Table 8.7 Reasons for rejecting the contractor's programme – and procedure (FIDIC and NEC4).

Engineer's / Project Manager's review and acceptance of the contractor's revised programme	FIDIC Yellow Book 2017 (clause 8.3)	NEC4 (clause 31.3)
Period for responding to the contractor's revised programme	14 days[a)]	2 weeks
Reasons for requesting a revised programme:		
o It fails to comply with the contract	Yes	Yes
o Inaccurately shows actual progress	Yes	Yes
o Is not consistent with the contractor's obligations	Yes	Yes
o Is not a reasonable depiction of the contractor's plans	No	Yes
Default provisions for Non-Objection to, or acceptance of the revised programme	Yes[b)]	Yes[c)]

a) FIDIC Yellow Book 2017, clause 8.3. Whilst clause 8.3 does not expressly oblige the Engineer to respond within 14 days (or within 21 days of receiving the contractor's 'Initial programme', unless he does so within that period, he will be deemed to have issued a Notice of Non-Objection.
b) FIDIC Yellow Book 2017, clause 13.3.2 (Variation by Request for Proposal)
c) NEC4, clause 31.3. Should the Project Manager fail to respond to the contractor's revised programme within 14 days, the contractor notifies him accordingly. If the Project Manager fails to respond within 1 week after the contractor's notice, then the contractor's revised programme is deemed to be "accepted".

Neither the engineer nor project manager can arbitrarily object or refuse to accept the revised programme, but the grounds for rejection under both FIDIC Yellow Book 2017 and NEC4 are fairly wide. Essentially, the contractor must accurately show the project's current progress, along with maintaining all the other requirements that are either provided in the contract or in the Employer's Requirements (FIDIC Yellow Book 2017) or the Scope (NEC4). The reasons for not accepting the contractor's revised programme for both contracts are summarised in Table 8.7.

There are default provisions in both the FIDIC Yellow Book 2017 and NEC4 if the engineer or project manager fails to properly respond to the contractor's revised programme within a set period of time. In both cases, the contractor's revised programme effectively becomes accepted in the absence of the engineer's or project manager's timely dissent.

8.6.4 Monitoring Progress on the Programme

Programming software can be used to monitor progress of individual activities against the 'baseline' programme:

o Progress shown as a 'drop line'.
  – The drop line is drawn vertically downwards at the date on which progress is recorded.
  – If the drop line is shown around an activity to the left of the vertical (i.e. in the past), then progress of that activity is falling behind; whereas
  – If the drop line is to the right of the vertical (i.e. in the future), then progress is ahead of schedule.
  – If the drop line is vertical from the report date through an activity, then it is proceeding according to programme (i.e. it's on time).

o Within the same line for an activity, the baseline programme activity can be shown (often as a narrow, single-colour line) beneath the as-built activity so a simple, visual comparison between the timing and duration of the 'baseline' and 'as built' versions of that activity can easily be made.

Monitoring progress of individual activities is usually insufficient for monitoring progress of the project as a whole or part of that project (e.g. such as a group of related activities or a section), as the interrelationship of those activities must also be considered. The logic links between the relevant activities, along with any applicable constraints and float, must be considered in order for the Critical Path to be calculated.

In practice, if the activities on the programme (or schedule) have been linked logically, the chosen programming software will calculate and show the Critical Path. Unless the programmed sequence of works is relatively simple – and a quick visual check will identify any errors or impracticability – the planner should thoroughly check the calculated critical path for errors, particularly those that could create a problematic sequence of work.

The contract, or the Employer's Requirements or Scope, may require the contractor to provide a report to accompany the revised programme (see Section 8.4.4).

8.7 Interim Payments

In the United Kingdom, there is specific legislation that governs the making of interim payments for contracts for construction and engineering projects. This legislation imposes specific procedures for both the employer (and any supervising officer acting on his behalf) and the contractor to follow. The natural consequence is that in order to be compliant with the UK legislation, all contracts' conditions for interim payments must include these core procedures, so there is relatively little freedom for the parties to depart from the requirements of the legislation. Consequently, if parties want to carry out the relevant construction or engineering work in the UK and be subject to UK law, then no matter what form of contract they select, the interim payment conditions will all be very similar.

This legislation is the Housing Grants, Construction, and Regeneration Act 1996 (as amended)[117] (HGCRA 1996), and it provides that the interim payment provisions for all qualifying construction contracts[118] (i.e. those that provide for construction operations[119]) must be compliant. As well as providing for the carrying out of the physical construction or engineering work, a qualifying construction contract importantly also includes design[120] and the giving of specific but related advice.[121] Therefore, any conditions of engagement for designers (refer to Chapter 4) will have to comply with the HGCRA 1996 in terms of providing the same procedures for interim payments (see Section 8.7.4).

---

117 Housing Grants Construction and Regeneration Act 1996 c53, as amended by the Local Democracy, Economic Development, and Construction Act 2009 c20.
118 HGCRA 1996, s104(1).
119 HGCRA 1996, s105(1).
120 HGCRA 1996, s104(2)(a) '...architectural, design or surveying work'.
121 HGCRA 1996, s104(2)(b) '...provide advice on building, engineering, interior or exterior decoration or the laying-out of landscape in relation to construction operations'.

The HGCRA 1996 is often referred to as the 'primary legislation' in this regard. In the event that a construction contract does not fulfil a requirement (or requirements) of the HGCRA 1996,[122] then the relevant provisions of the 'secondary legislation', namely the Scheme for Construction Contracts (England and Wales) Regulations 1998[123] ('the Scheme'), will apply by default. In other words, the Scheme will apply in the following ways:

o If the construction contract fails to fulfil one of the HGCRA 1996s requirements, then the equivalent requirement as set out in the Scheme will be implied (or 'slotted in').
o If the contract does not fulfil any of the HGCRA 1996s requirements, then the entirety of the Scheme will apply.

The HGCRA 1996 provides the contractor with the right to stage or periodic payments.[124] The only exceptions are:

o If the work being carried out is not classed as 'construction operations'.[125]
o The duration of the work is either specified to be – or the parties agree that it is – less than 45 days.

The majority of construction and engineering projects are likely to fall within the definition of 'construction operations'.

The terms 'stage' and 'periodic' payments are generally understood to have the following meanings:

o Stage payments trigger the interim payment procedure when the contractor has completed a defined stage of the works (for example, the substructure up to a particular level, a structural steel frame, some concrete work, or completion of the building's external envelope).
o Periodic payments trigger the interim payment procedure at defined intervals (usually every month).

The contractor's entitlement to receive an interim payment may also depend upon him submitting an application for payment. Some traditional contracts (such as JCT SBC 2016) give the contractor the option of applying for payment, but the default obligation rests with the employer (or the supervising officer) to value the contractor's works at the end of the appropriate stage or period and then certify or notify the amount the employer has to pay, in any event (i.e. without the contractor having to submit an application for interim payment).

In addition to providing the contractor with an entitlement to regular payments, the HGCRA 1996 also provides that all qualifying construction contracts must include a payment mechanism that provides the following:

o A mechanism for determining when the 'due date' for each payment occurs[126];
o That the employer (or the supervising officer) must notify the contractor of the amount he considers is due at the due date, and the basis of its calculation not later than five days after the 'due date' (the 'payment notice')[127];

---

122 HGCRA 1996, s110(3).
123 The Scheme for Construction Contracts (England and Wales) Regulations 1998 SI 649, as amended.
124 HGCRA 1996, s109.
125 HGCRA 1996, within the meanings of s105(2).
126 HGCRA 1996, s110(1).
127 HGCRA 1996, s110A(1)(a) and s110A(2).

o When the payment must be made by the employer after the due date (the 'final date for payment')[128]; and

o If the employer intends to pay a lesser amount that shown in the payment notice, then he (or the supervising officer) must notify the contractor of that lesser sum (along with the basis of its calculation) within a period before the final date for payment (a 'pay less notice').[129]

The parties are free to agree on (i) the period between the due date and the final date for payment, and (ii) the period before the final date for payment within which a pay less notice must be given to the contractor.[130] However, the parties are not permitted to change the five-day period after the due date for issuing the payment notice.

Importantly, the HGCRA 1996 provides that the employer (or payer) must pay what is referred to as the 'notified sum' to the contractor.[131] If the employer (either by himself or via the supervising officer) complies with the HGCRA's requirements, then the notified sum is either the amount set out in the 'payment notice', or the lesser amount set out in a 'pay less notice' (if one was issued).

If the employer has not provided a payment notice (either complying with the HGCRA's requirements as to form or timing), then the contractor may issue his own payment notice ('contractor's payment notice').[132] Unless the employer subsequently issues a compliant 'pay less notice', then the amount in the contractor's payment notice becomes the 'notified sum' which the employer must then pay. The final date for payment is postponed by the same period as the 'contractor's payment notice' was issued after the (employer's) payment notice should have been issued. If the particular contract requires the contractor to submit an application for payment, then this will become the 'contractor's payment notice' in the event that the employer fails to properly issue a payment notice, so there is no need for the contractor to submit a second notice (the contractor's payment notice) after the employer's default in this regard.

The HGCRA 1996 provides further protection to the contractor in the event of non-payment. If the employer has failed to pay an amount due by the final date for payment, the contractor may give seven days' notice to the employer of his intention to suspend performance.[133] The right to suspend then ceases when the amount due is paid in full. Suspension can provide a powerful incentive for the employer to pay, as not only will progress of the works be on hold, but he must also reimburse the costs and expenses that the contractor reasonably incurs from the suspension.[134] Suspension can be risky for the contractor in situations where the amount due cannot be readily established; for instance where the employer has only paid part of the sum that the contractor claimed. For instance, if the contractor wrongly suspends his obligations, then he may be committing a serious breach of the contract and one which may lead to serious consequences, such

---

128 HGCRA 1996, s110(1)(b).
129 HGCRA 1996, s111(3).
130 HGCRA 1996, s110.
131 HGCRA 1996, s111(1).
132 HGCRA 1996, s110B(1), which must set out the sum the contractor considers is due and the basis of its calculation (complying with s110A[2]).
133 HGCRA 1996, s112 (Right to suspend performance for non-payment).
134 HGCRA 1996, s112(3A).

as termination by the employer. However, if the notified sum along with the employer's failure to pay it in full by the final date for payment can be established, then suspension remains a viable remedy for the contractor.

8.7.1 JCT DB 2016

Interim payments under JCT DB 2016 can either be stage payments (Alternative A) or periodic payments (Alternative B), and either of these alternatives complies with s110(1) of the HGCRA 1996, so no parts of the Scheme will apply. The relevant entries against clause 4.7.1 in the Contract Particulars denote whether Alternative A or Alternative B has been selected, and if no option is selected, then Alternative B (periodic payments) applies by default. Each alternative in the Contract Particulars sets out the following information:

o Alternative A (stage payments) – entry against clause 4.7.1:
  – Identifies which documents describe the relevant stages.
  – A brief description of each stage, and their cumulative value (clearly assuming that the stages are physically completed in the order described in the Contract Particulars) which must equate to the Contract Sum.
o Applying to both Alternative A (stage payments) and Alternative B (periodic payments) – entry against clause 4.7.2:
  – The first 'Interim Valuation Date' which also sets out that that subsequent Interim Valuation Dates occur on the same or nearest Business Day[135] in that month. For example, if the first Interim Valuation Date is 31 July, and 31 August occurs on a Monday but is the Summer Bank Holiday, then the subsequent Interim Valuation Date would occur on Friday 28th August, as it must be in the same month. If no entry is made against clause 4.7.2, then the first Interim Valuation Date occurs one month after the Date of Possession.[136]

Under Alternative A and Alternative B, interim valuations continue until the due date for the final payment,[137] and the contractor shall submit an 'Interim Payment Application' to the employer not later than seven days prior to the Interim Valuation Date.[138] This must include:

o The sum that the contractor considers to be due at the due date,
o The basis on which that sum was calculated,[139] and
o Any further information set out in the Employer's Requirements.[140]

The due date is easily established for Alternative B (periodic payments), and this occurs on a monthly basis, seven days after each Interim Valuation Date. If the contractor provides the Interim Payment Application later than seven days prior to the Interim Valuation Date, the due date occurs seven days after it is received by the employer.

---

135 JCT DB 2016, clause 1.1 (Definitions) – any day excluding Saturdays, Sundays, and public holidays.
136 JCT DB 2016 – Contract Particulars – entry against clause 2.3 (Date of Possession of the Site).
137 JCT DB 2016, clause 4.7.2 (the due date for the final payment – clause 4.24.5).
138 JCT DB 2016, clause 4.7.2.
139 JCT DB 2016, clause 4.7.3.
140 JCT DB 2016, clause 4.7.4.

The due date under Alternative A occurs in the same way as for Alternative B (i.e. on a monthly basis), but at first glance, this may not seem apparent. This is because it can appear that the contractor's entitlement to an Interim Payment only arises once the contractor has completed one of the agreed stages (as set out in the Contract Particulars), but this is not the case. The entry in the Contract Particulars (against clause 4.7.2) shows that the contractor is entitled to Interim Payments on a monthly basis, after an Interim Valuation Date. This apparent confusion may have arisen as the previous version (JCT DB 2011) stated that 'Where Alternative A applies, an Interim Application shall be made as at completion of each stage specified in the Contract Particulars for Alternative A'.[141] Therefore, under Alternative A, whilst the contractor remains entitled to monthly Interim Valuations, unless he completes one or more of the agreed stages and is not claiming any other sums,[142] he will not be entitled to any further payment because the Gross Valuation (see Table 8.8) will not include any further sums due[143] (as no later stages will have been completed under clause 4.12.1.1).

Following the due date, the employer must give a payment notice not later than five days[144] after, with the final date for payment occurring 14 days later.[145] If the employer intends to pay a lesser sum than set out in the payment notice, he must issue his pay less notice not later than five days before the final date for payment.[146] If the employer fails to issue a payment notice, then the Interim Payment Application fulfils the requirement of a contractor's payment notice, so subject to any pay less notice, the employer must pay the amount stated as due in the relevant Interim Payment Application.[147] In the event that the employer is late in making an interim payment, then the contractor has the following remedies:

o He is entitled to be paid interest on the late payment under clauses 4.9.6 and 4.9.7; and
o He may exercise the right of suspension after giving seven days' notice to the employer, providing that the failure to pay extends beyond that period, and the right to suspend ceases upon payment in full (clause 4.11).

The 'Gross Valuation' for interim payments under each alternative are split into the following three categories:

o Amounts subject to retention
o Amounts *not* subject to retention
o Deductions

Retention is effectively an employer's incentive for the contractor to complete the work on time and rectify all patent defects within a particular period after completion. It is usually calculated as a percentage (usually between 3% and 5%, depending upon a project's

---

141 JCT DB 2011, clause 4.8.2.
142 For example, the contractor has not claimed any sums under clauses 4.121.2 to 4.12.1.4 or under clause 4.12.2.
143 JCT DB 2016, clause 4.12.1.1 (the cumulative value of stages completed) – see Table 8.8.
144 JCT DB 2016, clause 4.7.5.
145 JCT DB 2016, clause 4.9.1.
146 JCT DB 2016, clause 4.9.5.
147 JCT DB 2016, clause 4.9.3.

Table 8.8 JCT DB 'Gross Valuation' under Alternatives A and B.

Element of the "Gross Valuation" for Interim Payments under Alternatives A or B	Alternative A (Stage payments)	Alternative B (periodic payments)
Amounts subject to any Retention:		
– Cumulative value of completed states	4.12.1.1	–
– Work properly executed (inc design)	–	4.13.1.1
– Value of Changes (variations)	4.12.1.2	Included in 4.13.1.1
– Materials on site	–	4.13.1.2
– Listed Items (clause 4.15)	4.12.1.3	4.13.1.3
– Fluctuations	4.12.1.4[a]	4.13.1[b]
– Acceleration quotation	4.12.1	4.13.1
Amounts not subject to any Retention:		
– Adjustments under clause 4.3[c]	4.12.2.1	4.13.2.1
– Contractor's suspension costs (4.11.2)	4.12.2.2	4.13.2.2
– Direct loss and / or expense (4.19)	4.12.2.3	4.13.2.3
– Insurance Options B or C	4.12.2.4	4.13.2.4
– Any other amount for Fluctuations	4.12.2.5[d]	4.13.2.5
Less the following deductions:		
– Amount for defects (2.35)	4.12.3.1	4.13.3.1
– Failure to comply with instructions (3.6)	4.12.3.1	4.13.3.1
– Evidence of insurance (6.12.2)	4.12.3.1	4.13.3.1
– Breach of Joint Fire Code (6.19.2)	4.12.3.1	4.13.3.1
– Allowable Fluctuations amount[e]	4.12.3.2	4.13.3.2

a) JCT DB 2016 - clause 4.12.1.4 refers to JCT Fluctuations Option C (if applicable)
b) JCT DB 2016 – clause 4.13.1 refers to "any applicable Fluctuations Provision"
c) JCT DB 2016 – clauses 4.12.2.1 and 4.13.2.1 refer to additions under clause 2.5.2 (early use by Employer/Insurance Option A), 2.20 (patent infringement), 3.12 (inspection – tests), 6.10.2 (Insurance Option A – Terrorism cover), 6.10.2 (Insurance Option A – Terrorism cover Pool Re Cover), 6.11.3 (Insurance Option A – Terrorism Cover), 6.12.2 (Evidence of insurance) and 6.20 (Joint Fire Code)
d) JCT DB 2016 – excluding any amounts added under 4.12.1.4
e) JCT DB 2016 – clause 4.12.3.2 and 4.13.3.2, excluding any amounts under clauses 4.12.1.4 or 4.13.1

value – the 'Retention Percentage'), that is deducted from certain amounts due to the contractor as interim payments. JCT DB 2016 deals with retention in two ways:

o As a deduction, calculated as the 'Retention Percentage'[148] taken from the relevant sums included in the 'Gross Valuation' for Interim Payments under Alternative A or B; or
o No amounts of retention are deducted if the contractor provides the employer with a Retention Bond under clause 4.17.

---

148 JCT DB 2016, clause 1.1 defines the 'Retention Percentage' as the percentage stated in the Contract Particulars entry against clause 4.18. If no percentage is entered, the Retention Percentage is 3% by default.

In the event that retention is to be deducted, clause 4.18 provides that retention is deducted from the relevant sums in all interim payments as follows:

o Retention, calculated as the Retention Percentage, is deducted from the relevant sums in interim payments where practical completion (see Section 9.3.1) has not been achieved, in respect of the following:
  – The whole of the Works[149]; or
  – If the Works are completed in sections (see Section 8.3.1), from the relevant sum that represents the value of the uncompleted sections[150]; or
o Half of the retention (calculated as half of the Retention Percentage) is deducted from the relevant sum in interim payments:
  – Where the whole of the Works have reached practical completion but the Notice of Completion of Making Good (under clause 2.36 – see Section 9.5.1) has not been issued; or
  – Where the relevant sum is the total value of each section that has reached practical completion but for which no Notice of Completion of Making Good has been issued; or
  – In the case of partial possession, the value of the Relevant Part[150] under clause 2.32 (Defects, etc. – Relevant Part) has not been issued.

Under clause 4.16, the employer is obliged to act as trustee[151] for the retention, but is not obliged to invest the retention money. This provides that the employer should not simply deduct retention as a matter of arithmetic when calculating the amounts due under interim payments, but should set aside the retention and, if the contractor so requests, pay the retention into a separate bank account.[152] As long as the employer remains solvent, his obligation to act as trustee for the retention should be enforced.[153]

The Gross Valuation is summarised in Table 8.8.

Clause 4.14 sets out the constituent aspects of each Interim Payment as being the 'Gross Valuation' (either under clause 4.12 for Alternative A or 4.13 for Alternative B), less the following:

o Any amount of retention (clause 4.16),[154]
o The cumulative amounts of an advance payment to be repaid to the employer by the contractor,[155] and
o The amount previously paid to the contractor by the employer (in previous interim payments).[156]

8.7.2 FIDIC Yellow Book 2017

As FIDIC contracts are intended for use internationally (i.e. outside the UK), then it is perhaps unsurprising that the payment terms in the FIDIC Yellow Book 2017 do not comply

---

149 JCT DB 2016, clause 4.18.1 (Retention – amounts and periods).
150 JCT DB 2016, clause 4.18.2. Also refer to 9.3.1 for the 'Relevant Part' where the employer takes 'partial possession' of a Section.
151 A trustee is a person having nominal title to property that he holds for the benefit of another.
152 JCT DB 2016, clause 4.16.2.
153 *PC Harrington Contractors Ltd v Co-Partnership Developments Ltd*, 88 BLR 44, (1998).
154 JCT DB 2016, clause 4.14.1.
155 JCT DB 2016, clause 4.14.2.
156 JCT DB 2016, clause 4.14.3.

with the HGCRA 1996. What is perhaps surprising is that there are no modifications in the 'Particular Conditions Part B – Special Conditions' that could be used to achieve compliance with the HGCRA 1996 for a project in the UK. Such modifications would not be extensive and could be applied without significantly altering the rest of the payment terms.

Interim payments under FIDIC Yellow Book 2017 are made on the ordinary basis that they are for cash flow purposes only, and any payment certificate does not indicate the engineer's acceptance, non-acceptance, consent, or approval to any part of the contractor's work (clause 14.6.3). It provides for periodic interim payments, albeit the value of the work carried out to date can be calculated on one of two alternative bases:

o Estimated by the contractor under clause 14.3(i), or
o Set out as instalments in a Schedule of Payments[157] (clause 14.4). Any Schedule of Payments (included within the completed Schedules[158] as set out in the Contract Agreement[159]) only provides one[160] (of the 10) elements of the contractor's application for interim payment.

The procedure for periodic interim payments under clause 14.3 occurs on a monthly basis, unless a different period of payment is set out in the Contract Data.[161] The procedure commences with the contractor submitting his 'Statement'[162] (his application for interim payment) to the engineer. The overall timetable for the periodic interim payments (i.e. from the Statement to the date upon which the employer is obliged to pay the contractor) is often longer than the equivalent periods under JCT DB 2016[163] (refer to Section 8.7.1) or NEC4[164] (refer to Section 8.7.3). Clause 14.6.1 provides that the engineer issues the 'Interim Payment Certificate'[165] or 'IPC' within 28 days after receiving the contractor's statement, and unless a different period is set out in the Contract Data,[166] clause 14.7(b) (Payment) provides that the employer shall make the interim payment to the contractor within 56 days after the engineer receives the contractor's statement.

Figure 8.2 shows FIDIC Yellow Book 2017s interim payment provision for periodic interim payments under clause 14.3, along with (in the grey boxes and dashed lines) the modifications that would ensure compliance with the HGCRA 1996.[167]

---

157 FIDIC Yellow Book 2017, defined in clause 1.1.73. If no Schedule of Payments is included in the Contract, then the contractor submits estimates (non-binding) of the payments he expects to receive over each period of 3 months, starting 42 days after the Commencement Date, and concluding upon issue of the Taking Over Certificate for the Works (clause 14.4).
158 FIDIC Yellow Book 2017, defined in clause 1.1.72 – as shown in point 2(f) therein.
159 FIDIC Yellow Book 2017, refer to clauses 1.5(h), 1.6 (Contract Agreement) and the definition in clause 1.1.10.
160 FIDIC Yellow Book 2017, clause 14.4 provides that (unless otherwise stated in the schedule) the Schedule of Payments is treated as the 'estimated contract value of the Works executed' – as described in clause 14.3(i) – and does not fulfil the requirements of subclauses 14.3(ii) to 14.3(x), inclusive.
161 FIDIC Yellow Book 2017, clause 14.3. See 'Particular Conditions – Part A Contract Data', entry against subclause 14.3.
162 FIDIC Yellow Book 2017, defined in clause 1.1.79.
163 JCT DB 2016 – the overall period from the contractor submitting his Interim Payment Application to the final date for payment is 21 days.
164 NEC4, clause 51.2 – the overall period from the assessment date to payment is three weeks.
165 FIDIC Yellow Book 2017, defined in clause 1.1.45.
166 FIDIC Yellow Book 2017, Particular Conditions Part A – Contract Data, entry against clause 14.7(a).
167 Includes the references to the relevant sections of the HGCRA 1996.

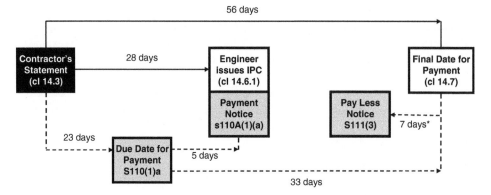

Figure 8.2 FIDIC Yellow Book 2017 interim payment (including HGCRA 1996 amendments).

The requirements for the contractor's 'Statement' under FIDIC Yellow Book 2017 are far stricter than those that apply to the contractor's application for interim payment[168] under JCT DB 2016. The contractor's statement must comply with clause 14.3 as follows:

o It must be in a particular format:
  – At least one hard (paper) copy and one copy in electronic format must be provided[169] unless the Contract Data requires any additional hard copies; and
  – The engineer must find its format acceptable.[170]
o It must show the sum that the contractor is claiming, and provide details and substantiation (in the form of supporting documents[171]) to allow the engineer to investigate that sum.[172] This includes:
  – The estimated contract value of the works executed,[173] either as (i) estimated by the contractor or as set out in the Schedule of Payments[174];
  – Variations[173];
  – Retention Money[175] (both amounts deducted[176] and due for release[177]);
  – Less amounts certified in previous IPCs[178];
  – Repayments for an Advance Payment (under clause 14.2)[179];

---

168 JCT DB 2016, clause 4.7.4 provides that the Employer's Requirements may specify further information is to be provided in addition to that required by clause 4.7.3.
169 FIDIC Yellow Book 2017, clause 14.3(b) – Particular Conditions Part A – Contract Data sets out the number of additional hard (paper) copies in the entry against clause 14.3(b).
170 FIDIC Yellow Book 2017, clause 14.3(a). If the criteria for the engineer to accept the Statement is not set out in the Employer's Requirements, the engineer (or if appointed, the Engineer's Representative) can set out the Statement's acceptance criteria in an instruction under clause 3.5 (Engineer's Instructions).
171 FIDIC Yellow Book 2017, clause 14.3(c).
172 FIDIC Yellow Book 2017, clause 14.3(c). In addition to investigating the amount claimed by the contractor, clause 14.3(c) seemingly invites the engineer to review the contractor's other contemporaneous documentation, such as the clause 4.20 monthly reports (see Section 8.4.6 above).
173 FIDIC Yellow Book 2017, clause 14.3(i).
174 FIDIC Yellow Book 2017, clause 14.4(a).
175 FIDIC Yellow Book 2017, clause 1.1.70 defines 'Retention Money'.
176 FIDIC Yellow Book 2017, clause 14.3(iii).
177 FIDIC Yellow Book 2017, clause 14.3(viii).
178 FIDIC Yellow Book 2017, clause 14.3(x).
179 FIDIC Yellow Book 2017, clause 14.3(iv).

- Provisional Sums (clause 13.4)[180];
- Plant and Materials (complying with clause 14.5)[181] – see below;
- For the contractor's use of utilities provided by the employer (clause 4.19)[182];
- Arising from fluctuations (or 'Changes in Cost' – clause 13.7) and Changes in Laws (clause 13.6)[183]; and
- Any other additions or deductions, either due under or in connection with the contract – such as under clause 3.7 (Agreement or Determination)[184] – see Section 9.7.2.1.

The contractor's compliance with all the requirements of clause 14.3 is likely to determine whether or not his statement is valid, so in the event that the project is governed by UK law, and a dispute arises over an application for interim payment, a court, adjudicator, or arbitrator may have to decide if it is. In doing so, they are likely to consider whether the statement is in 'substance, form and intent'[185] a statement complying with clause 14.3, and clause 14.3 sets out the requirements for satisfying each of those three characteristics in detail.

Clause 14.5 (Plant and Materials intended for the Works) anticipates that when procuring plant and materials intended for the works, the contractor may have to wait for a significant period between paying for the plant and materials and being paid for them by the employer. Provided that the following criteria are met, the contractor may include an amount for 'Plant and Materials' in his statement:

o The Contract Data includes a list of the relevant plant and materials, under the following entries:
- 'Plant and Materials for payment when shipped' – against clause 14.5(b)(ii), and/or
- 'Plant and Materials for payment when delivered to the Site' – against clause 14.5(c)(i).
If no entries are made in the Contract Data, then clause 14.5 does not apply.
o Under clause 14.5(a), that the contractor provides satisfactory evidence that the plant and materials:
- Have been procured,
- Comply with the contract (e.g. they have been successfully tested), and
- The costs of delivery (e.g. shipping or transport); and
o Evidence of either of the following:
- Under clause 14.5(b), evidence that the plant and materials are being shipped and are insured, along with any other documents or undertakings; or
- Under clause 14.6(c), evidence that the plant and materials have been delivered to site and are properly stored.

The engineer makes a determination (see Section 9.7.2.1) of the value of plant and materials and shall include 80% of this value in the IPC.

---

180 FIDIC Yellow Book 2017, clause 14.3(vii).
181 FIDIC Yellow Book 2017, clause 14.3(v). 'Plant' is defined in clause 1.1.66 and 'Materials' is defined in clause 1.1.53.
182 FIDIC Yellow Book 2017, clause 14.3(ix).
183 FIDIC Yellow Book 2017, clause 14.3(ii).
184 FIDIC Yellow Book 2017, clause 14.3(vi).
185 *Henia Investments Inc. v Beck Interiors Ltd*, EWHC 2433 (TCC), [2015], at paragraph 17.

FIDIC Yellow Book 2017 includes a provision for the employer to deduct retention in a similar way to JCT DB 2016; again this is an incentive for the contractor to complete the works (i.e. to achieve Taking Over)[186] (see Section 9.3.2) along with completing any outstanding works, rectification of defects, and testing (see Section 9.5.2). Similar to JCT DB 2016, clause 14.9 of FIDIC Yellow Book 2017 deals with retention in either of the following two ways:

o The Retention Money,[175] calculated as the 'percentage of retention' up to the 'limit of Retention Money' (both set out in the Contract Data[187]), which must be shown in the contractor's statement[188] as follows:
  – All of the Retention Money (i.e. the full 'percentage of retention') is included (i.e. shown as deductible by the employer)[189] in every contractor's statement before the issue of the Taking Over Certificate.
  – After the Taking Over Certificate is issued, the contractor's statement includes the first half of the Retention Money.[190]
  – After the final Defects Notification Period[191] has expired, the contractor's statement shall include the second half of the Retention Money.[192]
  – Unlike JCT DB 2016,[193] clause 14.3(iii) of FIDIC Yellow Book 2017 requires the contractor to effectively to apply for payment of the Retention Money.
o If the Special Provisions[194] apply, they include the substitute clause 14.9 that provides for a Retention Money Guarantee[195] instead of the Retention Money being deducted (as per clause 14.9 of the General Conditions).

Clause 14.6.1 (Issue of IPC) sets out that the engineer must issue the Interim Payment Certificate within 28 days after receiving the contractor's statement, and any 'supporting documents'. The inclusion of 'supporting documents' in clause 14.6.1 is perhaps surprising, given that they must already be provided as part of the statement under clause 14.3(c).

---

186 FIDIC Yellow Book 2017, clause 10 (Employer's Taking Over).
187 FIDIC Yellow Book 2017, Contract Data entries against clauses 14.3(ii) and 14.3(iii). Clause 14.9 (Release of Retention Money) provides that where completion by Sections applies, but no retention percentage is entered in the Contract Data against the percentage value of a Section, then the first half of the Retention Money is *not* released upon completion of that Section, but is released upon Taking Over of the Works – as per clause 14.9(a).
188 FIDIC Yellow Book 2017, clause 14.3.
189 FIDIC Yellow Book 2017, clause 14.3(iii).
190 FIDIC Yellow Book 2017, clause 14.9(a) for the Works, or clause 14.9(b) for a Section (see Section 8.3.2).
191 FIDIC Yellow Book 2017, clause 1.1.27 defines the 'Defects Notification Period' or 'DNP', also see Section 9.5.2.
192 FIDIC Yellow Book 2017, clause 14.9 (Release of Retention Money).
193 JCT DB 2016, clause 4.16 entitles the employer to 'deduct and retain' the retention until the events described in that clause. The contractor does not have to apply for the retention to be paid; instead, the employer's entitlement to deduct and retain the retention ceases.
194 FIDIC Yellow Book 2017, clause 1.1.78 defines the 'Special Provisions', which (if applicable) will be set out in Particular Conditions – Part B – Special Provisions.
195 FIDIC Yellow Book 2017, Annex F – Example Form of Retention Money Guarantee.

It therefore appears that the contractor may provide supporting documents after issuing the statement, but if he does so, it will have the following effects:

o The engineer is not obliged to issue the IPC until 28 days after receiving the supporting documents[196]; and

o Similarly, the employer is not obliged to make an interim payment to the contractor until 56 days after the engineer receives the supporting documents.[197]

The IPC must set out the amount that the engineer considers to be due,[198] along with any other amounts arising from clause 3.7 (Agreement or Determination) or arising under or in connection with the contract.[199] Importantly, the IPC must also set out each difference between the engineer's valuation (or consideration of the sum due) and the corresponding amount in the contractor's statement. In practice, this is likely to comprise either a spreadsheet showing the differences or a marked-up copy of the statement showing the same.

The final date for the employer to pay interim payments is set out in clause 14.7 and is 56 days from the date when the engineer receives the contractor's statement and supporting documents. The employer's obligation to pay the amount certified in the IPC depends upon:

o The contractor providing the employer with the Performance Security (see Sections 7.4.1 and 8.1.3) – clause 14.6(a); or

o The contractor appointing his representative (under clause 4.3 – see Section 8.2.2) unless he is already named in the contract – clause 14.6(b).

Clause 14.6.2 firstly permits the engineer to withhold an IPC if the amount payable to the contractor (after deducting retention and other permissible amounts) is less than the 'minimum amount of Interim Payment Certificate (IPC)' entered against clause 14.6.2 in the Contract Data. It is questionable if such a provision is permissible in the UK if the HGCRA 1996 applies to the contract (see Section 8.7). However, if by virtue of the contractor agreeing that he is only entitled to an interim payment that exceeds the minimum amount under clause 14.6.2, this provides valid grounds for a 'pay less notice'. Therefore, in practical terms, the employer would not be obliged to make any interim payment less than the minimum amount under clause 14.6.2, but the engineer:

o Could not avoid issuing the IPC; and

---

196 FIDIC Yellow Book 2017, clause 14.6.1, or until the contractor provides the 'Performance Security' (under clause 4.2 – see Section 8.1.2 above) or has appointed the 'Contractor's Representative' under clause 4.3.

197 FIDIC Yellow Book 2017, clause 14.7(b)(i).

198 FIDIC Yellow Book 2017, clause 14.6.1(a) – the corresponding amounts in the Statement, under clauses 14.3(i) to 14.3(v) and 14.3(vii) to 14.3(x). For the amount for 'Plant' and 'Materials' under clause 14.3(v), this can only be 80% of the value determined by the engineer.

199 FIDIC Yellow Book 2017, clause 14.6.1(b) – the corresponding amounts in the Statement under clause 14.3(vi).

o Would have to issue a valid pay less notice, citing clause 14.6.2 as the grounds for paying less than the amount shown in the IPC.

The remainder of clause 14.6.2 clarifies that the IPC cannot be withheld for any other reason, but the engineer may withhold (or in other words, make the appropriate reductions for) amounts in the following circumstances:

o Works not in accordance with the contract – clause 14.6.2(a);
o Any services or other obligation that the contractor has failed to perform – clause 14.6.2(b); and
o To account for an error or discrepancy in the contractor's statement – clause 14.6.2(c).

The engineer is only obliged to notify the contractor for the circumstances in clause 14.6.2(b), but he must provide his calculations of the amounts withheld (or reduced) under all subclauses. If the HGCRA 1996 applies to the contract, then the matters in subclauses 14.6.2(a) to (c) could be set out in either the IPC or a pay less notice. The engineer may also correct or modify any previous payment certificate under clause 14.6.3, in a subsequent payment certificate. This clause also requires the contractor to notify the engineer to refer a dispute for a 'determination' under clause 3.7 (Agreement or Determination – see Section 9.7.2.1). However, if the HGCRA 1996 applies to the contract, the contractor will have the right to refer a dispute to adjudication at any time (refer to Section 9.7.4.1).

Figure 8.2 provides an illustration of how to modify FIDIC Yellow Book 2017s interim payment provisions so they comply with the HGCRA 1996. In maintaining the same periods for the engineer to issue the IPC and the employer to make payment after the contractor provides his statement, the following amendments are required:

o To comply with s110(1)(a) of the HGCRA 1996, the 'due date for payment' should occur 23 days after the contractor provides his statement, in order to preserve the 28 days period for the engineer to issue the IPC;
o The IPC also becomes the payment notice in accordance with s110(1)(a) of the HGCRA 1996; and
o In order to comply with s111(3) of the HGCRA 1996, the employer has until seven days prior to the final date for payment to issue a pay less notice.

As with JCT DB 2016, the contractor's statement will negate the need to provide a separate 'contractor's payment notice' under s110B(1) of the HGCRA.

It appears that clause 16.1(c) provides the contractor with the right of suspension if the employer does not pay, akin to that provided by s112(1) of the HGCRA 1996, but the contractor may only do so if he gives the employer 21 days' prior notice[200] of his intention to do so.

Clause 14.8 provides for the contractor to receive 'financing charges' (in addition to any principal sum owed) for late payments, calculated as 3% compound interest (at monthly rests) unless another percentage is stated in the Contract Data (the entry against

---

200 HGCRA s112(2) provides that 'at least seven days' notice of intention to suspend performance' should be given, so clause 16.1(c) complies with that timescale.

clause 14.8). The contractor does not have to comply with clause 20 (Claims for Payment and/or EOT – see Section 8.10.4.1) when claiming for financing charges.

8.7.3 NEC4

NEC4's interim payment procedure is set out in clause 51 (Payment) and, for qualifying projects in the UK, is made compliant with the HGCRA 1996 by adopting the Secondary Option Clause Y(UK)2 (HGCRA 1996). It is very similar to the equivalent procedure in JCT DB 2016, insofar as:

o The contractor must apply for payment (an 'application for payment'[201]) before each 'assessment date', and stating the amount he considers is due on that date (clause 50.2).
o The project manager decides the first assessment date, which cannot be later than the 'assessment interval' (as set out in Contract Data Part One) after the 'starting date'.[202]
o Subsequent assessment dates occur at the end of each assessment interval until the issue of either a Defects Certificate or a termination certificate.[203]
o The project manager issues his certificate within seven days after the assessment date (clause 51.1).
o Clause Y2.2 provides the following:
 – The payment due date occurs seven days after the assessment date[204] (effectively providing that the date for the project manager's certificate and the due date are likely to occur simultaneously[205]).
 – The project manager's certificate is the 'notice of payment' – which must state the amount due at the payment due date and provide the basis for its calculation (the 'notified sum' – which may be zero).
o Clause 51.2 provides that the client makes the interim payment to the contractor within three weeks of the assessment date.[206] Accounting for the first week between the assessment date and the payment due date,[207] this effectively means that the final date for payment is 14 days from the payment due date.
o Clause Y2.3 provides that:
 – Any party giving a pay less notice must do so not later than seven days prior to the final date for payment; and
 – Unless a pay less notice is issued, the paying party (usually the client) pays the notified sum.

---

201 NEC4, clause 50.2 'application for payment' is not a defined term.
202 NEC4, clause 50.1.
203 NEC4, Option Y(UK)2, clause Y2.2.
204 This is analogous to JCT DB 2016 whereby the 'due date' occurs seven days after the Interim Valuation Date (clause 4.7.3).
205 This slightly differs from JCT DB 2016 as the clause 4.7.5 'payment notice' can be issued up to five days after the 'due date'.
206 This is expressed in a similar way (an overall period) to clause 14.7 of the FIDIC Yellow Book 2017, whereas JCT DB 2016 expresses the final date for payment as a period starting at the 'due date' (clause 4.9.1).
207 NEC4, clause Y2.2.

However, NEC4's interim payment procedure also has its own characteristics:

o Clause 50.4 provides that if the contractor does not submit his application for payment before the assessment date, he will either not be entitled to any payment or may have to repay sums to the client because the amount due is either:
  – The amount due at the previous assessment date; or
  – The amount that the project manager assesses at the (current) assessment date, but only if this is less than the amount due at the previous assessment date.
o Clause 50.5 permits the client to retain 25% (one quarter) of the 'Price for Work Done to Date'[208] if the Contract Data[209] does not identify a programme, and the project manager has not received the contractor's first programme for acceptance.[210] This presents a significant financial setback to the contractor and reinforces the importance that NEC4 places on the programme and on the contemporaneous management of the contract in general.

The contractor's application for payment must set out the amount due at the assessment date,[211] which is calculated as follows (clause 50.3):

o The 'Price for Work Done to Date' (see Table 8.9 for Main Option clauses A–E); along with

Table 8.9   NEC4 – 'Price for Work Done to Date' (Main Option Clauses A–E).

Main Option Clause	Clause	Definition of Price for Work Done to Date
Option A: Priced Contract with Activity Schedule	11.2(29)	Total of the Prices[a] for all completed[b] activities, either (i) in a group, or (ii) not in a group.
Option B: Priced Contract with Bill of Quantities ('BoQ')	11.2(30)	Total of the following, completed by the contractor: – Quantity of work in the BoQ multiplied by the rate – Proportion of lump sum in the BoQ
Option C: Target Contract with Activity Schedule	11.2(31)	The total Defined Cost[c] (i.e. the costs in the Schedule of Cost Components less Disallowed Cost[d]) forecast to the next assessment date, plus the Fee[e]
Option D: Target Contract with Bill of Quantities		
Option E: Cost Reimbursable Contract		

a) NEC4 – Option A (Priced Contract with Activity Schedule); the 'Prices' are defined in clause 11.2(32).
b) NEC4 – Option A (Priced Contract with Activity Schedule), clause 11.2(32) defines a 'completed activity' as one without notified Defects, which would delay following work if corrected.
c) NEC4 – Options C, D, and E – clause 11.2(24) defines 'Defined Cost'.
d) NEC4 – Options C, D, and E – clause 11.2(26) defines 'Disallowed Cost'.
e) NEC4 – clause 11.2(10) defines the Fee. In addition, clause 52.1 provides that all the contractor's costs that do not fall within the definition of 'Defined Cost' are deemed to be included within the Fee.

---

208 NEC4, 'Price for Work Done to Date' is defined in the following clauses for each of the Main Options: Option A (Priced Contract with Activity Schedule) – clause 11.2(29), Option B (Priced Contract with Bill of Quantities) – clause 11.2(30), Option C (Target Contract with Activity Schedule), Option D (Target Contract with Bill of Quantities), and Option E (Cost Reimbursable Contract) – clause 11.2(31),
209 NEC4, Contract Data Part One – '3 Time'.
210 NEC4, clause 31.1.
211 NEC4, clause 50.2.

o Other adjustments:
  - Amounts to be paid by the client to the contractor, and
  - Amounts to be retained[212] by the client or paid by the contractor.

If Option X16 (Retention) is selected, the client may retain an amount from sums otherwise due to the contractor in similar ways to JCT DB 2016 and FIDIC Yellow Book 2017. Unless Option X16 states that a retention bond applies,[213] the amount (the retention) that the client may retain[214] is calculated as the retention percentage applied to the 'Price for Work Done to Date', after deducting the 'retention free amount' (both of which are set out in Contract Data Part One[213]). The client retains the retention as follows:

o Until the earlier of completion of the whole of the works or the date on which the client takes over the whole of the works,[214] half the retention amount is released in the next assessment[215]; then
o The remaining half of the retention is retained until the Defects Certificate[216] is issued[215].

The 'Price for Work Done to Date' will vary depending upon which of the Main Option clauses have been selected. Table 8.9 summarises each of these.

Under Main Options C, D, and E, clause 50.9 provides that the contractor must notify the project manager when he has compiled the relevant Defined Cost and maintain records to show that its assessment is correct. Within 13 weeks, the project manager reviews the contractor's records and, if they are sufficient, notifies him that the Defined Costs are correct or that they contain errors. If the contractor's Defined Costs records are not sufficient, the project manager notifies the contractor that more records must be provided. The aim of clause 50.9 is to encourage the finalisation of Defined Costs.[217]

Clause Y2.5 provides the contractor with the same right of suspension in the event of non-payment as is provided in s112(1) of the HGCRA 1996 and confirms that this is a 'compensation event'. In addition, clauses 51.3 and 51.4 entitle the payee (usually the contractor) to interest on late payments.

8.7.4 Consultants' Conditions of Engagement

If a project is subject to UK legislation, then the payment terms in the consultant's conditions of engagement must similarly comply with the HGCRA 1996. For the four conditions of engagement considered in Chapter 4, Table 8.10 summarises the payment terms and their compliance with the HGCRA 1996.

---

212 NEC4, For example, under Option X16: Retention.
213 NEC4, Contract Data Part One – relevant entries against Option X16 (Retention) and clause X16.3 refers to the retention bond.
214 NEC4, Option X16 (Retention), clause X16.1.
215 NEC4, Option X16 (Retention), clause X16.2. Retention released under clause X16.2 is included in the assessment (under clause 50.1) after Completion or after the Defects Certificate (as the case may be).
216 NEC4, clause 44.3 – see Section 9.5.3.
217 NEC4, *Managing an Engineering and Construction Contract,* vol. 4, (2017), p. 48.

Table 8.10 Conditions of Engagement complying with HGCRA 1996.

Designer's Conditions of Engagement	Payment Terms' compliance with the HGCRA 1996	Payment Terms
RIBA Standard Professional Services Contract 2020 ('RIBA PSC')	Yes	Clauses 5.10–5.24
FIDIC Client/Consultant Model Services Agreement, 5th ed., 2017 ('FIDIC MSA')	No	Clauses 7.1, 7.2
Association of Consulting Engineers (ACE) Professional Services Agreement 2017 ('ACE PSA')	Yes	Clauses 7.3–7.7
NEC4 Professional Services Contract (June 2017 with January 2019 amendments) ('NEC4 PSC')	Yes	Clause 51 Option Y(UK)2

8.8 Variations

Understanding and managing variations in design and build contracts is very important as it is rare for engineering or construction projects to proceed to completion without any changes being made. How those changes were generated, and how they are managed in terms of their effects on the project's programme and cost, are of major importance to the employer and contractor – and are often considered from completely opposite standpoints. The employer may not want the project to overrun or overspend and, consequently, may not accept that a variation has occurred or may argue that its effect on programme and budget are minimal. Conversely, the contractor may seek to maximise his entitlements to extensions of time and additional payments. Considering these often opposing viewpoints, it is hardly surprising that variations (and their valuation) comprised around 40% of construction disputes in the UK between 2012 and 2018.[218]

The three design and build contracts all deal with the valuation of variations in two different ways. They either allow the parties to agree on the effects that the variation has on adjusting the contract price or the time for completing the works (i) before carrying out the variation, or (ii) after it has been undertaken. Option (i) gives the employer an opportunity to obtain further information from the contractor to consider before deciding whether he wants to instruct the variation.

8.8.1 General Points

A key part of understanding the concept of variations is to understand the following aspects:

o Is the variation recognised under the terms of the contract?

---

218 However, the valuation of variations as a disputed matter fell from 42% (in 2015 and 2018) to 26% in 2022. RIBA/NBS National Construction Contracts & Law Surveys (2012, 2013, 2015, 2018, and 2022).

o Does the variation result in an adjustment of contract sum (or contract price)?
  – An increase in the contract sum because of the variation requires (i) carrying out a greater quantity of work, (ii) similar work that is more expensive, or (iii) additional new work; or
  – A reduction in the contract sum because the variation requires (i) less work to be carried out, (ii) similar work that is cheaper, or (iii) an omission of work.
o Does the variation result in a change to the programme?
  – Additional time (or an extension of time) – for example, if the contractor has to carry out more work; or
  – Less time, because the contractor has less work to carry out.
o Does the variation provide any other entitlements to the contractor (such as loss and expense)?

8.8.2 Is the 'Variation' Recognised Under the Terms of the Contract?

To begin to answer this question, it is important to look at what has been changed, thereby ascertaining whether (i) the employer has instructed or imposed changes to the Scope of work or the conditions under which they are carried out, or (ii) those changes which have been introduced by the contractor.

In Sections 3.4 and 3.5, this book briefly summarised that the Employer's Requirements (or Scope[219]) is the employer's document that tells the contractor *what* he wants, and the Contractor's Proposals (or the Scope provided by the contractor for its design[102]) is the contractor's document that tells the employer *how* he is going to build what the employer wants. If the changes are obvious, then the employer will expect that he will have to pay extra for the varied work and possibly allow the contractor a longer period to build it.

Obvious examples of the employer changing what he wants could include:

o Instructing the contractor to carry out additional work (increased quantities), as the employer can clearly see the extra work being undertaken.
o Substituting a cheaper element of work for one that is more expensive.
o Extensive layout changes.

The employer may instruct other changes, the effects of which may be less obvious:

o Substituting similar products or elements of work.
o Minor layout or configuration changes.
o Omission of work.
o Altering the contractor's method of working.
o Restricting the choices available to the contractor.

The employer may initially perceive that changes of this latter nature may not have any or only a minor effect on the contract price or the programme. However, he may not appreciate that even minor changes in layout could require the structural or building services elements to be altered. Similarly, it may not be obvious that if an item of work is omitted, ancillary work may have already been completed so the overall saving is not as great as the

---

219 NEC4.

employer may expect. It is therefore important that the contractor explains the effects that changes may have; if the employer is not an experienced client, he should engage appropriate advisors who understand the implication of a change.

Section 8.4.2 set out that generally, save for a couple of exceptions, the contractor is free to choose his method of working. In addition, the rule of 'minimal performance' will apply to the contractor's work (see Section 3.1.2). This does not mean that the contractor can get away with doing less than the contract requires; it simply means he has to do the minimum to satisfy his obligations under the contract. In the case of *Sunlife Europe Properties Ltd v Tiger Aspect Holdings*,[220] Edwards-Stuart J described the rule in the following terms:

> '...where a contract can be performed in more than one way, the performing party can choose the way that is least onerous to him. He is under no obligation to perform the contract in some other more time-consuming or expensive manner, simply because the other party would like it, if there is a more economic and compliant means of doing so.'[221]

Therefore, if the employer instructs the contractor to change his method (or imposes a restriction which causes the contractor to change his method) from one that he has chosen as the most economical to satisfy the contract's requirements, then this is usually a variation under the contract. Common examples could be:

o Instructing a specific change in the method of working (e.g. changing from excavation by machine to excavation by hand) where either method complies with the contract.
o Changing (reducing) the project's working hours.
o Altering the project's logistics:
  – Changing the contractor's access routes to or from the site;
  – Making a change which requires site operations to be altered (such as working on an area used for accommodation or material storage earlier than programmed);
  – Imposing restrictions for access to the works (such as tower crane or hoist usage).

Another instance in which a variation can occur that is related to the contractor's chosen method is where the contract allows the contractor a choice of methods but the employer subsequently restricts or removes that choice. A variation may also arise if the employer restricts or removes the contractor's choice of materials or goods to be incorporated into the work. One example involves a contract in which the contractor was allowed to carry out filling works using either (i) imported fill or (ii) material arising from the preceding demolition work that had been crushed to an appropriate size. The contractor felt that the most economical way of carrying out the filling work was to use crushed material arising from the demolition work. The employer subsequently instructed the contractor to only use imported material. The court found in favour of the contractor – that the reduction in choice imposed by the employer in specifying imported fill amounted to a variation and the contractor was entitled to additional payment.[222] In a similar case, the court decided that

220 EWHC 1161 (TCC), [2013].
221 EWHC 1161 (TCC), [2013] at paragraph 28.
222 *English Industrial Estates Corporation v Kier Construction Ltd*, 56 BLR 93, (1991).

a variation occurred when the contract allowed the contractor the choice of laying water pipes that were manufactured by either Stanton or Staveley, but the engineer instructed that Staveley pipes be used.[223] The contractor based his original price on Stanton pipes, so was entitled to (i) additional payment for change to Staveley pipes, and (ii) the additional work because the Staveley pipes had larger joints.

If a change originates from the contractor, then it is far less likely that the employer will have to pay extra for it or allow additional time for it to be carried out. Changes in the contractor's design often arise as his design is developed, as coordination issues may arise which result in work having to be adapted or redone. The employer will often reject these changes on the ground that they have arisen through necessary 'design development' – after all, the contractor having single-point responsibility to the employer for developing the project's design is probably the reason why the employer procured the project on a design and build basis (see Section 3.1.2). It is therefore important that the contractor ensures that his design is fully coordinated; and engaging a 'lead designer' (see Section 4.2.6) is an important step to successfully incorporate and coordinate all elements of the design (including the architectural, structural, and building services elements, along with design from the relevant subcontractors).

The employer may reject the contractor's claims for variations on the ground that the resulting work is 'indispensably necessary'[224] to complete the work. The following examples are relevant to design and build contracts:

o If an obvious item of work (such as a roof or a floor) is not specified in the employer's requirements but the contractor is nevertheless obliged to complete the design and construct the works to that design, then it is unlikely that the missing item of work will be a variation.
o If the scope of work is greater than the contractor first anticipated,[225] carrying out work that the contractor deems additional to meet that scope of work will not be a variation. Therefore, it is important that the contractor effectively deals with the risk[226] of having to carry out additional works that he may not be able to quantify when pricing the project.

Generally, the contractor is not obliged to carry out variations after he has completed the works, as his obligations for carrying out any physical works will cease (save for correcting any defects within the designated period after completion).

8.8.3 JCT DB 2016

Under JCT DB 2016, a variation is called a Change,[227] and usually occurs when the employer issues an instruction under clause 3.9 (Instructions Requiring Changes) or 3.11

---

223 *J Crosby & Sons Ltd v Portland UDC*, 5 BLR 21 at 12, (1967).
224 *Keating on Construction Contracts*, 10th ed., (2016 and 2019), at 4–045.
225 *Atkins Ltd v Secretary of State for Transport*, EWHC 139 (TCC), [2013].
226 For example, in the case of refurbishment projects, where it may be difficult to ascertain the precise quantity and nature of the work when pricing the project, the contractor may include a 'risks and contingency' premium (see Section 3.6.3) or insist that 'provisional sums' (see Section 3.6.4) are included within the appropriate part of the contract for such works.
227 JCT DB 2016, clause 1.1 (Definitions) simply defines 'Changes' as 'see clause 5.1'.

(Instructions on Provisional Sums). A comprehensive definition of 'Change' is provided in clause 5.1, which sets out the criteria for a Change to occur:

o It must arise from a change in the Employer's Requirements; and
o The following changes[228] must be necessary:
  – Changes to the design,
  – Changes to the works' quality, or
  – Changes to the quantities.

The only exception to the above is if the Employer's Requirements are changed as a result of the employer instructing the contractor (under clause 3.13) to remove any work that does not comply with the contract.[229]

Clause 5.1 helpfully gives two subcategories of changes, which are as follows:

o Clause 5.1.1 – those affecting the physical works themselves. This includes additional work, substituted work, and omissions of work.
o Clause 5.1.2 – the changes in the conditions under which the contractor must undertake the work. This includes restrictions or limitations of site access, working space, and working hours, and having to carry out the works in a specific order. Clause 5.1.2 restrictions will still create a change, even if there are no changes to the physical works (i.e. under clause 5.1.1).

It is very clear that changes can only arise if the Employer's Requirements have changed. The employer has no ability to instruct a change to the Contractor's Proposals, but there should be no need for him to do so because the Contractor's Proposals should satisfy the Employer's Requirements (see Section 3.3.2 above). If the employer considers that the Contractor's Proposals do not satisfy the Employer's Requirements, then his course of action is not to instruct under clause 3.9 (Instructions Requiring Changes), but to instruct under clause 3.13 (Work Not in Accordance with the Contract). The employer is free to issue instructions requiring changes (under clause 3.9) but the contractor may reasonably object to the instruction on the following grounds:

o If the clause 3.9 or 3.11 instruction is likely to alter or modify the contractor's design, he may object to a change.[230]
o The contractor may also give his reasonable objections to the instruction on the grounds that it may compromise the safety of the contractor's design under the CDM Regulations. The employer shall then vary the instruction by removing the grounds for the contractor's objection and, until he does, the contractor is not obliged to comply with the clause 3.9 instruction.

---

228 JCT DB 2016, clause 5.1.1 – the changes are described as an 'alteration' or 'modification'.
229 The use of 'rectification' in clause 5.1.1 could arguably be a general description of the purpose of instructions issued under clause 3.13, as clause 3.13 only expressly obliges the contractor to 'remove' work that is not in accordance with the Contract. However, the corollary of a clause 3.13 instruction to 'remove' non-compliant work is that the contractor must replace it with work that is compliant to satisfy his obligation under clause 2.1.1 'carry out and complete the Works…' (underlining added) – see Section 9.3.1 (Meaning of 'Completion').
230 JCT DB 2016, clause 3.9.1.

JCT DB 2016 includes optional procedures for the parties to agree on the effects of the variation in terms of the price for carrying out the variation, any extension of time,[231] and loss and expense[232] prior to the variation being instructed. If selected in the Contract Particulars, paragraph 2 (Valuation of Changes – Contractor's Estimates) of Part 1 of the Supplemental Provisions provides that if the contractor considers that an employer's instruction under clause 3.9 (Instructions Requiring Changes) means it will need to carry out a 'Valuation' (under clause 5.2 – see Section 8.9.1), then the contractor must provide the following information to the employer within 14 days[233] of the clause 3.9 instruction:

o Financial information, including:
  – The adjustment to the Contract Sum, and
  – Any loss and/or expense.
o Practical information, including:
  – Additional resources, and
  – A method statement.
o The length of any required extension of time.

The employer then has a further 10 days to agree on the contractor's estimate. If no agreement is reached, the employer has two options, either:

o Instruct compliance with the clause 3.9 instruction, and the Supplemental Provision shall not apply; or
o Withdraw the clause 3.9 instruction. If the instruction is withdrawn, the contractor is entitled to be paid any design costs incurred when preparing the estimate.

There is a horrible 'sting in the tail' for the contractor if the Supplemental Provisions apply,[234] and if he fails to provide his estimate within 14 days in paragraph 2.2. He will not be entitled to any addition to the Contract Sum in any interim payment, and must wait until Final Payment for it – which, at the earliest, may be between 8 and 14 months after practical completion! In the author's experience, the Supplemental Provisions of JCT DB 2016 are often selected, but are rarely operated by either party.

Variations (or 'Changes') cannot be instructed under JCT DB 2016 after the 'Practical Completion Statement' is issued under clause 2.27.1, as this signifies that the contractor has completed the Scope of the Works (including any changes). The only instructions the contractor is obliged to carry out are those to rectify defects, shrinkages, or other faults under clause 2.35.1 (Schedules of Defects and Instructions).

8.8.4 FIDIC Yellow Book 2017

FIDIC Yellow Book 2017 uses the term 'Variation', for which a basic definition is given in clause 1.1.88. In the same way as JCT DB 2016, the detail of a variation is provided in the

---

231 Refer to Section 8.10.3.
232 Refer to Section 8.11.2.
233 JCT DB 2016 Supplemental Provisions Part 1, paragraph 2.2 – the parties may agree to a period other than 14 days, or within a period that is reasonable in the circumstances.
234 JCT DB 2016, Supplemental Provisions Part 1, paragraph 2.6.

variation clause 13 (Variations and Adjustments). From reading clause 1.1.88 and clause 13 together, the following bring about a variation:

o It must be by instruction[235] from the engineer.
o The engineer should state that it is a variation[236] in that instruction.
o The engineer's instruction must be issued prior to the Taking Over Certificate being issued.[237]
o The engineer may initiate[238] the variation, or it may be proposed by the contractor.[239]
o A 'Variation' is 'any change to the Works',[240] but does not permit the employer to omit work for the purpose of having it carried out by a third party or by himself (save where the contractor fails to remedy a defect[241]).

The FIDIC Yellow Book 2017 definition of a 'Variation' appears to be very wide, as it means 'any change to the Works'. The lack of a precise definition of 'Variation' (e.g. similar to that in clause 5.1 of JCT DB 2016) may be a source of disputes. The 'Works' are defined[242] clearly to include the 'Permanent Works', but also the 'Temporary Works', so in terms of the latter, a Variation may extend to changing the contractor's method of working where the engineer instructs a Variation to the Temporary Works.[243]

A Variation to the Permanent Works appears more straightforward. Clause 13.1 of FIDIC Yellow Book 2017 expressly refers to 'omissions', so by implication, it seems obvious that a Variation would also include 'additions' to the Works. The plain and ordinary meaning of 'change to the Works' would also include altering the specification of the Permanent Works, along with making design changes.

What is less clear is whether a variation includes changes in the way that the works are carried out (see clause 5.1.2 of JCT DB 2016 in Section 8.8.2 above). Arguably, if the employer imposes any restrictions on the contractor (e.g. changing the site's access routes, limitation of working hours, or instructing that the works be carried out in a particular sequence) then they may not be treated as Variations because they are not 'changes to the Works' themselves. However, such examples are likely to be grounds for the contractor to become entitled to an extension of time under clause 8.5(c).[244] It appears that the engineer can either deal with any extensions of time either under clause 20.2 (Claims for Payment and/or EoT), or under his determination[245] of any extension of time and adjustment to the contract price (see below).

---

235 FIDIC Yellow Book 2017, clause 1.1.88.
236 FIDIC Yellow Book 2017, clause 1.1.88 and clause 3.5 (Engineer's Instructions).
237 FIDIC Yellow Book 2017, clause 13.1.
238 FIDIC Yellow Book 2017, clauses 13.1, 13.2, and 13.3.1.
239 FIDIC Yellow Book 2017, clause 13.3.2 (Variation by Request for Proposal).
240 FIDIC Yellow Book 2017, clause 1.1.88. 'Works' is defined in clause 1.1.89, which means the Permanent Works and the Temporary Works, which are defined in clauses 1.1.65 and 1.1.82, respectively.
241 FIDIC Yellow Book 2017, clause 11.4 (Failure to Remedy Defects).
242 FIDIC Yellow Book 2017, clause 1.1.89.
243 FIDIC Yellow Book 2017, clause 1.1.82 defines 'Temporary Works', but it is thought that a Variation cannot instruct the contractor what plant and equipment to use as 'Contractor's Equipment' (defined in clause 1.1.15) does not form part of the Temporary Works.
244 FIDIC Yellow Book 2017, clauses 8.5(c) – 'any delay, impediment or prevention caused by…the Employer' and 4.15 (Access Route).
245 FIDIC Yellow Book 2017, clause 3.7 (Agreement or Determination).

Clause 13.3 sets out the procedure for variations. Firstly, the engineer gives a notice[246] to the contractor, setting out the change that he requires, along with any requirements for recording costs that the contractor should satisfy.[247] Secondly, it is clear that the contractor proceeds to carry out the engineer's instruction without delay, unless he submits a reasoned objection under clauses 13.1(a) to 13.1(e), which provide the following grounds for objection:

o The engineer's instruction may compromise the contractor's design[248] (preventing the works being fit for purpose) or may not be able to be undertaken safely[249] (similar to clause 3.5.1 of JCT DB 2016 – see Section 8.8.3);
o The instructed works are 'Unforeseeable'[250] with regard to the nature and Scope of the Works set out in the Employer's Requirements[251];
o 'Goods'[252] intended for the varied works cannot be readily obtained[253]; and
o The variation may adversely affect the contractor's ability to satisfy the Schedule of Performance Guarantees.[254]

Thirdly, under clause 13.3.1 and within 28 days (or some other period agreed on with the engineer) the contractor shall provide the following particulars to the engineer:

o A method statement including the resources that the contractor has used or intends to use[255];
o A programme for the varied work,[256] which must include any consequential amendments to the programme provided under clause 8.3; and
o The proposed adjustment of the contract price, including any costs for omitting work.[257]

Given that unless he objects, the contractor must proceed with the varied work almost immediately, and that the engineer's determination of the effects on the programme and contract price may not be due for another 60 days,[258] FIDIC's variation procedure appears to place some significant risks with the contractor. However, the contractor's cash flow risk may be alleviated as clause 13.3.1 provides that for Interim Payment Certificates that are due between the contractor commencing the variation and the engineer's determination,

---

246 FIDIC Yellow Book 2017, complying with clause 1.3.
247 FIDIC Yellow Book 2017, clause 13.3.1. See also Section 8.10.4.1.
248 FIDIC Yellow Book 2017, clause 13.1(e).
249 FIDIC Yellow Book 2017, clause 13.1(c).
250 FIDIC Yellow Book 2017, defined in clause 1.1.87 as something an experienced contractor would not foresee at the Base Date (28 days before the last tender submission date [clause 1.1.4]). JCT DB 2016 adopts a similar concept – the 'Base Date' is effectively the cut-off date for changes in legislation, etc. which may entitle the contractor to adjust the Contract Sum.
251 FIDIC Yellow Book 2017, clause 13.1|(a).
252 FIDIC Yellow Book 2017, clause 1.1.44 defines 'Goods' as any of the Contractor's Equipment, Materials (including those supplied to the Contractor), Plant, and Temporary Works.
253 FIDIC Yellow Book 2017, clause 13.1(b).
254 FIDIC Yellow Book 2017, clause 13.1(d).
255 FIDIC Yellow Book 2017, clause 13.3.1(a).
256 FIDIC Yellow Book 2017, clause 13.3.1(b).
257 FIDIC Yellow Book 2017, clause 13.3.1(c).
258 Allowing for the 28 days for the contractor to provide the clause 13.3.1 particulars and the 42-day time limit for the engineer's determination under clause 3.7.3.

the engineer must use provisional rates or prices for the varied work. However, there does not appear to be an equivalent obligation regarding time, so it seems that the engineer does not have to apply a provisional extension of time before making his determination under clause 3.7.

Clause 13.3.2 of the FIDIC Yellow Book 2017 gives the opportunity for the engineer to find out the likely effects that a variation may have on the contract price and programme. Firstly, the engineer notifies the contractor (providing him with details of the proposed change) and asks the contractor for his response, which is either:

o An objection citing the same grounds as in clauses 13.1 (a)–(e); or
o His clause 13.3.2 'proposal', containing the information required by clauses 13.3.1(a)–(c).

If the contractor submits a clause 13.3.2 proposal (and not an objection), the engineer responds[259] by either providing his consent or, seemingly, his comments.[260] Clause 13.3.2 states that the contractor shall not delay any work for the period during which the engineer is considering the proposal. The words 'any work' appear to mean the works (including previously instructed variations) but not the work that is the subject of the engineer's proposal request, because it can only be instructed as a variation once the engineer consents to the contractor's proposal.

Clause 13.3.2 only seems to operate in a straightforward manner if the engineer 'consents' to the contractor's proposal. He proceeds to issue a variation instruction, thereby obliging the contractor to get on with the varied work without delay and provide any further particulars that the engineer may reasonably require.

However, clause 13.3.2 seems unhelpful in the following respects:

o Curiously, the contractor's clause 13.3.2 'proposal' is not accepted in the same way that a 'Contractor's Estimate' (under JCT DB 2016s Supplemental Provisions – see Section 8.7.2) or a quotation for a Compensation Event (under clause 62.1 of NEC4 – see Section 8.7.4) can be accepted. The engineer may effectively ignore the contractor's proposal and proceed with the agreement/determination procedure.[261]
o If the engineer does not give consent to the clause 13.3.2 proposal (instead providing his 'comments'), there is no apparent procedure telling the contractor what to do with his proposal, either in line with the engineer's comments or otherwise. For example, the contractor does not need to revise his proposal in order to obtain the engineer's consent in the same way as he needs to revise his design under the design review procedure in clause 5.2.2(b) (refer to Section 3.5.2).

However, if the contractor incurs any cost when preparing a clause 13.3.2 proposal which does not receive the engineer's consent, he can recover this cost via clause 20.2 (Claims for Payment and/or EoT).

---

259 FIDIC Yellow Book 2017, clause 13.3.2 – the engineer's 'response' is another Notice, which he gives as soon as practicable after receiving the contractor's proposal.
260 FIDIC Yellow Book 2017, clause 13.3.2 provides that in either case of the engineer providing consent, or not providing consent to the contractor's clause 13.3.2 proposal, he can do so 'with or without comments'.
261 FIDIC Yellow Book 2017, clause 3.7 (Agreement or Determination).

Clause 13.2 contains a similar procedure whereby the contractor can propose a variation of his own volition. The contractor's clause 13.2 proposal contains the same information required for a clause 13.3.2 proposal (see above), and similarly the engineer either provides his consent or comments. However, the important differences between the two proposal clauses 13.2 and 13.3.2 are as follows:

o The contractor's clause 13.2 proposal should only be submitted to the engineer if (in the contractor's opinion) it would provide an additional benefit to the project, including the following:
 – It may result in earlier completion (by accelerations); or
 – It may improve the project's value, including increasing its operating efficiency, or reducing its operating or maintenance costs; whereas
o The engineer may request the contractor provide a clause 13.3.2 proposal on a purely speculative basis.

FIDIC Yellow Book 2017 is quite clear that the engineer may only instruct variations before he issues the Taking Over Certificate (clause 13.1), adopting a similar position to JCT DB 2016 (see Section 8.8.3) as the works are now complete.[262]

8.8.5   NEC4

NEC4, like its predecessors (NEC to NEC3), deals with variations in a novel way compared to construction contracts published by the JCT, FIDIC, or ICE. It groups together all events and matters for which the contractor would be entitled to changes in the price for the works, or the period in which he must complete the works, and calls them 'Compensation Events'. Whilst appearing slightly unusual to practitioners who may be more used to a JCT or FIDIC contract, where variations, extensions of time, and loss and expense matters each have their separate section within the contract's conditions, NEC's concept of the 'Compensation Event' is simple and clear.

The operation of NEC4's Compensation Event procedure retains many of the characteristics of both JCT's and FIDIC's variation, extension of time, and loss and expense procedures, but includes some important and fundamental differences that go to the heart of its operation. For the majority of this book, how each design and build contract deals with each procedural stage is captured in a single section, but for clarity, and to highlight the relatively unique nature of NEC4's Compensation Event procedure, this is covered in Section 8.8.5.1 (NEC4 – Compensation Events) and Section 8.8.5.2 (NEC4 – Compensation Event procedure) below. The single heading of 'compensation events' reflects NEC's ethos in dealing with project changes. Book Four (*Managing Change*) of NEC's *Managing Reality* series summarises[263] the overriding principles for compensation events as:

o The Contractor should not be '…out of pocket for things that happen that are outwith its control'; and
o 'Compensation events are events that are at the *Client's risk* in the contract.'

---

262 FIDIC Yellow Book 2017, clause 10.1(a) (Taking Over the Works and Sections) confirms that the Works have been completed and have passed the 'Tests on Completion'.
263 *Managing Reality*, Book Four – *Managing Change*, 3rd ed., (2017), '1.1. Introduction', p. 4.

8.8.5.1 NEC4 – Compensation Events

A simple way to understand compensation events is to compare them to similar or equivalent matters under a JCT DB 2016 and FIDIC Yellow Book 2017 contract. Table 8.11 tabulates the various compensation events from clause 60 of NEC4 and makes this comparison.

In addition to the compensation events under clauses 60.1 and Y2.4, Main Options B (Priced Contract with Bill of Quantities) and D (Target Contract with Bill of Quantities) include additional compensation events dealing with the differences in the final quantities of work[264] and the project manager's instruction to correct the Bill of Quantities.[265]

In terms of variations, the most obvious similarity to a 'Change' under JCT DB 2016 or a 'Variation' under FIDIC Yellow Book 2017 would be a project manager's instruction to change the Scope under clause 60.1(1). As with JCT and FIDIC, the contractor is obliged to carry out that instruction immediately.[266]

Given the 'Accepted Programme's' importance as a management tool, it is unsurprising that NEC4 emphasises the need for the client to comply with it. Compensation events under clauses 60.1(2), 60.1(3), and 60.1(5) all encourage the client, along with anyone working on his behalf ('Others'), to comply with the Accepted Programme. Similarly, and in keeping with the NEC ethos of dealing with matters contemporaneously, the compensation event under clause 60.1(6) encourages the project manager to promptly respond to communications.

Compared to JCT DB 2016 and FIDIC Yellow Book 2017, NEC4 may seem a little simplistic when dealing with matters such as 'force majeure' and those covered by insurance, such as the compensation event in clause 60.1(19). Both JCT DB 2016 and the FIDIC Yellow Book 2017 set out such matters in more detail, and in separate clauses. However, the same point could be raised for JCT DB 2016, which uses the 'impediment, prevention, default' Relevant Event (clause 2.26.6) to the same effect as many causes of compensation events: clauses 60.1(2), 60.1(3), 60.1(5), 60.1(9), 60.1(14), 60.1(16), and 60.1(18).

8.8.5.2 NEC4 – Compensation Events Procedure

NEC4's compensation event procedure shares many of the characteristics found in JCT DB 2016 and FIDIC Yellow Book 2017, in terms of the contractor giving prior notifications (JCT and FIDIC) and providing quotations for the varied work (very similar to FIDIC). The contractor provides some prior warning or notification to the employer (or supervising officer) if an event may occur that could affect project. 'Affect' in this context means an event which is likely to adversely affect the employer in terms of:

o Something that may disrupt, delay, or prolong the progress of the works – resulting in the project (or a section of it) being completed later than was either planned or agreed; or

o An event that may result in the employer having to make additional payments to the contractor – either for carrying out a variation, or paying financial compensation to cover the contractor's costs of dealing with disruption, delays, or prolongation to its works.

---

264 NEC4, Options B and D, clauses 60.4 and 60.5.

265 NEC4, Options B and D, clause 60.6, to either change how the method of measurement is applied or resolve an inconsistency or ambiguity.

266 NEC4, clause 27.3 provides that the contractor 'obeys' a valid instruction given by the project manager or the supervisor.

Table 8.11 NEC4 Compensation Events cross-references.

NEC4 Compensation Event – clause 60.1 (for all Main Option Clauses A, B, C, D, and E)	Equivalent matter under:	
	JCT DB 2016[a)]	FIDIC Yellow Book 2017[b)]
(1) Project Manager's Instruction ('PMI') to the Contractor to change the Scope	Change 5.1 EoT 2.26.1/2.26.2 L&E 4.21.1/4.21.2	Variation 13.3.1, 1.9, 4.7.3(ii), 8.5(a)
(2) The Client prevents access to parts of the site by the 'access date' or relevant dates on the Accepted Programme	EoT 2.26.6 L&E 4.21.5	2.1, 4.15, 11.7
(3) The Client fails to provide something by the Accepted Programme's date	EoT 2.26.3, 2.26.6 L&E 4.21.5	8.5(e)
(4) PMI not to start or stop work	Change 5.1 EoT 2.26.1/2.26.2 L&E 4.21.1/4.21.2	8.10, 8.12
(4) PMI to change a Key Date	N/A	Yes[c)]
(5) The Client (or other contractors) do not do their work in accordance with the Accepted Programme	EoT 2.26.6, 2.26.7[d)] L&E 4.21.5	4.6, 8.5(e), 8.6
(5) The Client (or other contractors) do not work in accordance with the Scope	EoT 2.26.6 L&E 4.21.5	8.5(e), 8.6
(5) The Client (or other contractors) undertake work that is not set out in the Scope	EoT 2.26.6/2.26.7 L&E 4.21.5	8.5(e)
(6) The PM[e)] fails to respond to a communication within the period for reply	EoT 2.26.6 L&E 4.21.5	N/A
(7) PMI dealing with an object of archaeological value or that is of interest found on the Site	EoT 2.26.4	4.23[f)]
(8) The PM changes a previous decision	Change 5.1 EoT 2.26.1/2.26.2 L&E 4.21.1/4.21.2	Variation 13.3.1 8.5(a)
(9) The PM's reason(s) for not accepting something is/are not set out in the contract	EoT 2.26.6 L&E 4.21.5	N/A
(10) The Contractor is instructed to search for a defect but there is not one	EoT 2.26.2.3 L&E 4.21.2.2	11.8
(11) Testing causes an otherwise avoidable delay	EoT 2.26.6 L&E 4.21.5	7.4, 7.6, 10.3, 12.2, 12.4
(12) The Contractor encounters unforeseen physical conditions within the site, which are not weather conditions	N/A	4.12.4
(13) The Contractor encounters adverse weather conditions within the meaning of the contract	EoT 2.26.8	8.5(c)
(14) The Client causes an event for which he is liable	EoT 2.26.6 L&E 4.21.5	17.2
(15) PM certifies 'taking over' before 'Completion' and the 'Completion Date'	N/A	10.2
(16) Client fails to provide necessary facilities, equipment, and materials for tests	EoT 2.26.6 L&E 4.21.5	8.5(e)

(Continued)

Table 8.11 (Continued)

NEC4 Compensation Event – clause 60.1 (for all Main Option Clauses A, B, C, D, and E)	Equivalent matter under:	
	JCT DB 2016[a]	FIDIC Yellow Book 2017[b]
(17) The PM corrects an assumption he made about a previous Compensation Event	N/A	N/A
(18) Client commits a breach of the contract which is not a Compensation Event	EoT 2.26.6 L&E 4.21.5	8.5(b), (e)
(19) An event occurs (which is not another Compensation Event) which prevents the Contractor from completing the Works[g] that is outside of each party's control, or that the contractor could not have reasonably foreseen	EoT 2.26.9[h], 2.26.10, 2.26.11, 2.26,12, 2.26.13, 2.26.14	1.13, 8.5(d), 13.6, 18.4[i]
(20) The PM does not accept a contractor's quotation for a proposed instruction	N/A	13.3.2
(21) Any additional Compensation Events set out in the Contract Data Part One occur	N/A	N/A
Options B and D – differences between the final quantities of work and the quantities in the Bill of Quantities: – If it does not arise from a change in the Scope; – If it causes the Defined Cost to change (see Table 8.12 in Section 8.9.3.1); or – If the Bill of Quantities for that rate exceeds 0.5% of the total of the Prices. (All Clause 60.4) – If the difference in quantity delays Completion or a Key Date (clause 60.5). – If the Project Manager corrects a mistake in the Bill of Quantities arising from a departure from the method of measurement, ambiguity, or inconsistency (clause 60.6) including between the Bills of Quantities and another document (clause 60.7).	N/A	N/A
Y2.4 Contractor suspension	EoT 2.26.5 L&E 4.11.2	EoT 16.1

a) DB 2016 – references to 'Change' mean a variation, 'EoT' means extension of time, and 'L&E' means direct loss and/or expense.

b) FIDIC Yellow Book 2017, clauses in this column are reference to 'Claims' for an extension of time and/or additional costs under clause 20.1 (Claims) unless otherwise stated.

c) FIDIC Yellow Book 2017, Supplemental Provisions for 'Milestones' must apply, and the Engineer may instruct a Variation in the Milestone (see clause 8.5(a)).

d) JCT DB 2016, clause 2.26.7 ascribes an EoT (Relevant Event under clause 2.26) to Statutory Undertakers failing to carry out their work, but no entitlement to L&E (Relevant Matter under clause 4.21).

e) Including those working on the Project Manager's behalf – such as the Supervisor.

f) FIDIC Yellow Book 2017, clause 4.23 (Archaeological and Geological Findings). The Contractor is only entitled to an extension of time (under clause 8.5(b)) and additional costs (clause 20.2) for following the Engineer's instructions after notifying him of the discovery of the relevant remains or items.

g) NEC4 clause 60.1(19) – the event can also prevent the Contractor from completing by the Accepted Programme's planned completion date.

h) JCT DB 2016 – EoT under clause 2.26.9 is for one of the 'Specified Perils' which is an event covered by the Employer's Joint Names Policy for insuring the Works and Existing Structures (and their contents) owned by him under Insurance Option C (see also clause 6.9).

i) FIDIC Yellow Book 2017 – clause 18.1 (Exceptional Events), the consequences of which are set out in clause 18.4.

Most employers, and particularly those who are public bodies embarking on infrastructure projects, will have a project budget, which ideally should not be exceeded. The purpose of a contractor giving a prior warning or notification to the employer is not (as some contractors often believe) intended to prevent a contractor from becoming entitled to additional time or money, but to give the employer an opportunity to either avoid or mitigate any adverse effects of the event.

For JCT DB 2016 and FIDIC Yellow Book 2017, the requirement for the contractor giving prior warning or notice is used for events which may adversely affect the works' progress and entitle the contractor to loss and expense or additional costs. However, with NEC4, the prior warning requirement also applies to variations, because as Table 8.11 shows, NEC4's compensation events include variations, along with matters giving rise to extensions of time and loss and expense (or additional costs).

In keeping with its core principles of mutual trust and cooperation[267] and its emphasis on monitoring and managing the project,[268] NEC4 places the obligation to notify a compensation event on both the project manager and the contractor. Similar to the procedure in FIDIC Yellow Book 2017 (clause 13.3.1 – see Section 8.8.3), NEC4 requires the contractor to submit 'quotations' for the project manager to assess. In the context of 'variations', notifying a compensation event will occur as follows:

o Clause 61.1 provides that the project manager must notify the contractor of a compensation event when issuing an instruction[269]; or

o Clause 61.3 obliges the contractor to notify the project manager of a compensation event if firstly, the contractor 'believes' the event is a compensation event, and secondly, if the project manager has not already notified the contractor in this regard.

The notification obligations in clauses 61.1 and 61.3 complement each other in both the practical and contractual senses. It is fairly obvious that an instruction to increase the quantity of work will constitute a compensation event (as the prices will increase) so it makes practical sense for the project manager (or supervisor) to notify the contractor accordingly. If the effects of a project manager's (or supervisor's) instruction are less obvious and may not have been immediately apparent to the person issuing the instruction, then clause 61.3 permits the contractor to notify of the compensation event. The contractor's obligation to notify under clause 61.3 would arise if the supervisor had issued an instruction but had not informed the project manager, who consequently failed to notify of a compensation event under clause 61.1. If the project manager notifies the contractor of a compensation event,[270] clause 61.2 provides that he also instructs the contractor to submit quotations.

Clause 61.3 contains perhaps one of the most 'infamous' provisions found in any of the design and build contracts – namely, the 'time bar' clause, which says that unless the contractor notifies the project manager that an event (or expected event) is a compensation

---

267 NEC4, clauses 10.1 and 10.2.
268 The importance of the 'Accepted Programme' is set out in Sections 8.6.2 and 8.6.2.
269 NEC4, clause 61.1. The 'instruction' can be given by the project manager or the supervisor, but only the project manager notifies the contractor of a compensation event.
270 NEC4, clause 61.1.

event within eight weeks of becoming aware of the event, then he will not be entitled to any changes in the prices, a key date, or the completion date. Clause 61.3 is considered to be an exclusion clause, intended to limit the client's liability to the contractor if the contractor fails to comply. In order to be effective, exclusion clauses must be very clearly worded, contain no ambiguities, and if such an ambiguity is present, then the clause will be interpreted against the party who drafted it, as Lord Bingham of Cornhill said in *Dairy Containers Ltd v Tasman Orient Line CV*[271]:

> 'The general rule should be applied that if a party, otherwise liable, is to exclude or limit his liability or to rely on an exemption, he must do so in clear words; unclear words do not suffice; any ambiguity or lack of clarity must be resolved against that party.'

The wording of clause 61.3 seems clear when read alongside the NEC4 contract as a whole and, in particular, considering the way in which NEC4 is written. Clause 61.3 sets out:

o What the contractor must do,[272]
o The period of time in which he must do it,[273] and
o The sanction imposed upon him if he fails to comply.[274]

The operation of the 'time bar' clause is contentious, and it has a number of perfectly valid arguments for and against. A contractor falling foul of the time bar provision will naturally feel aggrieved if this prevents him from recovering additional payment and time, for what would otherwise be a perfectly legitimate compensation event under the contract (for example, a project manager's instruction under 60.1(1) that adds significant work to the Scope). Conversely, a client may feel aggrieved if he was prevented from effectively managing a compensation event because the contractor failed either to issue an early warning (see Section 8.4.7 above) or to notify the project manager. These circumstances may arise where the client is a public body and faces deadlines or specific timescales for obtaining additional funding for the project.

Time bar clauses have a number of advantages and disadvantages as follows:

o Advantages
  – Encourage the parties to deal with matters as they happen (or relatively soon after), instead of leaving them to later stages or end of the project.
  – Promote the resolution of issues or conflicts during the course of the project.
  – Contractor's notification obligations assist the client with managing the compensation event (such as taking steps to mitigate delay or further time to allocate additional funding to the project).

---

271 1 WLR 215, [2005].
272 NEC4, clause 61.3. The contractor must notify the project manager that an event has occurred that has resulted or may result in a compensation event.
273 NEC4, clause 61.3. The contractor must notify the project manager within eight weeks after becoming aware that the event has happened.
274 NEC4, clause 61.3. The sanction is clear in that the contractor will not be entitled to any change in the Prices, the Completion Date, or a Key Date.

o Disadvantages
  - Perceived unfairness of the risk of losing a legitimate entitlement on a 'technicality' (failure to comply with an administrative procedure).
  - Can result in a lack of clarity whereby the contractor formats all communications as 'notices' to reduce the risk of failing to comply with the time bar clause.
  - Can require additional administrative resources for the contractor, the project manager, and the client.

The exceptions to the contractor having to comply with the eight week 'time bar' are as follows:

o The project manager (or supervisor) issued an instruction or a certificate or has changed a previous decision which caused the event[275]; or
o It was reasonable for the contractor not to 'believe' that the event is a compensation event.

Whilst not avoiding compliance with clause 61.3, the contractor may claim that he did not reasonably become 'aware' of the event until later, but must still notify within eight weeks of when he did become aware. Clause 61.3's requirements for the contractor to 'believe' and become 'aware' that an event is a compensation event are subjective. It is thought that the contractor would have to genuinely 'believe' that an event was not a compensation event, and had genuinely not become aware the event, in order to justify a failure to notify under clause 61.3. There may be some discrete arguments that a contractor could genuinely become aware of an event or genuinely believe that an event is a compensation event; and these may centre on circumstances such as that the contractor's belief or awareness may be triggered by the later consequences of the event instead of the occurrence of the event itself. Any deliberate avoidance by the contractor of either believing that the event is a compensation event or becoming aware of the event (or even its subsequent consequences) is unlikely to assist the contractor in justifying any failure to comply with clause 61.3.

Upon receipt of the contractor's clause 61.3 notification (of a compensation event), the project manager has one week (or such longer period as the contractor agrees) to issue one of the following replies[276]:

o The project manager confirms that the event (in the contractor's clause 61.3 notification) is a compensation event and instructs the contractor to submit quotations; or
o He may reject the contractor's clause 61.3 notification, either because it was late[277] or because the notified event:
  - Is not one of the compensation events listed in clause 60.1;
  - Is either unlikely to or is not going to occur, or will not affect the defined cost, completion date or a key date; or
  - Was the contractor's fault.

---

275 NEC4, clause 61.3. The NEC4 version of clause 61.3 is less subjective than its NEC3 predecessor. Clause 61.3 of NEC3 provided that the contractor must notify a Compensation Event within eight weeks or would not be entitled to any changed Prices, Key Dates, or changes to the Completion date, 'unless the Project Manager should have notified the event to the Contractor but did not'.
276 NEC4, clause 61.4.
277 NEC4, The contractor failed to provide the clause 61.3 notification within eight weeks.

Clause 61.4 also places a 'time bar' on the project manager. If he fails to respond to the contractor's clause 61.3 notification within either one week, or a longer period agreed upon with the contractor, the contractor notifies the project manager. If a response is not forthcoming for another two weeks[278] then the project manager's failure to respond in time is firstly considered to be his acceptance that the notified event is a compensation event and, secondly, is treated as his instruction to the contractor to submit quotations.

If an event is either yet to occur or has occurred and, in both instances, it is impossible for the contractor to accurately forecast the effects of the event, then the project manager's clause 61.2 instruction[279] includes his assumptions,[280] upon which his assessment of the compensation event (under clause 63) will be based. If the project manager's assumptions prove to be incorrect, he corrects them,[281] which is compensation event under clause 60.1(17).

The contractor submits his quotation (or a revised quotation) within three weeks[282] of the project manager's instruction. The three weeks period can be extended with the project manager's and contractor's agreement before the project manager's reply is due.[283] Similar to the 'detailed particulars' provided by the contractor in clause 13.3.1 of FIDIC Yellow Book 2017, clause 62.2 of NEC4 required the contractor's quotation to include[284]:

○ His proposed changes to the prices (this can be an addition or a reduction in the case of an omission); and
○ Any changes to the accepted programme to show alterations to the completion date and any key dates.

Unless the duration is extended (see above), the project manager must reply to the contractor's quotation within two weeks of receiving it, and his clause 62.3 reply must set out:

○ That he accepts the contractor's quotation[285];
○ That a revised quotation is to be submitted by the contractor, provided that the project manager gives his reasons why[286]; or
○ That the project manager will assess any changes to the prices, completion date, or any key dates.

As with the project manager's response to the contractor's clause 61.3 notification, another 'time bar' applies to his reply to the contractor's quotation, albeit the project manager is given a greater opportunity to issue his clause 62.3 reply. If he fails to provide his clause 62.3 reply within two weeks (or such extended period as is agreed), then the

---

278 NEC4, clause 61.4. The two weeks period after the contractor's notification under clause 61.4 (i.e. of the project manager's failure to respond to the contractor's clause 61.3 notice) is not adjustable by agreement.
279 NEC4, clause 61.2, the project manager's instruction to the contractor to submit quotations.
280 NEC4, clause 61.6.
281 NEC4, via a notification under clause 61.6.
282 NEC4, clause 62.3.
283 NEC4, clause 62.5.
284 NEC4, clause 62.2.
285 NEC4, The project manager's acceptance of the contractor's quotation is a 'notification' (refer to clause 13.7).
286 NEC4, clause 62.4.

contractor notifies the project manager of that failure. The contractor's clause 62.2 quotation[287] is treated as being accepted by the project manager, if he fails to reply within a further two weeks following the contractor's clause 62.6 notification.

It appears that NEC4 maintains a similar position to JCT DB 2016 and the FIDIC Yellow Book 2017 insofar as the project manager cannot instruct any compensation events to vary the Scope after the completion date. Clause 11.2(16) defines the 'Scope' as including valid instructions, and clause 11.2(2) defines 'Completion' as when the contractor has completed the Scope; implying that once the Scope of work is complete, no further instructions to vary can be issued. Clause 61.7 also provides that no compensation event is notified after the Defects Certificate is issued.[288] This should not be interpreted as permitting the project manager or supervisor to issue instructions to vary the Scope after completion, but allowing the contractor to notify compensation events occurring during the 'defect correction period';[289] for example, those occurring under clauses 60.1(10),[290] 60.1(11),[291] and 60.1(16).[292]

Clause 63.6 sets out that the only rights that the contractor and client possess in terms of changing the prices, the completion date, or any key dates are provided by the compensation event procedure. An initial reading may appear as if NEC4 intended to impose an 'exclusive remedies' clause by including clause 63.6. This would mean that the parties' rights to claim any common law damages for breach of contract by the other would be excluded, and their sole rights would be confined to the compensation event procedure. However, there are a number of points to consider regarding clause 63.6:

o In legal terms[293]:
  – In order to effectively take away any common law rights or remedies that a party may have, the clause in question should contain clear words to that effect; and
  – A lack of clarity, or ambiguity in the clause is resolved against the party seeking to rely on it.
  Comparing the wording of NEC4 clause 63.6 with, say, clause 44.4 of the MF/1 (1988) contract[294] which in addition to restricting the parties' 'rights', also restricted their 'obligations and liabilities', a party's liability to the other for common law damages may not be affected.[295]
o Clause 63.6 obviously refers to compensation events, which include any breach by the client at clause 60.1(18). Therefore, the compensation event procedure would appear to permit the contractor to make a claim that he could otherwise make as a common law damages for the client's breach, under the compensation event procedure in any event.

---

287 NEC4, clause 62.6 or if the contractor has provided more than one quotation, the quotation nominated by the contractor in his clause 62.6 notification.
288 NEC4, issued by the supervisor (clause 44.3).
289 NEC4, Refer to Contract Data Part One (4 – Quality Management).
290 NEC4, clause 60.1(10) – supervisor instructs the contractor to search for a defect but none are found.
291 NEC4, clause 60.1(11) – unnecessary delays are caused by a test or an inspection.
292 NEC4, clause 60.1(16) – client's failure to provide necessary facilities, etc. for tests and inspections.
293 *Dairy Containers Ltd v Tasman Orient Line* CV, 1 WLR 215, [2005].
294 *Strachan & Henshaw Ltd v Stein Industrie (UK) Ltd and GEC Alsthom Ltd*, 87 BLR 52, (1998).
295 However, a number of NEC4 commentaries suggest that similar wording in clause 25.3 effectively excludes the client's rights to recover amounts other than the 'additional cost' (as stated in that clause) if the contractor's work fails to achieve the condition to meet a Key Date.

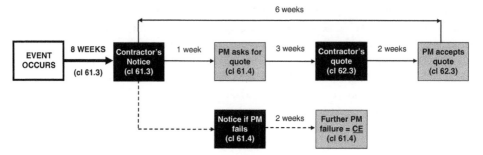

Figure 8.3 NEC4 Compensation Event Procedure.

o If Secondary Option X18 (Limitation of liability) is selected, then arguably, clause 63.6 may not restrict the contractor's liability to the client for matters such as consequential losses.[296]

Figure 8.3 summarises the compensation event procedure,[297] which includes the quotation and assessment stages (covered in Sections 8.9.3.1 and 8.9.3.2).

Clause 66.1 provides that the compensation event is 'implemented' if the project manager:

o Accepts the contractor's clause 62.3 quotation (or it is deemed to have been accepted); or
o Notifies the contractor of his assessment under clause 64.3.

Once a compensation event is implemented, unless the contract provides otherwise, it cannot be revised (refer to Section 8.9.3).

8.9 Valuation (or Assessment) of Variations

The assessment of variations for this Section 8.9 only considers the method of ascertaining the change in value of the physical works arising from a supervising officer's instruction. Whilst NEC4's compensation event procedure also deals with extensions of time (changes to the completion date) and any resulting additional costs, these aspects will be covered alongside their equivalent provisions under JCT DB 2016 and FIDIC Yellow Book 2017 in Section 8.10 (Extensions of Time) and Section 8.11 (Loss and Expense).

Whilst the three design and build contracts share some common procedures for valuing variations, they all deal with valuation in different ways. JCT DB 2016 places the obligation of valuation on the employer and, on the face of it, requires relatively little input from the contractor unless the Supplemental Provisions are being operated (see Section 8.7.2). JCT's general approach should be contrasted with the ways that FIDIC and NEC4 require far greater input from the contractor, who submits 'particulars' (FIDIC) or 'quotations' (NEC4) for the engineer or project manager to assess. As JCT generally does not use a 'quotation' procedure, it contains fairly prescriptive 'Valuation Rules'[298] for the employer to follow when valuing a variation (change).

---

296 NEC4, Option X18 (Limitation of liability), clause X18.2.
297 NEC4, Figure 8.2 also shows the procedure for 'assessment' of Compensation Events – discussed further in Section 8.9 (Valuation (or assessment) of variations).
298 JCT DB 2016, clauses 5.4–5.7.

As a general rule, JCT DB 2016, FIDIC Yellow Book 2017, and NEC4 all place an initial reliance on the contracts' pricing document when valuing variations, albeit the way the pricing document is employed will depend upon the contract in question, or in the case of NEC4, which of the Main Option clauses[299] is selected. The level of detail shown in the pricing document is often pivotal to the ease of valuing variations. If the pricing document consists of a fairly comprehensive schedule of rates, or a bill of quantities, then the supervising officer can assess the variation's value with more precision. However, if the pricing document does not contain much pricing detail (for example, in the case of JCT DB 2016, the CSA may only contain a pricing summary that is similar to NRM2's 'condensed' format – see Section 3.6.1), it may be virtually impossible for the employer to identify an appropriate rate or price from within the summary total for a particular building element (e.g. substructure). The contractor may be obliged to provide further information, either as required by the contract[300] or by the supervising officer,[301] and in some cases (for example, where the valuation of a variation is significantly disputed) the contractor may need to provide evidence of the rates and prices that comprise its Contract Sum (JCT) or Contract Price (FIDIC) even if that level of detail is not shown in the contract's pricing document. If additional pricing information is required, then the employer (along with any consultants) should not misuse or over-analyse any additional pricing information, particularly if it does not form part of the contractor's pricing document and is evidence of how the contractor has calculated its price for an element of the works. They should consider the following:

o The employer may have been happy to let the contract (often for many millions of pounds in value) be based upon a fairly rudimentary pricing document, which may not show rates and prices for individual items of work.
o The contractor's pricing information may not show a linear or consistent build-up to the rates or prices in the pricing document, as he may have 'moved' money around or made commercial adjustments (refer to Section 3.6.5).

8.9.1 JCT DB 2016

It is perhaps unsurprising that JCT DB 2016 contains the most detailed method for valuing variations, given that its design and build contract is based upon the traditional JCT SBC 2016 contract that, in general, operates a bill of quantities as its pricing document. Like its traditional counterpart, JCT DB 2016 relies upon the values of work set out in the pricing document as the principal means of valuing changes (variations) and provisional sums.

JCT DB 2016s pricing document is the Contract Sum Analysis (CSA), but as pointed out in Section 3.6.1, the CSA's format depends upon the stipulations set out in the Employer's Requirements. JCT DB 2016 no longer contains provision to use a bill of quantities, so unless the employer prescribes the way in which and level of pricing detail he requires, it is likely that the contractor will simply provide a CSA similar to NRM2's condensed format[302]

---

299 NEC4, Main Option Clauses are: Option A (Priced Contract with Activity Schedule), Option B (Priced Contract with Bill of Quantities), Option C (Target Contract with Activity Schedule), Option D (Target Contract with Bill of Quantities), and Option E (Cost Reimbursable Contract).
300 JCT DB 2016, clause 4.7.4.
301 FIDIC Yellow Book 2017, clause 13.3.1.
302 NRM2, Template for pricing summary for elemental bill of quantities (condensed), Appendix D, p. 272.

(refer to Section 3.6.1). If the CSA only contains summary prices for each element of work (such as for substructure, superstructure, etc.), the employer may face difficulty in determining the appropriate 'value' to use, particularly if the change relates to a discrete part of that element of work. For example, if the employer instructs the contractor to provide additional steel reinforcement for the substructure but the CSA only shows an overall price for the substructure, it will be difficult to ascertain just the value of the steel reinforcement (within the overall substructure price) to use when valuing the change.

The JCT DB 2016 'Valuation Rules' (in clauses 5.4–5.7.1) relate to the valuation of the physical works and any costs directly arising from their execution. However, clause 5.7.2 of the Valuation Rules specifically exclude the addition of direct loss and/or expense, confining such cost to the Relevant Matters and their calculation in clauses 4.19–4.23 (see Section 8.11).

They split the valuation of a change into the following three categories:

o If the work can be valued by measurement;
o If it can be valued on a 'daywork' basis; or
o If the change can neither be valued by measurement nor by daywork, then a fair valuation shall be made.

As a general rule, work can be valued by measurement if it can be quantified before the work is undertaken (for example, by measuring a revised construction drawing). Simple examples for a construction project would be to instruct the contractor to provide a longer section of fencing (an addition) or provide the same length of fencing but of a different specification (a substitution). This should be contrasted with additional work that cannot realistically be quantified beforehand – for example, instructing the contractor to carry out investigative works, such as to carefully excavate down to reveal the position of existing, buried services. This type of work would typically be valued on a 'daywork' basis (under clause 5.5).

Clause 5.4 deals with valuation by measurement, and sets out the following procedure:

o If the changed work[303] is of 'similar character' to work for which a value is included in the CSA, then it shall be valued consistently with the values in the CSA, but making appropriate adjustments to that value, if the changed work is carried out under different conditions[304] (also see clause 5.6) or there is a significant change in its quantity.
o If the change results in omitting work, the omission is valued using the values in the CSA (because the Contract Sum will already include a value for such work).
o If the changed work is not of 'similar character', then the employer must make a 'fair valuation'.

---

303 JCT DB 2016, clause 5.4.2, the 'Change' comprises additional or substituted work (i.e. a Change under clause 5.1.1.1).
304 JCT DB 2016, clauses 5.4.2 and 5.6. For example, the CSA may include for work of 'similar character' to the additional or substituted work (the Changed work), but it is carried out at low level, whereas the changed work requires access and special lifting. The valuation under clause 5.4.2 would be based upon the values for the changed physical work in the CSA, plus an appropriate adjustment for the value for providing access and lifting.

There are several important terms that are central to how clause 5.4.2 operates. Firstly, work of 'similar' character is not the same as work of an 'identical' character.[305] Therefore, a slight change in specification may not entitle the contractor to a change in the unit value (or unit rate) of the work, provided that it is carried out under similar conditions to the similar work for which there is a value in the CSA.

Secondly, the only circumstances in which the CSA's values for the work of 'similar character' are changed are as follows:

o If the conditions change under which the work is to be undertaken; some examples being:
  – The work may be carried out at height,
  – Carried out in an area with restricted access, or
  – May have to be installed or erected outside normal working hours (e.g. at night or during a weekend).
The values in the CSA for such work are likely to be increased to reflect the additional costs of hoisting, lifting, and/or additional labour for 'out of hours' working.
  – Conversely, if the work is carried out under different, but easier conditions (such as being erected at ground level, or in an area with ample access or during normal working hours), then the values in the CSA may be reduced accordingly.
o If the quantity of that type of work significantly changes:
  – If the quantity of work significantly reduces, then the values in the CSA may be increased if the contractor is unable to take advantage of a 'bulk buying' discount; whereas
  – If the quantity significantly increases, then the rates, prices, or values in the CSA may be reduced to reflect a discounted price.
  – A significant change in quantity (increase or decrease) may also require a change in the conditions under which the work is undertaken. Working to smaller areas may be more labour-intensive, whereas working to a larger area may allow the contractor's operatives to work more efficiently, and may permit the work to be carried out by machines.

Clause 5.4.2 does not permit the values in the CSA to be adjusted or form the basis of an adjusted value, price, or rate[306] where the changed work is not of similar character to the work which is included in the CSA. Instead, it requires a 'fair valuation' to be made. This seemingly precludes the common quantity surveying practice of using a pro rata adjustment of the CSA's values.[307] The Court considered the meaning of a 'fair valuation' in the case of *Weldon Plant v Commission for the New Towns*[308] as follows:

> '...a fair valuation had to include each of the elements which are ordinarily to be found in a contract rate or price: elements for the cost of labour, the cost of plant, cost of

---

305 *Oxford Dictionary of English*, 3rd ed., (2010): 'similar' – 'having a resemblance in appearance, character or quantity, without being identical'.
306 Often referred to as a 'star rate' or a 'derived rate'.
307 A simple example would be changing the thickness of a brick wall from 1 brick to 2 bricks thick. It is unlikely that this change would be of similar character (as the wall's character and appearance would be changed), so in the absence of a provision requiring a valuation by 'fair rates and prices', a pro rata valuation is made, doubling the original rate (or price) to reflect the doubling of the wall's thickness.
308 *Weldon Plant Ltd v Commission for the New Towns*, EWHC 496, [2000], in which the contractor (Weldon) was appealing against the effects of an arbitrator's award.

materials, cost of overheads and profit. In my judgement a fair valuation has not only ordinarily to include something on account of each of these elements, but also it would not be a fair valuation within the meaning of the contract[309] if it did not do so.'[310]

With regard to the matter of and addition for overheads, and the principle of the contractor not being penalised by a 'fair valuation', HH Judge Humphrey Lloyd QC continued:

'...the arbitrator did not deal with the addition which in my judgement has to be made in order to ensure that the contractor obtains a contribution from the costs of the business it undertakes towards its fixed or running overheads. For reasons that I have already given it would not be fair if the valuation did not include an element on account of such contribution. It would mean that such contribution would have to be found elsewhere, presumably from the contractor's margin for profit or risk. In my view a valuation which in effect required the contractor to bear that contribution itself would not be a fair valuation, in accordance with the principles of clause 52(1)[311] which are intended to secure that the contractor should not lose as a result of having to execute a variation (except, as I have stated, to the extent that its costs etc are of its making).'

Clauses 5.4.1 and 5.4.4 also respectively provide that allowances should be made for the following additions or omissions:

o The value of any design work, and
o Site facilities (including administrative functions) and any temporary works.

A fair valuation should also be made under clause 5.7.1 if any of the work (be that additions, substitutions, or omissions of work) or other liabilities directly arising from a change cannot reasonably be valued either by measurement or dayworks (see below) or a change in conditions under clause 5.6.

If the changed work (additional or substituted work) cannot be valued by measurement under clause 5.4, then the work is valued in accordance with the 'Dayworks' clause 5.5. JCT DB 2016 relies upon various definitions of the prime cost of daywork, depending upon whether (and if so, how many and what discipline of) specialist trades are used. The four definitions of daywork referred to in clause 5.5 (and its footnote) are as follows:

o The RICS 'Definition of Prime Cost of Daywork carried out under a Building Contract'[312] ('the RICS definition');

---

309 *Weldon Plant Ltd v Commission for the New Towns.* In this case, the contract was the ICE Conditions (6th ed.).
310 *Weldon Plant Ltd v Commission for the New Towns*, EWHC 496, [2000], at paragraph 16.
311 JCT DB 2016, clause 5.4.2 is similar to clause 52(1) of ICE 6th ed., which states: 'Value of ordered variation: (1) The value of all variations ordered by the Engineer in accordance with clause 51, shall be ascertained by the Engineer after consultation with the Contractor in accordance with the following principles. (a) Where work is of similar character and executed under similar conditions to work priced in the Bill of Quantities it shall be valued at such rates and prices contained therein as may be applicable; (b) Where work is not of a similar character or is not executed under similar conditions...the rates and prices in the Bill of Quantities shall be used as the basis for valuation so far as may be reasonable failing which a fair valuation shall be made...'.
312 Currently the third edition (2007).

o As agreed between the RICS and
- the Electrical Contractors Association,
- the Electrical Contractors Association of Scotland, and
- the Building and Engineering Services Association.

The basis for calculating the daywork rates are set out in the definitions. In the case of the RICS definition, this sets out how the relevant hourly rate is calculated[313] and provides two options for expressing the hourly rate, as follows:

o 'Option A' by applying the Percentage Additions as stated in the Contract Particulars to the prime cost of such works in the RICS definition; or
o 'Option B', which is by proving 'all inclusive' hourly rates.

The prime cost of works and the percentage additions (for Option A) or the 'all inclusive' rates (Option B) are set out in the document referenced in the Contract Particulars, against the entry for clause 5.5. This is usually the CSA, or a clearly referenced section of it (see Section 3.6.1).

As an alternative to operating the Valuation Rules, the parties can agree to select paragraph 2 (Valuation of Changes – Contractor's Estimates) of Part 1 of the Supplemental Provisions so, instead, the contractor provides the employer with an estimate of the value of change (see Section 8.8.2).

JCT DB 2016 also provides for price adjustments for 'Fluctuations' if the option is selected in the Contract Particulars.[314] The parties can select either that no Fluctuations Provisions applies, or one of the following:

o JCT Fluctuations Option A (Contribution, levy, and tax fluctuations),[315]
o JCT Fluctuations Option B (Labour and materials cost and tax fluctuations),[316]
o JCT Fluctuations Option C (Formula adjustment),[316] or
o Another fluctuations option selected by the parties (which must be set out in the Contract Particulars).

8.9.2   FIDIC Yellow Book 2017

FIDIC's procedure for assessing the change to the contract price arising from a variation (along with Provisional Sums – see clause 13.4) relies upon more input from the contractor than under JCT DB 2016, as he must provide his proposals for adjustment of the contract price under clause 13.3.1(c). The contractor's clause 13.3.1 proposal may also set out any 'modifications' to the programme, but adjustment (extensions) of time is covered under Section 8.10 (Extensions of Time).

The engineer then proceeds to either agree or 'determine' (see Section 9.7.2.1) the adjustments to the contract price on the following bases, depending upon whether the contract contains a 'Schedule of Rates and Prices'.

---

313 RICS. (2007). *Definition of Prime Cost of Daywork,* 3rd ed.; see Sections 3.1 to 3.5, along with Sections 4 (Materials and Goods), 5 (Plant), and 6 (Incidental Costs, Overheads and Profit).
314 JCT DB 2016, Contract Particulars, entry against clause 4.2, 4.12, and 4.13 (Fluctuations Provisions).
315 JCT DB 2016, JCT Fluctuations Options A is set out in Schedule 7 to the contract.
316 JCT Fluctuations Options B and C are no longer included in the JCT DB 2016 contract, but can be downloaded from the JCT website.

If the contract includes a schedule of rates and prices:

o If the same items of work as are contained in the schedule of rates and prices are included in the variation, then those same rates or prices shall be used to value the variation.

o If the variation's work is similar to (but not the same as) the work in the schedule of rates and prices, then the rate or price for similar work shall be used in the valuation.

o A new rate or price may be used for valuing the variation if the rate or price in the schedule of rates and prices is not appropriate because:

 – The variation's work is not of similar character to work in the contract, or

 – Is not carried out under similar conditions to work in the contract, then the new rate or price will be derived from those in the schedule of rates and prices, and adjusted appropriately to reflect the prevailing circumstances of carrying out the variation.

o If the schedule of rates and prices does not contain any appropriate rates or prices from which the 'new rate or price' can be derived, then the variation shall be valued using 'Cost Plus Profit'.[317]

Clause 13.3.1's procedure for valuation is similar to JCT DB 2016s 'Valuation Rules' (see Section 8.9.1), with the exception that rates for a variation where the work is not of 'similar character' may be 'derived' from the rates and prices in the contract, whereas JCT skips this method and goes straight from applying rates in the CSA to applying a 'fair valuation'. Given that FIDIC refers to a Schedule of Rates and Prices (whereas JCT simply refers to the CSA, which may not break down the contractor's pricing in as much detail), it implies that more pricing detail may be shown in the Schedule.

FIDIC's definition of 'Cost Plus Profit' is the equivalent of JCT's 'fair valuation', as it includes:

o The 'Costs' (as defined in clause 1.1.19), comprising the contractor's reasonable expenditure in performing the contract (e.g. costs of labour, plant, materials [or goods] and subcontract costs), along with overheads and taxes[318] (or similar costs) but excluding profit; and

o The profit percentage that is either set out in the Contract Data,[319] or if no percentage is stated, is 5% as provided by clause 1.1.20.

Clause 14.1 provides that Provisional Sums are also valued in accordance with clause 13.3.1, with the additional provision that the engineer can instruct the contractor to obtain quotations from suppliers or subcontractors for their element of the work. The engineer may then instruct which suppliers or subcontractors the contractor shall select. If he fails to respond within seven days of receiving a quotation, the contractor may select the supplier or subcontractor.

If the contract does not contain a schedule of rates and prices, then variations are to be valued using the 'Cost Plus Profit' method.

---

317 FIDIC Yellow Book 2017, 'Cost Plus Profit' is defined in clause 1.1.20 and means the 'Cost' (as defined in clause 1.1.19) plus the agreed percentage for profit, or if no percentage is agreed, clause 1.1.20 provides that the 'profit' is 5% (5%).

318 FIDIC Yellow Book 2017 clause 14.1(b) states that the contractor shall pay all 'taxes, duties and fees' whereas JCT DB 2016 states that the Contract Sum is exclusive of VAT (Article 2: Contract Sum).

319 FIDIC Yellow Book 2017, Particular Conditions Part A – Contract Data, entry against clause 1.1.20.

Using 'dayworks' as a method of valuing variations is not strictly part of the 'valuation rules' in clause 13.3.1. Valuing work on a daywork basis can only occur in the following circumstances set out in clause 13.5:

o The contract has to include a Daywork Schedule[320];
o The variation is of a 'minor and incidental' nature; and
o The engineer instructs that the variation be undertaken on a daywork basis.

The engineer may also instruct the contractor to obtain quotations from its subcontractor or suppliers, and a similar procedure to that for valuing Provisional Sums (clause 14.1) applies. The contractor shall keep and submit 'accurate statements' (presumably in the form of 'daywork sheets'), for the engineer's approval.

As described in Section 8.8.3, the operation of FIDIC's variation (and valuation) procedure can take almost 90 days, so clause 13.3.1 provides that until such time as the variation's value is agreed or determined, the engineer must value the variation using provisional rates or prices when including its value in interim payments.

FIDIC Yellow Book 2017 also permits the contract price to be adjusted in the following circumstances:

o Clause 13.6 – Adjustments for Changes in Laws; and
o Clause 13.7 – Adjustments for Changes in Cost, provided that there are 'Schedule(s) of Indexation' included in the contract.

8.9.3   NEC4

As with JCT DB 2016 and FIDIC Yellow Book 2017 (where variations are assessed by the supervising officer), the project manager assesses the compensation event under NEC4. The project manager's assessment can be carried out in the following ways:

o Under clause 63.2, the project manager and the contractor can effectively 'agree' on the change in the prices arising from the compensation event by agreeing to lump sums or rates; however, there is no provision for them to 'agree' on the length of time by which the completion date or any key dates are delayed.[321]
o If the contractor provides a clause 62.1 quotation within the allotted period from the project manager's request,[322] the project manager assesses this in accordance with clause 63, unless the following occur (as provided in clause 64.1):
  – The project manager decides that the contractor's clause 62.1 quotation does not properly assess the compensation event, and no revised clause 62.1 quotation has been instructed; or
  – If the contractor fails to include a programme (showing the relevant changes) with his clause 62.1 quotation; or
  – The project manager has not accepted the contractor's programme[323] when he provides the clause 62.1 quotation.

---

320 FIDIC Yellow Book 2017, defined in clause 1.1.26.
321 NEC4, clause 63.5 sets out the method for assessing delays to the Completion Date.
322 NEC4, three weeks under clause 62.3, or within any period extended by the project manager under clause 62.5.
323 NEC4, The project manager does not accept the contractor's programme for one of the reasons in clause 31.1.

In which case, the project manager assesses the compensation event himself.

o Clause 64 provides that the project manager makes his own assessment of a compensation event if the contractor:
  – Fails to provide his clause 62.1 quotation within the time allowed[324]; or
  – If the contractor has defaulted regarding the programme,[325] by either failing to provide an accepted programme,[326] failing to submit a revised or altered programme for acceptance[327]; or the contractor's submitted programme falls foul of one (or more) of the reasons that the project manager may refuse to accept the programme.[328]

The project manager assesses the effects of a compensation event as follows:

o Changes to the prices (where not agreed under clause 63.2):
  – Additions (or increases) are dealt with under clauses 63.1;
  – Omissions (expressed as reductions in the prices) are covered under clauses 62.3 and 62.4; and
o Delays to the completion date and/or to key dates are covered under clause 63.5.

The project manager must notify the contractor of his assessment within the same period as the contractor was given for providing his quotation.[329] The period for his assessment is stated to begin as soon as the '...Project Manager's assessment becomes apparent', so in practice, this is likely to start when one of the four events in clause 64.1 occurs or where he notifies the contractor that he will make his own assessment.[330] It appears that the project manager is obliged to provide details of his assessment,[331] although they may not necessarily need to accompany his clause 64.3 notification, but must be provided within the relevant period. If he fails to do so, the contractor notifies the project manager, giving him a further two weeks in which to assess the compensation event, failing which the contractor's quotation (or if more than one quotation was submitted, the contractor notifies the project manager which one should be used) is deemed to be accepted.[332]

Clause 66.1 provides that once the contractor's quotation is accepted (either by the project manager notifying the contractor of his acceptance,[333] or because he failed to reply in time,[334] or after the project manager notified the contractor of his own assessment[335]),

---

324 NEC4, clause 64.1.
325 NEC4, clause 64.2.
326 NEC4, clause 11.2(1) definition of 'Accepted Programme'.
327 NEC4, Some examples: the contractor may have failed to provide a first programme for acceptance within the period set out in the Contract Data (clause 31.1) or within the 'period for reply', or a period longer than the interval stated in the Contract Data (clause 32.2).
328 NEC4, The project manager does not accept the contractor's programme for one of the reasons in clause 31.3 – see Section 8.6.3 and Table 8.7.
329 NEC4, clause 64.3. Usually, the contractor provides his quotation within three weeks of the project manager's instruction to do so (clause 62.3) unless the project manager extends the period under clause 62.5.
330 NEC4, clause 64.3. Also see *Managing Reality*, Book Five – *Managing Procedures*, 'Pro Forma 21: Project Manager's Reply to a Quotation for a Compensation Event', p. 59.
331 NEC4, clause 64.3.
332 NEC4, clause 64.4.
333 NEC4, clause 62.3, the project manager's acceptance of the contractor's clause 62.2 quotation.
334 NEC4, clause 62.6.
335 NEC4, clause 64.3 provides that the project manager notifies the contractor of his assessment within a period commencing from when it was apparent that the project manager's assessment was required, and

then the compensation event is 'implemented'. Implementing a compensation event almost irreversibly[336] changes the prices, the completion date (and any key date),[337] and clause 66.3 provides that a revised assessment of these changes cannot be made unless the following occur:

o The project manager corrects an assumption[338] on which the compensation event's assessment was based; or
o The project manager's assessment of the final amount (under clause 53) does not become 'conclusive' evidence for the reasons set out in that clause, although the scope for altering any implemented compensation event appears limited (see Section 9.6.3).

The applicable methods of assessment for each are described in the following sections:

o 8.9.3.1 – Assessing changes to the Prices, and
o 8.9.3.2 – Assessing changes to the Completion Date or a Key Date.

8.9.3.1 Assessing Changes to the Prices

NEC4's principles for compensation events are set out in Section 8.8.4 and can be summarised as follows:

o No compensation event is the fault of the contractor because they all arise from employer risk events; and
o The contractor should not suffer a disadvantage (either commercially or in terms of time for completion) as a result of a compensation event.

In the event that the project manager and the contractor have not agreed on lump sums or rates for assessing changes to the prices, the project manager assesses the increases or decreases (the changes) in the prices by the extent to which the compensation event has affected the defined cost and the resulting fee adjustment (clause 63.1). This method applies to assessment of compensation events for all of NEC4's Main Option clauses.

NEC4's companion, *Managing Reality*, explains the basis for assessing compensation events by using Defined Cost. The theory is that the contractor is 'no better or worse off'[339] if the client instructs a change for which he bears the risk. It further explains that the contractor's risk position (when providing quotations for compensation events that change the prices) is the same as when he was tendering for the project.

For such an important and widely applicable term, the definitions of 'Defined Cost' are a little scattered throughout NEC4, and some common-sense interpretation is required. Firstly, the correct definition for the applicable Main Option clause should be identified. These are summarised in Table 8.12.

---

lasts for the same period of time as the contractor had to provide his clause 62.2 quotation (either two weeks under clause 62.2, or such extended period under clause 62.5).

336 Akin to the effect of the Final Statement in JCT DB 2017 (clause 1.8 – Effect of Final Statement).
337 NEC4, clause 66.2.
338 NEC4, clause 61.6, project manager's correction of an assumption, which is a compensation event under clause 60.1(17).
339 NEC4, *Managing Reality*, Book Four – *Managing Change*, 3rd ed., (2017), p. 45.

Table 8.12 Applicable definitions of 'Defined Cost'.

Definition Clause	Defined Cost Definition	Applicable Main Option Clauses[a]
11.2(23)	The cost of the components in the Short Schedule of Cost Components is the Defined Cost	A and B
11.2(24)	The cost of the components in the Schedule of Cost Components excluding the Disallowed Cost[b]	C, D, and E

a) NEC4 – Main Option Clause F (Cost reimbursable contract) is not considered in this book.
b) NEC4 – Main Option Clauses C, E, and E – 'Disallowed Cost' is defined in clause 11.2(26).

Table 8.13 Explanation of actual and forecast Defined Cost.

Category of Defined Cost (clause 63.1)	Explanation/interpretation
'actual Defined Cost'	This is the Defined Cost of work that the contractor has already undertaken ('the work done') by the 'dividing date'. It is not the value of work carried out to date – as set out in the rates or prices contained in the Short Schedule of Cost Components, the Schedule of Cost Components, or any Bill of Quantities.
'forecast Defined Cost'	This is a forecast of the Defined Cost of the work that the contractor must carry out after the 'dividing date'. It is not the value of the work yet to be carried out – as calculated from rates and prices contained the Short Schedule of Cost Components, the Schedule of Cost Components, or the Bill of Quantities, but the forecast Defined Cost of such work as is necessary to give effect to the compensation event.

Clause 52.1 provides another definition of 'Defined Cost', but this focuses more on allocating the contractor's recoverable costs into either type of Defined Cost and the Fee.[340] The categories of Defined Cost in clause 63.1 are:

o Actual Defined Cost, and
o Forecast Defined Cost.

Unfortunately, NEC4 does not define what is meant by the 'actual' or 'forecast' categories of defined cost, but it is not difficult to arrive at a common-sense interpretation, when clause 63.1 is read alongside the explanation provided in NEC4's companion – *Managing Reality*[341] – as Table 8.13 summarises.

In very basic terms, the two categories of defined cost are either the actual or forecast costs of the physical work (such as the costs of labour, plant, materials, and subcontractors) that the contractor has carried out or will carry out. The 'Fee' is intended to cover the contractor's ancillary costs such as overheads (such as administration, marketing,

---

340 NEC4, clause 52.1 gives a definition of 'Defined Cost' as essentially the costs incurred by the contractor, which are not included in the Fee, which is defined in clause 11.2(10) and is the 'fee percentage' stated in Contract Data Part Two (under '1 General') which is applied to the 'Defined Cost'.
341 NEC4, *Managing Reality*, Book Four – *Managing Change*, 3rd ed., (2017), pp. 24–25.

recruitment, and training) and profit, along with any applicable taxation and bonds or guarantees.[342]

The second step for the project manager is to establish when the 'dividing date' occurred, which will either be:

o The date of an instruction from the project manager or supervisor, or
o The date on which the contractor issued the project manager with his notification of a compensation event under clause 61.3.

In practical terms, the date of the project manager's or supervisor's instruction is likely to be an instruction to change the Scope[343] or change a previous decision.[344] As this will often be the earliest date that the contractor hears about the change, it is highly unlikely that he will have started to do any of the necessary work resulting from the instructed change in the Scope or a previous decision. Conversely, if the contractor has notified a compensation event under clause 61.3, there is a greater likelihood that some of the compensation event's effects have already taken place (for example, the contractor may have already incurred delay cost – perhaps for a period of up to eight weeks[345]).

Drawing NEC4's overriding principles for a compensation event, along with the categories of defined cost together, an illustration of how the change to the prices are assessed can be given:

Example

(1) For a highways project let under NEC4 Main Option Clause B (Priced Contract with Bill of Quantities), the project manager instructs a change in the Scope, adding 100 no. pre-cast concrete bollards – 300 mm diameter and 750 mm high.
(2) In the Bill of Quantities, the contractor has priced the same bollard at £205 each, yet the actual cost of installing each bollard is £240. On this basis, and if there are no compensation events, the contractor will lose £35 for each bollard that he installs. This would be the case if the contract was let under the JCT SBC/Q 2016 form, where the contractor takes the risk for the sufficiency of this rates and prices, even in the event of a variation – so would only be entitled to recover £205 for installing each bollard.
(3) However, applying clause 61.3, it is clear that the 'dividing date' is the date of the project manager's instruction, so at that date, the contractor has not started the compensation event work. Therefore the assessment of the change in the prices will only use 'forecast Defined Costs'. NEC4's principle provides that the contractor should not suffer a commercial disadvantage when undertaking compensation events, so the compensation event would be assessed as the forecast Defined Cost of £240 per bollard plus the Fee.

It is vital that the project manager consistently applies the concept of 'Defined Cost' (either actual or forecast) to his assessments. Assessing the compensation event on a basis

---

342 NEC4, *Managing Reality*, Book Four – *Managing Change*, 3rd ed., (2017), p. 49.
343 NEC4, A compensation event under clause 60.1(1).
344 NEC4, A compensation event under clause 60.1(8).
345 NEC4, The contractor is only entitled to a change in the Prices or Completion Date/Key Date if he notifies a compensation event within eight weeks after becoming aware that the event has happened.

that is not consistent with clause 61.3, even if intended to be for the parties' convenience, is likely to result in an imbalance between the risk positions that the client and the contractor have accepted under NEC4. For example, if the project manager and the contractor agree on rates or lump sums[346] to assess the change in prices for one compensation event but revert to the 'Defined Cost' approach when assessing another, their risk positions may vary as follows:

o For the compensation event assessed on the basis of agreed rates and prices:
  – The contractor may benefit from the project manager agreeing to a rate or price which is particularly lucrative, providing a greater profit margin than he would obtain via the 'Defined Cost' method of assessment; or
  – The contractor may agree to rates or prices which provide less of a profit margin, or even result in a loss.
  – The party believing that they have 'lost out' may seek to recover their losses (in the case of the contractor) or additional monies paid (in the case of the client) when assessing subsequent compensation events.
  – The project manager (or a cost consultant assisting him) may misuse or misapply the agreed rates and prices when assessing subsequent compensation events. He may wrongly try to compare the price assessed using the 'Defined Cost' method with an agreed rate or price, even when the two methods are based upon entirely different bases and concepts of risk.

Managing Reality claims[347] that the 'Defined Cost' approach dispenses with arguments that often occur under other contracts (such as JCT SBC 2016, JCT DB 2016, or FIDIC Yellow Book 2017) when valuing works using unit rates or prices. In particular, arguments over the applicability or appropriateness of adjusting a rate or price (in a bill or quantities or priced schedule of work) to give a new rate that reflects the changes in quantities, changes in specification, or changes in conditions under which the variation is (or will be) carried out.

In the author's view, and from a quantity surveying perspective, the opposite of *Managing Reality*'s claim is often true. The reasons for this are as follows:

o Unlike JCT DB 2016 or FIDIC Yellow Book 2017, the rates and prices for work in NEC4's activity schedule[348] or bill of quantities[349] are not used as the basis for assessing the change in the prices, as NEC4 clearly assesses the change in the prices using the defined cost method. Therefore, under NEC4, the contractor only bears the pricing risk of the rates or prices in the activity schedule or bill of quantities, and that risk is not transferred when assessing compensation events. Contrast this with the contractor's position under JCT DB 2016 or FIDIC Yellow Book 2017, where the contractor always bears the pricing risk for changes where his rates and prices form the basis of valuing the works.
o Where a rate or a price is to be adjusted, the basis of that rate or price remains, so ordinarily, any argument is confined to the rate 'adjustment' only. The rate adjustment is usually

---

346 NEC4, clause 63.2. See also the 'risk allowances' for cost and time to be included under clause 63.8.
347 NEC4, *Managing Reality,* Book Four – *Managing Change*, 3rd ed., (2017), p. 45.
348 NEC4, Option A (Priced Contract with Activity Schedule).
349 NEC4, Option B (Priced Contract with Bill of Quantities).

assessed as additional costs required to carry out the work. For example, if the work is carried out at height, then the additional elements are likely to include the costs of providing access and any labour for extra handling. The amounts in respect of physically carrying out the item of work itself (i.e. the elemental costs for labour, materials, plant, or subcontractor work) are likely to be unchanged.

However, NEC4's defined cost method is likely to create more potential matters for dispute for the parties to argue over, as following elements can be disputed:

o The forecast defined cost of the original work: For example, the contractor may claim that the relevant price in his activity schedule or bill of quantities represents the forecast defined cost. However, this may not be the case, as the contractor may have under-priced the work in order to provide a competitive tender. As a matter of course, the project manager should investigate this in order to ascertain an accurate value for the original work's forecast defined cost.
o The forecast defined cost for the changed work:
o The project manager and contractor may be in dispute over the appropriate rates or prices in the Schedule of Cost Components or the Short Schedule of Cost Components (which are often expressed as hourly rates) to use for the resources required (items of plant and personnel) in order to carry out both the original and the changed work; and
o The durations for which the resources in the Schedule of Cost Components or Short Schedule of Cost Components (plant and people[350]) are required to carry out the original and the changed work. A quantity surveyor will be able to measure the quantities of both the original and changed work, and then apply the appropriate rates of output per unit of the original and changed work. For example, Spon's pricing books include the outputs (or durations of required resource per unit of work) for most items of work,[351] to which the rates and prices in the Schedule of Cost Components or Short Schedule of Cost Components can be applied.

It is often far less complicated to adopt what is a more traditional approach, which is:

o Use one source of cost data[352] from which to value variations, and
o The contractor takes the pricing risk for those rates and prices.

This ensures that any disputes over valuing work using new or derived rates are usually confined to the adjustments to that rate or price, rather than a wholesale dispute over all elements of it. The rate or price can be directly applied to the quantity of work, so arguments over the duration or output for that work can similarly be avoided.

This traditional approach fits neither with NEC4's ethos for the client taking the risk for all compensation events nor with its defined cost method. This is not to say that the methods of valuation for JCT or FIDIC contracts are better or worse, but parties to NEC4 contracts should be aware that *Managing Reality*'s claim that, seemingly, NEC4 inherently reduces the disputes over valuation of variations (when compared to JCT or FIDIC) may

---

350 NEC4, Option B, clause 63.13, the project manager and contractor can agree on a new rate for 'People' but not for any other category of resource.
351 Spon's *Architects and Builders Price Book* includes 'Labour hours' for most items of work, and its *Civil Engineering and Highway Price Book* includes 'Gang hours'.
352 Such as the rates and prices in a bill of quantities or a schedule of rates.

Table 8.14 Summary of NEC4 Main Option Clauses' assessment of compensation events.

Main Option Clauses for assessing compensation events	Option A	Option B	Option C	Option D
Are the Prices (Options A and B) or Defined Cost (Options C and D) reduced from changes to the client's Scope proposed by the contractor.	Yes 63.12	Yes 63.12	No 63.13	No 63.13
Prices are reduced by multiplying the compensation event's effect by the 'value engineering percentage'.[a)]	63.12	63.12	–	–
Changed Price assessments are in the form of changes to:	Activity Schedule 63.14	Bill of Quantities 63.15	Activity Schedule 63.13	Bill of Quantities 63.15
– For work 'not yet done'	–	Changed rate, quantity or lump sum, or new rate[a)]	–	Changed rate, quantity or lump sum, or new rate[b)]
– For 'work done'	–	New lump sum item	–	New lump sum item
New rate for 'People Rates'	63.16	63.16	–	–

a) NEC4 – 'Value engineering percentage' – see Contract Data Part One (6 – Compensation events) for Options A or B.
b) NEC4 – Option B, clause 63.15 – if there is no rate in the Bill of Quantities, a new rate is calculated by reference to the 'method of measurement' (set out in Contract Data Part One, 6 – Compensation Events), unless the project manager and contractor otherwise agree.

not be entirely correct, and actually may require a greater quantity surveying involvement than would otherwise be required under the other design contracts.

The methods of assessing the change to the prices (or defined cost) for each of NEC4's Main Option clauses A to D are set out in Table 8.14.

Option E (Cost Reimbursable Contract) assesses a change to the prices in accordance with clauses 63.1–63.4 of NEC4's Core Clauses.

If selected, Option X1 (Price Adjustment for Inflation) is used to change the prices for Options A, B, C, and D on a similar 'fluctuations' basis as the Fluctuations Option in JCT DB 2016 and the Adjustment for Changes in Cost (clause 13.7) in the FIDIC Yellow Book 2017.

Under Option C (Target Contract with Activity Schedule) or Option D (Target Contract with Bill of Quantities), the project manager must assess the 'contractor's share'.[353] He does this on two occasions; firstly providing a 'preliminary assessment' at completion, and then a 'final assessment' with assessing the 'final amount due'. As both assessments are essentially

---

353 NEC4, refer to Option C, clauses 54.1–54.4, and Option D, clauses 54.5–54.8 for the 'Contractor's share'.

tasks that take place on and after completion, they are covered in Chapter 9 (Concluding the Contract), in Section 9.6.3.

8.9.3.2 Assessing Changes to the Completion Date and Any Key Dates

The general principles for the contractor firstly claiming and then the employer (or his representative) assessing and granting extensions of time are covered in more detail in Section 8.10 (Extensions of Time). However, given that NEC4 sets out how to assess compensation events in terms of (i) changes to the prices and (ii) to the completion date and to any key dates, then it seems appropriate to address the procedure for point (ii) in this section.

The importance of the contractor's accepted programme cannot be understated, as it is central to the method of assessing changes to the completion date and changes to any key dates, and to how NEC4 requires those changes to be expressed.

NEC4's requirements for the contractor's programme (and the accepted programme) are set out in clause 31, and these are summarised in Section 8.6.2 and Table 8.5. In terms of the matters (completion date and key dates) for which changes (due to compensation events) are assessed, the contractor's programme must show the following:

o Completion date and 'planned completion'; and
o Key dates, and the 'planned date' on which the condition for satisfying the key date is likely to be met.

The use of 'planned' dates is central to NEC4's ethos of good management practice and are considered independently of the contractual dates (namely, the completion date and any key dates), because in practice, the 'actual' dates when the contractor completes the works or hits a key date (when compared to the relevant dates stipulated by the contract) may be significantly different. As seen in Section 8.4.5, the contractor is obliged to complete the works and the work for any key dates 'on or before' the completion dates or key dates set out in Contract Data Part One.[354]

NEC4's close monitoring and management of the programme (see Section 8.6.2) enables the client to be kept abreast of when the contractor is likely to complete the works and key dates from the following:

o The contractor's revised and accepted programmes; and
o The benefit of the project manager's advice regarding the same,[355] that is, accepting or rejecting the contractor's revised programme.

This not only allows the client to form an objective view of progress (so as not to solely rely on what the contractor is telling him) but also to manage himself and any responsibilities he has for the project. If the contractor is likely to complete the project (or stages of it) early, then by virtue of the NEC4's management procedures such as revised programmes, early warnings, and notifying compensation events, the client may be able to arrange for his own early occupation or use of the project. Similarly, if the contractor is likely to satisfy the condition for meeting a key date early, and the client is responsible for necessary works that

---

354 NEC4 – clause 30.1 (completion date) and clause 30.3 (Key Date). If Option X5 (Sectional Completion) is selected, the 'completion date' will be for each section of the works.
355 See Section 8.6.3 and the reasons for the project manager not accepting the contractor's programme.

follow on from that, then he can make arrangements for his own work (or that of others) to be brought forward. Conversely, if the contractor is likely to complete the works (and satisfy the key date conditions) later than the dates in Contract Data Part One, then the client has a greater opportunity to mitigate any delay and its effects.

The contractor's programme is NEC4's key document for monitoring progress, so it is unsurprising that the method of assessing delays to the completion date and to any key dates also involves the contractor's programme. Clause 63.5 provides that delays to either the completion date or a key date are established by the following steps:

o Firstly, the 'dividing date'[356] is established (see Section 8.9.3.1 above).
o Secondly, the following periods of time (as at the 'dividing date') are established:
  – That the compensation event has caused either 'planned completion' to be later than the planned completion on the accepted programme; or
  – Similarly, the 'planned date for meeting a condition' for a key date to be delayed beyond those respective 'planned' dates shown on the latest accepted programme.
o Thirdly, and in order to ensure that the assessment of delay is accurate as at the 'dividing date', it must allow for:
  – Any events (that have hindered or assisted the works' progress) that are already accounted for in the latest accepted programme[357] (see clause 62.2).
  – Likewise, it must allow for any such events that occurred between the date of the last accepted programme and the 'dividing date'.
o Finally, the same relevant period of delay to either 'planned completion' or to the 'planned date for meeting a condition' for a key date is then applied to the completion date or the key date. NEC4's 'User Guide' neatly captures this principle as follows:

> 'By taking this approach, if planned Completion is delayed the Completion Date is delayed by the same period. If planned Completion is not delayed the Completion Date is unchanged. The same principle applies to the planned achievement of a Condition required by a Key Date…'[358]

The actual method of assessing the delays is not proscribed, but it is likely to be one of the accepted methods (such as critical path analysis), which are covered in more detail in Section 8.10 (Extensions of Time), unless the parties agree on the method beforehand.

As with many aspects of NEC4, it assumes that everyone is carrying out their functions in accordance with the contract, particularly at the required times. Therefore, it is virtually essential to have an up-to-date accepted programme.

8.10   Extensions of Time

The three design and build contracts covered by this book deal with matters such as extensions of time and additional costs (or loss and expense – covered in Section 8.11) in different

---

356 The 'dividing date' is either (i) the date of an instruction from the project manager or supervisor; or (ii) the date when the contractor issued the project manager with his notification of a compensation event under clause 61.3.
357 For example, any events that were the subject of an Early Warning (clause 15).
358 NEC4, *User's Guide – Managing an Engineering and Construction Contract,* (2017), p. 64.

ways. JCT DB 2016 and FIDIC Yellow Book 2017 contain specific procedures that are separate from those dealing with other matters, such as variations. This can give the impression that extensions of time and additional costs (loss and expense) are more 'contentious' than variations or other important events occurring under the contract, and that they are often (and certainly in the case of FIDIC) considered to be 'claims' items. However, the parties should remember that the purpose of extension of time provisions is to preserve the employer's right to delay damages in the event that he has caused the progress of works to be adversely affected. An effective extension of time clause will firstly protect the contractor from delay damages for a period of employer delay, whilst preserving the employer's right to claim delay damages for any period of contractor delay.

It is often the case that the parties leave claims until the later stages of the project, despite the almost universal advice from the legal community that claims should be dealt with contemporaneously. It is almost as if 'claims' is a dirty word, and in many instances the parties either shy away from dealing with extensions of time or additional costs until the last minute. This often causes further contention because they are having to argue whether a claim is permissible (because one party considers the claim was made too late), before even addressing the matters of extension of time or additional costs themselves.

In comparison, NEC4's approach is a breath of fresh air. It includes all matters that entitle the contractor to additional payment or additional time for completion[359] under the heading 'compensation event' (see Sections 8.8.5 and 8.8.5.1), and it encourages the parties to deal with all such matters (including claims) contemporaneously and within the same period of time.

In the event of delays, all three design and build contracts provide for extensions of time to be granted to the contractor if any employer risk event or a neutral event occurs. Plainly, the contractor should not be, and is not, entitled to extensions of time if he is at fault for a delay.

JCT DB 2016 and FIDIC Yellow Book 2017 provide separate clauses and procedures that govern extensions of time, whereas NEC4 deals with extensions of time under the general heading of 'compensation events' (which also includes variations – see Sections 8.8.5, 8.8.5.1, 8.8.5.2, 8.9.3, 8.9.3.1, and 8.9.3.2). JCT DB 2016 and FIDIC Yellow Book 2017 deal with variations separately (see Sections 8.8.3 and 8.9.1 along with Sections 8.8.4 and 8.9.2, respectively).

The extension of time procedures for JCT DB 2016 and FIDIC Yellow Book 2017 focus on the following steps:

o Firstly, the contractor notifies the employer (or the supervising officer) that a delay to the completion date has occurred or is likely to occur.
o Secondly, the contractor must identify which grounds in the contract entitle him to an extension of time.
o Thirdly, the contractor provides further information in support of his claim, such as records and additional particulars.
o Fourthly, the contractor may have to show the delaying event's effects on a programme.

---

359 NEC4, 'compensation events' are summarised in Table 8.11.

o Finally, the employer or supervising officer must assess the contractor's information and decide upon (i) whether the contractor is entitled to an extension of time, and (ii) the length of that extension.

In broad terms, these five steps are also captured in NEC4's 'compensation event' procedure (see Section 8.8.5.2). The extension of time procedures in JCT DB 2016, FIDIC Yellow Book 2017, and NEC4 are largely administrative, and (with the slight exception of NEC4 – see Section 8.9.3.2) do not address the more technical task of *how* to ascertain an extension of time. Further guidance is provided in Section 8.10.1.

8.10.1 Ascertaining Delaying Effects

The methods of assessing delays will vary depending upon the necessary sequence of carrying out a project's Scope of works. Two differing examples of a sequence are as follows:

o A 'linear' sequence, where each subsequent activity (or a significant number of them) must necessarily follow its preceding activity. Examples of 'linear' engineering projects could include a road, an airport runway, or a railway. An example of a 'linear' construction project could be the construction of a warehouse unit.
o A 'non-linear' sequence could involve a project with a definite completion date but where the Scope of work comprises many activities (or subsets of activities) that are often repetitive in nature. They can largely be undertaken independently from one another but may be constrained by the available resources (usually labour). An engineering example of a 'non-linear' project could involve repairing potholes on a highway. A construction example of a non-linear project could be refurbishing rooms in a hotel.

It is almost universally accepted, both in and outside of the UK, that if the nature of a project's activities are generally linear, that some form of critical path analysis should be employed in order to ascertain the effects of a delaying event.

In Section 8.6.1, some terminology used in critical path analysis was included in the notes, but it is worth repeating here.

o Programmes:
   – 'Baseline'. This is usually the contractor's initial programme, or a programme against which progress and delays can be measured. Some planning software applications show the 'baseline' programme as narrower bars underneath the bars for the 'updated' programme activities, so a simple, visual comparison between the two can be made.
   – 'Updated'. Initially the 'baseline' programme is updated to reflect current progress and new events (such as variations and progress) to become the first 'updated' programme, and each subsequent updated programme becomes the current updated programme.
   – 'Rescheduled'. After inputting the relevant information (such as variations, progress, or delays) into the previous version of the programme (beginning with the baseline programme), the planning software can then reschedule the programme, so the Gantt chart shows any changes to the activities' sequence, timing, and completion that arise from the inputted information.

o Critical path:
  – The Society of Construction Law, Delay, and Disruption Protocol (SCL):[360]

  > '*The longest sequence of activities through a project network*[361] *from start to finish, the sum of whose durations determines the overall project duration. There may be more than one critical path depending upon workflow logic. A delay to progress of any activity on the critical path will, without acceleration or re-sequencing, cause the overall duration to be extended, and is therefore referred to as a "critical delay".*'

  – The Chartered Institute of Building (CIOB) Guide to Good Practice in the Management of Time[362]:

  > '*The longest sequence of logically linked activities from the start of a schedule*[361], *to its finish that determine the earliest possible achievement of a key date, sectional completion date, or the completion date.*'

o Float[363]:
  – SCL: '*The time available for an activity in addition to its planned duration.*'
  – CIOB[362]: '*The degree of flexibility available for the scheduling of an activity…The amount by which an activity can be delayed without causing delay to subsequent activities or the completion date.*'
o Concurrent delay:
  – SCL: '*True concurrent delay is the occurrence of two or more delay events at the same time, one an Employer Risk Event, and the other a Contractor Risk Event, and the effects of which are felt at the same time.*'
o Mitigation measures:
  – Acceleration (see Section 8.5),
  – Re-sequencing the programme, and
  – Recovery (achieving better progress than planned or anticipated).

The widespread use of sophisticated programming or planning software (now commonplace with construction and engineering projects – see Section 8.6.1), has made delay analysis much easier to depict and understand. Provided that the user generates a logical sequence of work (or 'network of activities'), and links the relevant activities appropriately, most planning software can automatically show the critical path of activities, the float, and any critical delay. This is referred to as critical path analysis (CPA).

---

360 Second edition (2017).
361 'Project network' and 'Schedule' are other names for a Gantt chart programme.
362 Second edition (2018).
363 This also includes 'Free Float' – which the CIOB define as 'The period by which an activity may be delayed without delaying the start of any of its successor activities' and 'Total Float', the CIOB definition of which is 'The amount of time an activity can be delayed without delaying a key date, sectional completion date or the completion date, without contravening a constraint'. The CIOB define a 'constraint' as 'A restriction on the ability of a critical path network schedule to obey the activity and / or resource logic of the schedule'.

The extent of difficulty in ascertaining the effect of a delaying event can vary significantly, depending upon the complexity of either the project (as a whole), the quality of availability records, or the interrelationship of the affected activities. Examples at each end of this 'scale of difficulty' are as follows:

o A simple example would be if the employer fails to provide the contractor access to the site for four weeks (so that he is unable to carry out any physical work), then on the basis that (i) the work can be progressed from one activity to the next in a single series of steps, (ii) the work is continuous,[364] and (iii) it is realistic to expect the project would otherwise have been completed on time, it seems highly likely that the entire project will be delayed by four weeks.

o A more complex example could be one in which a change (variation) to the specification of some steelwork-mounted plant has been instructed. The structural loading for the steelwork may need to be changed, so this could require more time to re-design the structural steel members and their connections, along with procuring, fabricating, and installing the new steelwork. Whilst it sounds like progress could be delayed, the following may not be immediately clear and may require investigating to ascertain the following:

- If a delay to the steelwork activity will similarly affect the progress of subsequent activities;
- If the delay can be accommodated within the overall duration of the steelwork activities because the contractor has made sufficient time-risk allowances (float) within his programme for such delays;
- If the delay can be naturally accommodated within the overall steelwork programme because it occurs at the same time as other steelwork activities (i.e. concurrently); or
- If the contractor is able to mitigate any delays.

There are essentially four CPA methods, which are summarised in Table 8.15.[365]

Critical path analysis functions best when supported by plenty of factual information. If the contractor keeps records of progress (such as photographs, time-lapse video, allocation sheets for labour and plant, delivery tickets, marked-up programmes and drawings, substantiation for payment applications, etc.), then the programme's depiction of progress should be fairly accurate. If the majority of the relevant logic links between activities can be verified from the project's drawings and specification, then the CPA exercise becomes less subjective, and there should be less room for disputes.

For non-linear projects, it is often necessary to analyse the resources used for each individual task (or subset of tasks if the mini-sequence is repetitive). A 'line of balance diagram'[366] is often used in tandem with an appropriate CPA, where the focus is on the quantity and distribution of resources.

---

364 The contractor has planned for neither any 'breaks' nor any 'float' when progressing the work.
365 Also see the *SCL Delay and Disruption Protocol*, 2nd ed., (2017), at paragraph 11.5.
366 *CIOB, Guide to Good Practice in the Management of Time in Major Projects*, 2nd ed., (2018), defines 'Line of balance' as 'An illustration of sequence of repeated tasks across a number of work areas identifying resource, location and time used to optimise the use of resources across all trades working in defined areas' on p. 203.

Table 8.15 Critical Path Analysis (CPA) methods.

CPA method	Description of method	Required resources
Impacted As Planned	o The relevant delay is inserted into the original (or baseline) programme to calculate the revised completion date. o Critical path and delay impact may be determined prospectively.	– Original contract (or baseline) programme – Record of the delay(s) duration
Time Impact Analysis[a]	o The 'baseline' programme is updated periodically[b] to include actual progress and the insertion of delay events (either actual or forecast durations). The programme is then updated (or re-scheduled) after each period (or 'window') to calculate the updated completion. o The critical path is determined contemporaneously, and the impact of delay can be determined prospectively or retrospectively.	– Logic-linked baseline programme – Updated progress information (including delay events and accurate records of progress)
As-Planned vs. As-Built	o A comparison is made between the original or baseline programme (the 'As-Planned programme'), and a programme showing the actual sequence and timing of all work activities (the 'As-Built programme'). The differences between the programmes shows the extent of delay and (on further analysis) indicates who is responsible for individual delays. o The critical path can be determined contemporaneously, but the delay impact is usually determined retrospectively.	– Original or 'baseline' programme – Updated progress information (including delay events and accurate records of progress)
Collapsed As-Built[c]	o An accurate 'as-built' programme with detailed logic links is compiled, from which delaying events are removed, so the programme can be 'collapsed' to show the net effect of the delay events and any contractor delays relative to the contract's completion date(s). o The critical path and impact of delay are determined retrospectively.	– Logic-linked as-built programme – Updated progress information (including an accurate record of progress) – Does not require a baseline programme

a) May also be referred to as 'Time slice analysis', as there are notable similarities between the two methods.
b) For example, in 'windows' of one month.
c) Also may be referred to as 'Retrospective Longest Path Analysis'.

8.10.2 Procedures for the Supervising Officer

The specific procedures for the supervising officer under each of the three design and build contracts are considered under each contract in Sections 8.10.3 (JCT DB 2016), Section 8.10.4 (FIDIC Yellow Book 2017), and Section 8.10.5 (NEC4).

Ordinarily, the supervising officer ought to observe the following principles when assessing extensions of time:

o Not adopting a 'wait and see' approach.[367] The supervising officer (or employer) is encouraged to try and deal with the extension of time as close to the delaying events as possible;
o Correctly applying the relevant contractual provisions[368];
o Acting in a fair, unbiased,[369] impartial,[370] rational, and reasonable manner[368]; and
o Carrying out an analysis that is calculated, logical, and methodical, not one that is impressionistic or general[368].

The supervising officer's use of planning software applications is not thought to be essential to him assessing extensions of time,[371] as older methods (impliedly not involving planning software) remain valid if planning software cannot be used accurately. However, given that in the majority of cases,[372] it is only practicable to show many of the programme's express requirements in the FIDIC Yellow Book 2017[373] and in NEC4[374] (refer to Section 8.6.2) by using planning software, it would be advisable for the supervising officer to either have access to the relevant software, or if not, to the services of someone who has.

8.10.3   Extensions of Time under JCT DB 2016

The way in which the time for the contractor to complete the works (or a section thereof) can be extended under JCT DB 2016, can be summarised as follows:

o Completion of the works (or a section) is likely to be delayed beyond the relevant completion date – so no extension of time is given for a delay that does not cause the completion date (or sectional completion date) to be postponed.
o The employer does not express an 'extension of time' in terms of a period of time (e.g. six weeks). Instead, the employer gives an extension of time by confirming (or 'fixing') a later date as the completion date for the work or section.[375]
o The delay events which entitle the contractor to an extension of time are called 'Relevant Events' (see Table 8.16).
o Before the completion date of the works (or a section),[376] the employer's obligation to consider an extension of time is only triggered once the contractor provides a compliant notice and particulars of the delay to the employer.[377]

---

367 *SCL Delay and Disruption Protocol,* 2nd ed., (2017), Core principle 4, p. 5.
368 *John Barker Construction Ltd v London Portman Hotel Ltd,* 83 BLR 31, (1996).
369 *Sutcliffe v Thackrah,* AC 727, [1974].
370 *Costain Ltd v Bechtel,* EWHC 1018, [2005].
371 *City Inn Ltd v Shepherd Construction Ltd,* BLR 269, [2008], paragraphs 27–29.
372 *SCL Delay and Disruption Protocol,* 2nd ed., (2017), Guidance Part B: Guidance on Core Principles (at para 1.43) encourages the contractor to provide its programme to the supervising officer in its 'native electronic form' (e.g. Oracle Primavera, Elecosoft Powerproject, or Microsoft Project) and not just as a .pdf.
373 FIDIC Yellow Book 2017, clause 8.3.
374 NEC4, clause 31.2.
375 JCT DB 2016, clause 2.25.1.
376 JCT DB 2016, clause 2.27 (Practical Completion).
377 JCT DB 2016, clause 2.25.1.

Table 8.16 JCT DB 2016 Relevant Events (clause 2.26) and Relevant Matters (clause 4.21).

Description of Relevant Event / Relevant Matter	Relevant Event clause	Relevant Matter clause
Changes, and instructions for Changes (variations)	2.26.1	4.21.1
Instructions for:		
– Resolving inadequacies (clause 2.12) or discrepancies (clause 2.13)	2.26.2.1	–
– Postponing work (clause 3.10) and Provisional sums (clause 3.11)	2.26.2.2	4.21.2.1
– Opening up / testing compliant work (clauses 3.12, 3.13)	2.26.2.3	4.21.2.2
Deferment of possession (clause 2.4)	2.26.3	–
Compliance with, and instructions under clause 3.15 (Antiquities)	2.26.4	4.21.3
Contractor's suspension – non-payment (clause 4.11)	2.26.5	4.11.2
Employer causing an impediment, prevention or default	2.26.6	4.21.5
Statutory Undertaker / statutory obligations	2.26.7	–
Exceptionally adverse weather conditions	2.26.8	–
Loss or damage caused by Specified Perils (clause 6.8)	2.26.9	–
Civil commotion, terrorism	2.26.10	–
Strike, or lock-out (industrial action)	2.26.11	–
Exercise of statutory power by a public body after the Base Date[a)]	2.26.12	–
Delayed receipt of permission/approved by a statutory body[b)]	2.26.13	4.21.4
Force majeure	2.26.14	–

a)   JCT DB 2016, defined in clause 1.1 – refer to the Contract Particulars.
b)   JCT DB 2016, clause 2.26.13, the contractor must take all practicable steps to avoid or reduce the delay.

o If the works (or a section) continue after the relevant completion date,[378] the employer may consider an extension of time of his own volition (i.e. without the contractor providing a notice and particulars of the delay).

o The contractor must always use his best endeavours (see Section 8.4.3) to prevent delay.

o Once an extension of time is given, it can only be reduced in the event that the employer instructs a 'Relevant Omission'.[379] However, relevant omissions cannot result in a negative extension of time, whereby the contractor is obliged to complete by a date that is earlier than the original relevant completion date.[380]

o The only way that the contractor becomes obliged to complete by an earlier completion date than the date set out in the contract particulars is by agreeing to a 'Pre-agreed Adjustment' with the employer (by an Acceleration Quotation[381] – see Section 8.5.1).

Table 8.16 summarises the relevant events and identifies which of those are 'Relevant Matters' that also entitle the contractor to additional payment (refer to Section 8.11.1).

---

378 JCT DB 2016, clause 2.25.5, in the event that the Works (or a Section) have not reached practical completion after the relevant Completion Date has passed.
379 JCT DB 2016, clause 2.23.3, the omission by instruction (under clause 3.9) of any work or obligation.
380 JCT DB 2016, clause 2.25.6.3.
381 JCT DB 2016, Schedule 2 – Supplemental Provisions, Part 2, paragraph 4 (Acceleration Quotation).

Prior to practical completion (either of the works or a section), the employer's obligation to consider if the contractor is entitled to an extension of time is triggered by *both* of the following two events (under clause 2.24):

o The contractor must firstly notify[382] the employer under clause 2.24.1, and his notification must include:

   – The cause(s) of delay, which include the 'material circumstances'[383]; and
   – Identify which (if any) relevant events have caused or are likely to cause the delay.

   Importantly, the contractor's clause 2.24 notice must be issued in the event that the contractor has caused delay (i.e. the delay is not covered by a relevant event). This supports the notice's purpose to inform the employer of delays so he has a greater opportunity to take steps to avoid or reduce their effects.

o Clause 2.24.2 requires the contractor to give written 'particulars' of the effects of each delay cited in the clause 2.24 notice. The contractor can provide the particulars with the clause 2.24.1 notification or, if it is not practicable to do so, he can provide them afterwards. However, the particulars must be provided to trigger the employer's obligation to decide whether or not the contractor is entitled to an extension of time (under clause 2.25.1).

Clause 2.24.2 also requires the contractor to include an estimate of any delay with the 'particulars', but, unlike the contractor's clause 2.24.1 notice and clause 2.24.2 particulars, the estimate is not required to trigger the employer's decision under clause 2.25.1.

Clause 2.24.3 encourages the contractor to revise the particulars and the estimate if there is any 'material change' in either, along with providing any further information that the employer may reasonably require. The former part of clause 2.24.3 is important, as it recognises that the progress of work may be 'fluid' in relation to the completion date or as measured against any programme. The contractor should constantly keep the employer informed of any changes to progress resulting from the delays notified under clause 2.24.1. It is also important for the employer and the employer's agent to appreciate that progress may change, and the contractor's estimate of delay may need revising. They should not accept the contractor's first estimate as his final, or come-what-may answer.

The employer's response to the contractor's notice and particulars comes in the form of his decision under clause 2.25.1 as to whether or not,[384] he considers that the contractor is entitled to an extension of time. The employer's clause 2.25.1 decision should be provided as soon as is 'reasonably practicable', but in any event:

o Given not later than 12 weeks after receipt of the contractor's clause 2.24.2 particulars; or
o If there are less than 12 weeks until the completion date (of the works or a section), then his clause 2.25.1 decision must be provided before the relevant completion date.

---

382 JCT DB 2016, clause 1.7 (Notices and other communications) states that any notice must be in writing. Clause 2.24.2 (when referring to the 'particulars' being sent separate to the clause 2.24.1 notice, also provides that they must be 'in writing', so the 'written' requirement in clause 2.24 is clear).

383 Material circumstances are the significant or relevant facts connected with or relevant to the delay event.

384 JCT DB 2016, clause 2.25.2 provides that the employer shall give a clause 2.25.1 decision to the contractor even if he decides that the contractor is not entitled to an extension of time.

The employer's clause 2.25.1 decision must state the following:

o If an extension of time is granted, and if it is:
o The extension of time granted for each relevant event set out in the contractor's clause 2.24.1 notice; and
o If the clause 2.25.1 decision is not the employer's first extension of time decision, and in the event that the employer has instructed a relevant omission, the earlier date fixed as the completion date owing to the relevant omission (but subject always to clause 2.25.6.3).

8.10.4 FIDIC Yellow Book 2017

The introductory part of this Section 8.10 briefly summarised the procedures with which each of the three design and build contracts operate when dealing with extensions of time and additional costs or loss and expense (see Section 8.11). It compares the approach taken by NEC4, that all matters capable of changing the contract price and completion dates (i.e. variations, extensions of time, and additional costs) are covered under the 'compensation event' umbrella, against the approaches taken by JCT DB 2016 and the FIDIC Yellow Book 2017, whereby variations have a separate procedure away from extensions of time and additional costs (or loss and expense). These latter procedures under JCT and FIDIC were deliberately referred to as 'claims' (the inverted commas were intentional), as both can give the impression that extensions of time and additional costs are somehow more contentious than variations.

FIDIC Yellow Book 2017 takes this a step further than JCT DB 2016, by including the 'Employer's and Contractor's Claims' section at clause 20, which applies to claims both for extensions of time and additional payment.[385] The 'Claims' clause 20 is firstly covered in Section 8.10.4.1 (as it also applies to Additional Costs in Section 8.11), and Section 8.10.4.2 deals with the extension of time procedures.

8.10.4.1 FIDIC's 'Claims' Procedure
When dealing with 'claims'[386], clause 20 of the FIDIC Yellow Book 2017 includes a similar procedure to that employed by NEC4 when dealing with compensation events (see Section 8.8.5.2). It imposes time bars on the parties' entitlement if they do not notify the other of claims, or comply with the procedure set out in clause 20.2 (Claims for Payment and/or EoT[387]). FIDIC's imposition of time bars involves more procedural stages than under NEC4, because a party is able to contest a claim if it applies in the particular circumstances, whereas clause 61.3 of NEC4 simply applies the time bar if the contractor fails to notify a compensation event within eight weeks. In relative terms, FIDIC's time bar in clause 20 seems to have fewer 'teeth' than its counterpart in NEC4.

FIDIC's claims procedure applies to both the employer and the contractor as follows[388]:

o The employer may claim:
  – A reduction in the contract price,

385 FIDIC Yellow Book 2017, clause 20.1(b) and clause 20.2 (Claims for Payment and /or EoT).
386 FIDIC Yellow Book 2017, 'claim' defined in clause 1.1.5.
387 FIDIC Yellow Book 2017, 'EoT' or 'Extension of Time' defined in clause 1.1.38.
388 FIDIC Yellow Book 2017, clause 20.2 (Claims for Payment and/or EoT).

- An extension to the Defects Notification Period[389] or 'DNP' (see Section 9.5.2);
- o The employer or the contractor is claiming an additional payment from the other (including payment of delay damages under clause 8.8 from the contractor to the employer – see Section 9.4.2); and
- o The contractor may claim an extension of time ('EoT').

The employer will only be entitled to his claims (for extending the DNP or recovering additional payments from the contractor) if he complies with clause 20.2.[390]

The first step is for the claiming party to notify the engineer (the 'Notice of Claim'), and they must comply with the time bar in clause 20.2.1 This provides that within 28 days of the issuing party either becoming aware (or when it should have become aware) of the matter[391] requiring such notice that the Notice of Claim must be issued. If the issuing party fails to comply with the 28-day time limit, then he will not be entitled to any of the relief described in the bullet points above,[392] unless:

- o The engineer fails to notify[393] the claiming party within 14 days after receiving the Notice of Claim.[394] If the engineer fails to so notify, then the Notice of Claim will deemed to be valid, unless the other party disagrees, and notifies the engineer accordingly[395]; or
- o After receiving the engineer's notice (under clause 20.2.2) within the 14-day period, the claiming party may either:
  - Disagree with the engineer, or
  - Provide justifying circumstances for the late submission[396] of the Notice of Claim with a 'Fully Detailed Claim' (see below).

The second step is for the claiming party to submit its 'Fully Detailed Claim' in accordance with clause 20.2.4. Unless the claiming party and the engineer agree otherwise, this must be submitted not later than 84 days after the claiming party first became aware (or ought to have become aware) of the matter for which it issued the Notice of Claim (or a maximum of 56 days after it issued the Notice of Claim). The Fully Detailed Claim includes:

- o A statement of the Claim's contractual or legal basis.[397] This element is essential, as failure to provide this statement within 84 days will invalidate the Notice of Claim,

---

389 FIDIC Yellow Book 2017, 'DNP' defined in clause 1.1.27 and see clause 11.3 (Extension to the Defects Notification Procedure).
390 FIDIC Yellow Book 2017, clause 20.2.7.
391 FIDIC Yellow Book 2017, clause 20.2.1 – the matter is described as an 'event or circumstance'.
392 FIDIC Yellow Book 2017, clause 20.2.1 provides that not only will the claiming party not be entitled to the relief flowing from the Notice of Claim, but also that any liability that the other party may have in connection with the claim matter will be discharged.
393 FIDIC Yellow Book 2017, clause 20.2.2 provides that the engineer must issue a 'Notice' (defined in clause 1.1.56) and complying with clause 1.3 (Notices and Other Communications).
394 FIDIC Yellow Book 2017, clause 20.2.2 (Engineer's Initial Response).
395 FIDIC Yellow Book 2017, clause 20.2.2, the other party's notification and disagreement with the 'deemed valid Notice of Claim' is then reviewed by the engineer under clause 20.2.5 (Agreement or Determination of the Claim).
396 It appears that in setting out such circumstances, the claiming party must accept that the Notice of Claim was submitted late.
397 FIDIC Yellow Book 2017, clause 20.2.4(b).

unless the engineer fails to notify the claiming party of such failure within 14 days. As with the engineer's failure to notify under clause 20.2.2 (see above),[393] his failure to notify receipt of the clause 20.2.4(b) statement of the legal basis within 14 days deems that the Notice of Claim is valid, unless the other party notifies their disagreement.[398] Similarly, if the engineer properly notifies the claiming party that the clause 20.2.4(b) statement has not been received, the claiming party can either disagree with the engineer's notice or provide the justifying circumstances for its late submission.[399]

o The matter[391] requiring the Notice of Claim – described in detail.
o Contemporary records,[400] which must be maintained by the contractor.
o Detailed particulars in support of the reduction in the contract price or extension to the DNP (claimed by the employer), an EoT (claimed by the contractor), or a claim for additional payment (claimed by either party).

The fully detailed claim is made when the matter giving rise to it has already occurred, and impliedly, has ceased (i.e. there has been cause, and the claim is the effect on either the contract price, DNP, EoT, or additional payment sought).

If the matter giving rise to the Notice of Claim is continuing (referred to as 'claims of continuing effect' under clause 20.2.6), then the claiming party may make an 'interim fully detailed claim'. In such cases, the claiming party's first 'fully detailed claim' (under clause 20.2.4) shall be considered as an interim claim,[401] and the claiming party shall submit further interim claims on a monthly basis,[402] each time setting out the accumulated effect[403] of the claim. The claiming party must then submit the 'final fully detailed claim' not later than 28 days after the end of the matter[404] for which the Notice of Claim was originally issued.

Clause 20.2.5 (Agreement or Determination of the Claim) provides that by notifying the parties, the engineer must either confirm his agreement to the claim or make a determination in accordance with clause 3.7 (Agreement or Determination). The engineer must issue either notification to the parties in 42 days (unless otherwise agreed[405]) in respect of:

o A fully detailed claim under clause 20.2.4,
o An interim fully detailed claim under clause 20.2.6(c), or
o A final fully detailed claim under clause 20.2.6(d).

If the engineer does not agree with the claim, his notice is an 'Engineer's Determination' (under clause 3.7.2), which must be fair and take appropriate account of all relevant

---

398 FIDIC Yellow Book 2017, clause 20.2.4.
399 FIDIC Yellow Book 2017, clause 2.20.4.
400 FIDIC Yellow Book 2017, clause 20.2.3 (Contemporary Records) kept by the claiming party (in this Section 8.10.4.1 – the contractor), and the contractor shall allow the engineer to inspect these records.
401 FIDIC Yellow Book 2017, clause 20.2.6(a).
402 FIDIC Yellow Book 2017, clause 20.2.6(c).
403 FIDIC Yellow Book 2017, clause 20.2.6(c) – the accumulated amounts of (i) additional payment, (ii) reduction in the Contract Price, (iii) the accumulated EoT, or (iv) the accumulated extension to the DNP.
404 FIDIC Yellow Book 2017, clause 20.2.6(d).
405 FIDIC Yellow Book 2017, clause 3.7.3 – 'time limit for determination' under clause 3.7 (Agreement or Determination) are proposed by the engineer and then agreed by the parties.

circumstances.[406] If the engineer does not notify within the 42-day (or other agreed) period, clause 3.7.3(ii) provides that the engineer is deemed to have made a determination rejecting the claim[407] (see Section 9.7.2.1). Clause 20.2.5 sets out that the engineer must agree or determine:

o Any additional payment that a party is claiming from the other[408] and
o Any extension to either[409]
  – The time for completion[410] (if the claiming party is the contractor) or
  – The defects notification period (if the claiming part is the employer).

The engineer may notify[411] the claiming party that he requires further information, which must be provided to him as soon as is reasonably practicable.[412] Whilst clause 20.2.5(iv) suggests that the 'time limits' for agreement or determination (under clause 3.7.3[413] – see Section 9.7.2.1) commence from when the engineer receives the further information, it appears that clause 20.2.5(ii) indicates that the engineer has to respond to the claim's legal or contractual basis[414] within 42 days of receiving the fully detailed claim (or interim fully detailed claim as the case may be), in any event.

The engineer must agree or determine the claim irrespective of whether he issued the following notices:

o Clause 20.2.2 (the claiming party's Notice of Claim was submitted late), or
o Clause 20.2.4 (the fully detailed claim or interim fully detailed claim was submitted late).

In the case of a late submission under either clause 20.2.2 or 20.2.4, the engineer's agreement or determination must consider if the Notice of Claim is valid, or if the late submission of a fully detailed claim (or interim claim) is justified. In making these considerations, the engineer may account for the following circumstances:

o If the non-claiming party is put at a disadvantage by the late submissions;
o The extent that the non-claiming party already knew of
  – The matters giving rise to the Notice of Claim or
  – The legal or contractual basis of the claim.

Clause 20.2.7 provides that the engineer should include amounts that have been 'substantiated' in each Payment Certificate. This indicates that the engineer may progressively consider the claim for agreement or determination over several months, and include

---

406 FIDIC Yellow Book 2017, clause 3.7.2 (Engineer's Determination), and must state it is a 'Notice of the Engineer's Determination'.
407 FIDIC Yellow Book 2017, clause 3.7.3(ii) provides that the engineer must provide a Notice of determination within 42 days.
408 FIDIC Yellow Book 2017, clause 20.2.5(a).
409 FIDIC Yellow Book 2017, clause 20.2.5(b).
410 FIDIC Yellow Book 2017, defined in clause 1.1.86, also see clause 8.2 (Time for Completion).
411 FIDIC Yellow Book 2017, clause 20.2.5(i).
412 FIDIC Yellow Book 2017, clause 20.2.5(iii).
413 Time limit is 42 days unless otherwise agreed.
414 FIDIC Yellow Book 2017, the 'statement' under clause 20.2.4(b).

'on account' sums in subsequent payment certificates whilst his task under clause 20.2.5 is proceeding.

8.10.4.2   Extensions of Time under FIDIC Yellow Book 2017

Clause 8.5 of the FIDIC Yellow Book 2017 provides that the contractor is entitled to extensions of time[415] on the following bases:

o If the cause of delay is a variation,[416] the contractor may be entitled to an extension of time without having to comply with clause 20.2 (Claims for Payment and/or EoT)[417]; or
o One of the following matters, for which the contractor must comply with the 'Claims' procedure in clause 20.2 (refer to Section 8.10.4.1):
   – An act of prevention by the employer[418];
   – Exceptionally adverse climatic conditions[419];
   – Unforeseeable shortages of materials[420]; or
   – Clause 8.5(b) – any other clause under the contract conditions that entitles the contractor to an extension of time.

Table 8.17 summarises the other conditions that entitle the contractor to an extension of time.

8.10.5   Extensions of Time under NEC4

Subject to demonstrating the effect of a delay (refer to Section 8.10.1), the contractor is entitled to extensions of time under NEC4 if a compensation event occurs (refer to Section 8.8.5.1), and the contractor must comply with the compensation event procedure (refer to Section 8.8.5.2). Assessing changes to the completion date or a key date is addressed in Section 8.9.3.2.

8.11   Additional Costs (Loss and Expense)

In contracts for construction work, there are a number of circumstances in which the contractor may become entitled to additional payment. An obvious example would be a change in the scope of works: either increasing the quantity of work or changing the specification for work from something cheaper to something more expensive. A less obvious example could be where price adjustments or fluctuations entitle the contractor to additional payment even when the scope of work has not been changed. However, in the context of this section, 'additional costs' or 'loss and expense' refers to the compensation that the contractor may be entitled to receive following the employer committing a

---

415 FIDIC Yellow Book 2017.
416 FIDIC Yellow Book 2017, 'Variation' defined in clause 1.1.88, and see clause 13 (Variations and Adjustments).
417 FIDIC Yellow Book 2017, clause 8.5(a).
418 FIDIC Yellow Book 2017, clause 8.5(e).
419 FIDIC Yellow Book 2017, clause 8.5(c).
420 FIDIC Yellow Book 2017, clause 8.5(d).

Table 8.17 Matters under clause 8.5(b) which entitle the contractor to an extension of time/EoT.

Clause 8.5(b) – Other clauses in the Contract Conditions entitling the contractor to an extension of time (EoT) or additional payment	Clause
Errors in the Employer's Requirements	1.9(ii)
Delays assisting the employer obtaining permits, permissions, etc.	1.13
Employer fails to give possession	2.1
Contractor instructed to cooperate with others[a]	4.6
Errors in setting out	4.7.3(ii)
Unforeseeable physical conditions	4.12.4
Access route unavailable	4.15
Archaeological and geological Findings	4.23
Employer's instructions to attend tests	7.4
Remedial work caused by the employer	7.6
Delays caused by authorities	8.6
Employer suspending the Works (under clause 8.9)/prolonged suspension	8.10, 8.12
Taking Over Parts	10.2
Contractor prevented from carrying out tests on completion	10.3
Right of Access after Taking Over	11.7
Engineer instructs the contractor to search for a defect and it is not remedied at the contractor's cost	11.8
Employer causes delay to tests after completion	12.2
Employer delays giving the contractor access for modifications or adjustments to the Works	12.4
Variation by Request for Proposal	13.3.2
Adjustments for Changes in Laws	13.6
Suspension by contractor	16.1
Liability for Care of the Works (delay in rectifying)	17.2
An Exceptional Event[b] occurs	18.4

a) FIDIC Yellow Book 2017, clause 4.6(a) to 4.6(c) describes the parties with whom the contractor may be instructed to cooperate.
b) FIDIC Yellow Book 2017, 'Exceptional Event' is defined in clause 1.1.37 and clause 18.1 (Exceptional Events) such as events beyond a party's control in clauses 18.1(i) to 18.1(iv), or other events in clauses 18.1(a) to 18.1(f).

breach of the contract – in other words – 'financial compensation'. This is the principle of common-law damages, which was illustrated in the case of *Robinson v Harman*:

> The rule of common law is, that where a party sustains a loss by reason of a breach of contract, he is, so far as money can do it, be placed in the same situation, with respect to damages, as if the contract had been performed.[421]

---

421 *Robinson v Harman*, 154 ER 163 at 855, (1848).

Therefore, in financial terms, the contractor is reinstated into the financial position he would have enjoyed if the employer had not been in breach. In short, the contractor is to be reimbursed, but not enriched, so he should not profit from the employer's breach.

The three design and build contracts do not expressly refer to the matters entitling the contractor to additional costs (or loss and expense) as 'breaches'. JCT DB 2016 refers to them as 'Relevant Matters',[422] and NEC4 refers to them as 'compensation events'.[423] FIDIC Yellow Book 2017 is less clear as it refers to the 'additional payment…under any Clause of these Conditions'.[424] On its face, this seems a wide interpretation, and it could include payments for variations (under clause 13) or payment of any sums that the engineer had failed to properly include in the Interim Payment Certificate.[425] However, in the context of this section, 'additional payment' will be treated to mean common-law damages.

8.11.1 Categories of Additional Costs (Loss and Expense)

The category of additional costs that the contractor seeks to recover will depend upon what losses or expenditure he has incurred or is likely to incur. For construction and engineering projects, it is common for the contractor to incur costs in the following categories (or heads of claim):

o Prolongation costs – incurred when completion of the works (or a section) is prolonged by a critical delay, for which the contractor is entitled to claim additional costs. Typical costs may include:
  – The contractor's site costs – such as site management (or similar personnel), accommodation, plant and equipment hire, attendant labour.
  – Financial costs – a contribution towards his overheads for the prolonged period, costs of financing retention deductions, additional insurance premiums for the prolonged period.
  – Subcontractor costs – the above categories of costs incurred by subcontractors (but claimed from the contractor).
o Disruption costs – incurred when completion of the work (or a section) may be completed within the programmed or allotted period, but additional resources were necessary to ensure the works' completion. In general, disruption costs will be time or resource related, so could include:
  – Additional labour, subcontractor costs (such as overtime payments).
  – Additional plant or equipment costs for completing an activity.
  – Additional site management (supervision).
  – Other logistical costs – for example, measures for accommodating the disrupted works within the overall programme of works (storage of materials, costs of later deliveries, postponing unaffected activities to accommodate additional labour to complete the disrupted activity, and temporary works).

---

422 JCT DB 2016, clause 4.21.
423 NEC4, clause 60.1.
424 FIDIC Yellow Book 2017, clause 20.2.
425 FIDIC Yellow Book 2017, clause 14.6.1. However, clause 14.8 (Delayed Payment) provides that the contractor does not have to comply with clause 20.1 (Claims) when seeking financing charges for late payment.

8.11.2 JCT DB 2016

This principle of common-law damages is reflected in the wording used in the JCT DB 2016 contract as it refers to 'direct loss and/or expense'.[426] This implies that the contractor is entitled to recover the amount he has lost, or any additional expenditure he has incurred that arises from one of the Relevant Matters in clause 4.21 (refer to Table 8.16). It does not imply that the contractor is to be rewarded or profit from the Relevant Matter. The grounds for the contractor becoming entitled to 'loss and expense' are as follows[427]:

o If the contractor has incurred or is likely to incur loss and expense (which he will not recover under any other clauses[428]); and
o If the progress of his works has been or is likely to be affected by:
  – The employer deferring giving possession of the site (under clause 2.4); or
  – The regular progress of the works (including progress of a section) or part of the same has been or is likely to be affected by one or more of the relevant matters (refer to Section 8.10.3 and Table 8.16).
  (The 'clause 4.19.1 matters')

The procedure for the contractor to claim loss and expense is similar to that which the contractor must follow when seeking an extension of time (refer to Section 8.10.3). Clause 4.20 provides that:

o The contractor notifies the employer as soon as it is reasonably apparent to him that one of the clause 4.19.1 matters may affect progress[429];
o Either with or soon after the contractor's notification, the contractor provides an initial assessment[430] of the following (which should include sufficient information to enable the employer to ascertain the loss and expense):
  – The loss and expense already incurred and
  – Any further amounts that are likely to be incurred.
o At monthly intervals thereafter, the contractor provides updates of the initial and subsequent assessments of the loss and expense.[431]
o The employer then notifies the contractor of the amount of loss and expense that he has ascertained,[432] as follows:
  – Within 28 days after receiving the contractor's initial assessment under clause 4.20.1 (and any other information under clause 4.20.2), and
  – Within 14 days after receiving the contractor's updates under clause 4.20.3.
  The employer's clause 4.20.4 notifications must provide sufficient detail for the contractor to identify any differences between the amounts claimed in the contractor's initial

---

426 JCT DB 2016, clause 4.19.1.
427 JCT DB 2016, clause 4.19.1.
428 JCT DB 2016, clause 4.19.2.
429 JCT DB 2016, clause 4.20.1.
430 JCT DB 2016, clause 4.20.2.
431 JCT DB 2016, clause 4.20.3. The updates should be in a form and manner which is reasonable for the employer to request – and (similar to the contractor's initial assessment under clause 4.20.2) provide sufficient information from which the employer can ascertain the loss and expense.
432 JCT DB 2016, clause 4.20.4.

assessment (under clause 4.20.2) or his updates (under clause 4.20.3) and the employer's ascertainment.

If the contractor's loss and expense consist of 'prolongation costs' (see Section 8.11.1), then prior to or at the same time as submitting his notice and initial assessment under clauses 4.20.1 and 4.20.2, he should submit the required notices and particulars that progress of his works is either being delayed or is likely to be delayed (under clauses 2.24.1 and 2.24.2) and that completion of the works (or a section) is likely to be delayed beyond the relevant completion date. In doing so, the employer cannot simply reject the contractor's claim for loss and expense in principle, on the basis that the contractor has not applied for an extension of time under clause 2.24.

However, if the contractor's loss and expense claim consists of 'disruption costs' (see Section 8.11.1), then there may be no requirement to submit the clauses 2.24.1 and 2.24.2 notice and particulars if the effects of the clause 4.19.1 matters will not delay completion of the works (or a section) beyond the relevant completion date.

In practical terms, the contractor's initial assessment and subsequent updates are likely to be set out within his Interim Payment Applications (under clause 4.7), as the contractor will obviously want to be reimbursed[433] as quickly as possible – as the monthly interval for the clause 4.20.3 updates correlates with the monthly basis for Interim Payments under both Alternative A (stage payments) or Alternative B (periodic payments – refer to Section 8.7.1).

8.11.3   FIDIC Yellow Book 2017

FIDIC Yellow Book 2017 refers to 'additional payment',[434] which this section considers to be additional payment under the 'Claims' procedure in clause 20.1 – that is, sums recoverable under what would otherwise be the principle of damages (described in Sections 8.11 and 8.10.4.2). In this context, the matters for which the contractor is entitled to additional payment are summarised in Table 8.17.

8.11.4   NEC4

Under NEC4, the contractor is only entitled to additional costs (changes in the prices) for a compensation event (refer to Sections 8.8.5 and 8.8.5.1), and he must comply with the compensation event procedure (refer to Sections 8.8.5.2). Assessments of additional costs are made in accordance with the procedure described in Section 8.9.3.1.

8.12   Testing and Defects

The three design and build contracts include differing procedures for rectifying defective (or non-compliant) work before completion. Contracts for construction projects generally

---

433 JCT DB 2016, The Gross Valuation for Interim Payments include loss and expense as amounts not subject to retention under Alternative A (stage payment) clause 4.12.3, or Alternative B (periodic payments) clause 4.13.3, both of which are 'Items included in adjustments' under clause 4.2.4.
434 FIDIC Yellow Book 2017, clause 20.1.(b).

include higher obligations for the contractor prior to completion, as there will be more focus on the aesthetic element of the building works (for example, if the project is a 'Grade A' office building, in the centre of a large commercial district). For construction projects (and those let under JCT DB 2016), the contractor not only has to ensure that the building can fully function as intended, but also that all defects (other than very minor ones) are completed and the building is clean and ready for occupation, letting, or to be used as intended.

Engineering projects usually place more focus on achieving the project's function, rather than on its appearance. The contractor must ensure that the works can function and achieve the required levels of performance, so any 'defects' must not impede the functionality or output generated by the completed works. Therefore, the contractor's obligations in this regard are concentrated on avoiding or rectifying 'functional' or 'performance' defects, rather than aesthetic ones.

This Section 8.12 considers the procedures for dealing with defects *before* completion. Chapter covers the meaning of 'completion' for each design and build contract (in Section 9.3), including the extent that defects must be rectified beforehand. Chapter 9 also covers the procedures for rectifying defects after completion (see Section 9.5).

The three design and build contracts contain varying degrees of rights and obligations for the employer and contractor in relation to matters of quality control. For a contract such as JCT DB 2016, which is predominantly used for construction projects, the relevant contract conditions envisage that a significant proportion of the works will be constructed from materials and goods that have already been manufactured. For example, the employer's requirements may specify materials or goods by giving a product reference (e.g. Ibstock bricks, Metsec structural framing system, or British Gypsum plasterboard products). In such cases, the manufacturers often guarantee that their products have satisfied a particular standard, so ordinarily, the standard of certain materials or goods can readily be accepted without needing to be tested. Of course, most construction projects also require the contractor (either by himself or via his subcontractors) to construct parts of the works, either by carrying out a particular process or from constructing or assembling that part from its constituent products. All of these elements will need to be tested. Some common examples where building elements need to be tested are:

o The project's structural elements. The contractor may have to carry out ground treatment works (such as ground compaction or piling), construct reinforced concrete foundations, and construct an in situ concrete frame or a steel frame.
o Building services. The contractor will have to demonstrate that the electrical, plumbing, and air conditioning systems function correctly. For more complex buildings, a building management system (BMS) may be installed to control the building services, along with its safety functions such as fire alarms, sprinklers, and fire zoning and ventilation. Testing of such systems is often referred to as 'commissioning'.
o The building as a whole upon completion to demonstrate compliance with any required environmental, social, and economic sustainability standards. In the UK, this may be the Building Research Establishment's Environmental Assessment Method (BREEAM).

In many cases, the proportion of manufactured goods and products for construction projects is relatively higher than for an engineering project. Consider a highways project, where the only manufactured (or 'off the shelf') elements may be street lighting, barrier systems, and drainage products, so the contractor may be constructing almost the majority of the project (such as the carriageway or bridges) from its constituent materials.

Clearly, matters of quality control and testing must be carried out progressively and, for projects such as highways, are carried out beyond the completion (or taking over) date. Some of these tests may be at stages of the project proscribed by regulations that are specific to the nature of the project (such as highways, power generation, water treatment, rail, and nuclear). The contractor will need to test the following:

o Civil engineering elements, such as
  – Ground treatments (ground compaction and California Bearing Ratio 'CBR' tests)
  – Piling
  – Concrete batch testing and samples
  – Reinforcement testing
o Foundations
o Superstructure
o Finished road surface (including settlement of the carriageway)
o Operation of plant including:
  – Output levels and efficiency
  – Safety of operation
  – Power or fuel consumption and emissions
o Other tests required by project-specific (often statutory) regulations.

Engineering projects for which the contractor is to design and install a significant amount of plant often require key components to be tested after manufacture (factory acceptance testing) and then tested once installed on site (site acceptance testing).

As FIDIC Yellow Book 2017 and (to a lesser extent) NEC4 focus on the engineering project, it is unsurprising that both contracts contain more extensive testing and quality assurance procedures than are found in contracts for construction projects (such as JCT DB 2016).

Whilst inspections and testing are an inherent part of construction and engineering projects, the time and resources required can often delay the progress of the works, so it is respectively incumbent on both the contractor and the employer (including his representatives) to make work available for inspection and testing, and carry out the same at the appropriate time and with as little disrupting or delaying effect on the works' progress as possible. Most consultants' best practice guides recommend that inspections and testing are carried out proportionately and efficiently:

o The RIBA *Good Practice Guide: Inspecting Works* recommends: '*Special attention will be required to the timing of predictive inspections and visits to witness the carrying-out of tests. It will be necessary to keep in close touch with the contractor to avoid visits being made too late, or having to instruct the postponement of work that would otherwise cover up work to be inspected. It may have been agreed, or the contract may require, that the contractor give notice before particular items of work are ready for inspection (although failure of the contractor to give such notice would not excuse the architect from inspecting work before it is covered up).*'[435]

---

435 Jamieson. (2009). *Good Practice Guide: Inspecting Works*. RIBA Enterprise, p. 75.

o The *Clerk of Works and Site Inspector Handbook* advises: '*The Clerk of Works should maintain systematic preliminary inspections and keep the architect informed of progress to completion and any difficulties likely to arise, and only when notified by the contractor that the works, or part of the works, are complete should the Clerk of Works then undertake a final inspection with the architect....*'[436]

In the event that the employer insists on carrying out inspections and testing either at inappropriate times (e.g. when work has been covered up or assembled) or unnecessarily (e.g. when there are little or no reasonable grounds for doing so), then the contractor is entitled to compensation under the three design and build contracts – particularly if the testing shows that the contractor's work complies with the contract.

8.12.1   JCT DB 2016

JCT DB 2016 contains two procedures for ensuring that the contractor constructs the works in accordance with the contract; one deals with inspection, testing, and rectification before practical completion of the works (or a section), and the second deals with the same after practical completion (i.e. during the Rectification Period). This Section 8.12.1 is only concerned with inspection, testing, and rectification before completion, as the contractor's obligations to rectify defects before 'practical completion', and managing defects during the Rectification Period are covered in Chapter 9 (Concluding the Contract).

Unlike NEC4, JCT DB 2016 contains no obligation for the contractor to search for defects, so the onus is on the employer or the employer's representative to check the quality of the contractor's work at the appropriate times. During the course of the works (i.e. before practical completion of the works or a section), clause 3.12 permits the employer to instruct the contractor to inspect and test the following:

o Any work that is covered up,
o Any goods or materials, or
o Any executed work.

Clause 3.12's inclusion of 'any executed work' seems wide enough to permit the employer to instruct inspections and tests of any work that is off site, as well as unfixed materials or goods that are on site or already incorporated into the works. It may not include workmanship, as this is specifically dealt with in clause 3.14,[437] and poor workmanship is more likely to be self-evident without requiring any testing. The contractor is not entitled to additional payment[438] for carrying out 3.12 inspections and testing in the following circumstances:

o Quite obviously, if the work being inspected or tested does not comply with the contract[439]; or

---

436 *Clerk of Works and Site Inspector Handbook*. (2018). RIBA Publishing, p. 205.
437 JCT DB 2016, clause 3.14. The employer may issue instructions as are 'reasonably necessary', and where this is so, the contractor is not entitled to any additional payment or any loss and expense.
438 JCT DB 2016, clause 3.12 expresses additional payment for inspections and testing being 'added to the Contract Sum'.
439 JCT DB 2016, whilst there is no definition of 'Contract' (as expressed in clause 3.12), it is fairly clear that clause 3.12 means the 'Agreement' (i.e. the Recitals, Articles, Contract Particulars, Conditions, Employer's Requirements, Contractor's Proposals, and Contract Sum Analysis).

o If the cost is already provided for in the Employer's Requirements or in the Contractor's Proposals. Whilst the requirement and responsibility for testing and inspections would naturally be set out in either of these documents, the most suitable document for setting out a specific cost or confirming that such a cost is included in the Contract Sum would be the Contract Sum Analysis.

The contractor will be entitled to additional payment if the inspection or test shows that the relevant work complies with the contract. If the inspection or testing shows that the relevant work is not in accordance with the contract, then the employer may:

o Instruct the contractor to remove the relevant work (including goods or materials) from site.[440]
o Instruct a change[441] (as otherwise instructed under clause 3.5.1) but with the following effects:
    – The instruction must be reasonably necessary (and is subject to the contractor's right to object – under clauses 3.9.1 and 3.9.4); and
    – The contractor is not entitled to additional payment or an extension of time.
o The employer must consider the criteria set out in Schedule 4 (Code of Practice) when issuing instructions to open up the works to carry out further tests or inspections.[442] If the instructions are reasonable[443] (i.e. the test or inspection is justified), then the contractor is not entitled to any additional payment, unless the test or inspection shows that the work complies with the contract; then the contractor is entitled to an extension of time[444] and loss and/or expense.[445]

In the event that the contractor fails to comply with the employer's instructions to remove or rectify the works (under clauses 3.13 and 3.14), the employer has the following remedies:

o Firstly, clause 3.6 (Non-compliance with Instructions), entitles the employer to engage and pay others (and recover any additional costs of doing so from the contractor) to carry out the work that he instructed the contractor to carry out, if the contractor fails comply with that instruction within seven days.
o Clause 2.35 (Defects) provides that the employer may 'otherwise instruct' that in lieu of the contractor rectifying defects, an appropriate deduction from the Contract Sum be made.
o Interim Payment (clause 4) places the financial onus on the contractor to carry out the works in accordance with the contract, as follows:
    – For Alternative A (stage payments), the presence of non-compliant work is likely to render one of the stages incomplete, so the contractor will not be entitled to payment until the relevant stage has been completed[446]; or

---

440 JCT DB 2016, clause 3.13.1.
441 JCT DB 2016, clause 3.13.2.
442 JCT DB 2016, clause 3.13.3.
443 JCT DB 2016 – For example, if the instruction is reasonable because the contractor has not kept relevant records (or ensured that a relevant subcontractor has not kept records) – see paragraph 6 of Schedule 4 (Code of Practice).
444 JCT DB 2016, a Relevant Event under clause 2.26.2.3.
445 JCT DB 2016, a Relevant Matter under clause 4.21.2.2.
446 JCT DB 2016, (Gross Valuation – Alternative A) clause 4.12.1.1.

- For Alternative B (periodic payments), the contractor is not entitled to be paid for work that is not 'properly executed', so the employer will not include a value for defective work within the interim valuation.[447]

o Finally (and the remedy with the most serious consequences for the contractor), the failure by the contractor to comply with a notice or an instruction to remove any 'work, materials or goods' that does not comply with the contract, and provided that such failure results in the works being 'materially affected', is a 'specified default' under clause 8.4.1.3, and it entitles the employer to notify the contractor of the specified default (under clause 8.4.1). If the contractor's failure continues for a period of 14 days from receiving the employer's clause 8.4.1 notice, then by a further notice, the employer may terminate the contractor's employment under the contract (see Section 9.1.1).

8.12.2 FIDIC Yellow Book 2017

The introduction to this Section 8.12 alluded to civil or process engineering projects often requiring far more comprehensive testing procedures than are found in contracts for construction projects. FIDIC Yellow Book 2017 includes separate testing, inspection, and quality control procedures for particular stages of a project as follows:

o Clause 7 (Plant, Materials, and Workmanship) provides the general quality control procedures during the course of the works:
- Samples (clause 7.2)
- Inspection (clause 7.3)
- Testing by the contractor (clause 7.4)
- Defects and rejection (clause 7.5)
- Remedial work (clause 7.6)
o Clause 9 (Tests on Completion[448]) sets out the tests that must be completed in order to permit 'Taking Over the Works and Sections' by the employer (under clause 10.1[449]).
o Clause 11 deals with defects after taking over (completion).
o Clause 12 covers tests after completion.

This Section 8.12.2 only considers the procedures under clauses 7 (Plant, Materials, and Workmanship) as procedures for testing and rectifying defects occurring on or after completion (or taking over) are covered in Chapter 9 (Concluding the Contract).

In relation to testing, clause 7 provides the following two obligations for the contractor:

o Clause 7.1(a) provides that the contractor carry out testing of plant (i.e. that forming part of the Permanent Works[450]) and materials[451] 'in the manner specified in the Contract'; and

---

447 JCT DB 2016 (Gross Valuation – Alternative B) clause 4.13.1.1.
448 FIDIC Yellow Book 2017, 'Tests on Completion' defined in clause 1.1.84. These tests must be set out in the Specification.
449 FIDIC Yellow Book 2017, clause 10.1(a) provides that 'Tests on Completion'.
450 FIDIC Yellow Book 2017, 'Plant' is defined in clause 1.1.66 and 'Permanent Works' is defined in clause 1.1.65.
451 FIDIC Yellow Book 2017, 'Materials' is defined in clause 1.1.53.

○ Clause 7.3 sets out that the contractor is obliged to firstly notify the engineer (the clause 7.3 notice) when any work is ready for inspection,[452] and must provide the following to assist with quality control and demonstrate that the works are in accordance with the contract:

- Samples of Materials,[453] including those provided by the manufacturer, or any additional samples[454] instructed by the engineer (clause 7.2);
- Full access for inspections at any relevant stage[455] (either on or off site), to be carried out by the employer's personnel[456]; and
- Any other inspections or duties that are specified in the Employer's Requirements or the contract conditions.[457]

The employer's personnel then either (i) promptly notify the contractor that there is no need to carry out an inspection or test, or (ii) without unreasonable delay, they carry out the required inspection or testing.[458] If there is any failure by the employer's personnel to either notify or inspect within the time set out in the contractor's clause 7.3 notice (or as otherwise agreed), then the contractor may proceed to package, transport, or cover up the relevant work. If the contractor fails to issue a clause 7.3 notice, then he must uncover and subsequently reinstate any work (at his own cost and risk) as required by the engineer.

Clause 7.4 obliges the contractor to provide all necessary equipment, apparatus, and personnel to carry out the testing, and it appears that unless the testing location is varied by the engineer (a variation under clause 13[459]), tests are carried out at a location notified by the contractor. If the tests are delayed by the contractor or by the employer, then each party is liable to the other[460] under clause 20.2 (Claims for Payment and/or Extension of Time).

Once the testing is complete, the contractor must promptly provide the results to the engineer, who 'endorses'[461] the test if it has been passed. If the test shows the work is non-compliant, then under clause 7.5, the engineer notifies that the relevant work is

---

452 FIDIC Yellow Book 2017, clause 7.3, which must be before work is packaged, stored, transported, or covered up. The contractor should also specify the time for the Employer's Personnel to carry out their inspection.

453 FIDIC Yellow Book 2017, 'Materials' are defined in clause 1.1.53, which fall within the definition of 'Goods' (clause 1.1.44) but excludes 'Plant' (defined in clause 1.1.58).

454 FIDIC Yellow Book 2017, additional samples are a Variation (see clause 13).

455 FIDIC Yellow Book 2017, clause 7.3(b).

456 FIDIC Yellow Book 2017, 'Employer's Personnel' is defined in clause 1.1.32, which includes the engineer, any appointed Engineer's Representative, and any other employees of the employer or the engineer.

457 FIDIC Yellow Book 2017, clause 7.3(c).

458 FIDIC Yellow Book 2017, clause 7.3. Inspection and testing also includes examining or measuring the relevant work.

459 FIDIC Yellow Book 2017, clause 7.4, if the testing shows that the relevant work is not in accordance with the contract, then the contractor is responsible for the costs and any delays arising from the clause 13 Variation (of the testing location).

460 FIDIC Yellow Book 2017, clause 7.4. The contractor will be entitled to 'Claims' (under clause 20.2) if there is a delay because the engineer instructed that the contractor should not proceed with testing (seemingly, in the engineer's absence). Similarly, the contractor will be liable for the Employer's 'Claims' (again, under clause 20.2) if he delays testing.

461 FIDIC Yellow Book 2017, clause 7.4 does not specify the form of either the engineer's endorsement, or 'test certificate' (provided instead of the endorsement), so it does not appear as if either constitutes a 'Notice' (as defined in clause 1.1.56).

defective,[462] and the contractor must promptly (although no period of time is specified) propose what remedial work is necessary for the relevant work to pass the testing process ('the clause 7.5 proposal'). The engineer must notify the contractor within 14 days of receiving the contractor's clause 7.5 proposal, identifying the extent to which the proposed work would not result in compliance with the contract so the contractor can submit a revised proposal. However, if the engineer does not do so, he is deemed to have issued a Notice of No-Objection.[463] Alternatively, and in the event that the engineer is not satisfied that the contractor's clause 7.5 proposal will cause the works to become compliant, he may, by issuing a further notice under clause 7.5(b), reject the proposed work,[464] and clause 11.4(a) shall apply.[465] This permits the employer to carry out the work himself, and recover the cost from the contractor as a claim under clause 20.2.[466]

Clause 7.6 (Remedial Work) provides that the engineer may instruct the contractor to carry out remedial work[467] in the following circumstances:

o To ensure that the relevant work complies with the contract after failing an inspection or test,[468] either:
  - Following the engineer's instruction to that effect under clause 7.5(a), or
  - At any time.[469]
o To ensure the 'safety of the Works' in the following circumstances:
  - If the contractor suffers a delay or incurs additional cost arising from an act by the employer (or the employer's personnel),[470] or from an Exceptional Event,[471] then the contractor may make a claim under clause 20.2; or
  - In circumstances where the contractor is liable for the cost, such as a contract default, an unforeseeable event, or an accident.

The engineer may issue such clause 7.6 (Remedial Work) instructions at any time prior to issuing the Taking Over Certificate.

If the contractor fails to comply with either (i) the engineer's clause 7.5 notice, or (ii) the engineer's clause 7.6 instruction (either as soon as is practicable or within either

---

462 FIDIC Yellow Book 2017, clause 7.5 (Defects and Rejection). The relevant work described in the engineer's notice may include design, workmanship, plant, or materials.

463 FIDIC Yellow Book 2017, 'No-objection' is defined in clause 1.1.55.

464 FIDIC Yellow Book 2017, clause 7.5(b) provides that the engineer may reject plant, materials, workmanship, or design.

465 FIDIC Yellow Book 2017, clause 11.4 (Failure to Remedy Defects).

466 FIDIC Yellow Book 2017, clause 11.4(a). The employer may carry out the work himself or engage others to do it on his behalf.

467 FIDIC Yellow Book 2017, clause 7.6(a) and clause 7.6(b) provide that remedial work also includes repair work, and replacement of plant and materials, or re-execution of any work (including plant and materials) that is not in accordance with the contract.

468 FIDIC Yellow Book 2017, clause 7.3 (Inspection) and/or clause 7.4 (Testing by the Contractor).

469 FIDIC Yellow Book 2017, clause 7.6.

470 FIDIC Yellow Book 2017, clause 7.6 (i).

471 FIDIC Yellow Book 2017, clause 7.6(ii). Clause 1.1.37 defines 'Exceptional Event' and clause 18.4 (Consequences of an Exceptional Event) will apply.

28 days[472] or the period set out in the instruction), then the employer has the following remedies:

○ Engage others (i.e. employ and pay them) to carry out the remedial work set out in the engineer's clause 7.6 instruction[473]; or

○ Unless the contractor provides a reasonable excuse as to why he has not complied with either the clause 7.5 notice or clause 7.6 instruction (and in the latter case, he must comply within 28 days[474]), or subsequently complies within 14 days of receiving the Engineer's Notice to Correct under clause 15.1, the employer may (by issuing a second notice under clause 15.2.2) terminate the contract (see Section 9.1.2).

8.12.3   NEC4

The quality control provisions in NEC4 remain largely the same as under NEC3, albeit with few minor changes. This Section 8.12.3 only covers quality control matters during the course of the works (i.e. prior to the completion[475]), so the correction of a defect after completion is covered in Chapter 9 (refer to Sections 9.2.3 and 9.5.3).

As mentioned earlier in the book, NEC4 refers to the client and the Scope instead of NEC3's references to the employer and the Works information, and amends the title of Section 4 from 'Testing and Defects' (NEC3) to 'Quality Management'. This change is to accommodate the new clause 40 entitled 'Quality Management System'.

The importance of the Scope cannot be understated in relation to testing and defects under NEC4, as it sets the contractor's obligations in terms of the quality management system (clause 40), along with governing the tests, inspections, and procedure for rectifying defects (clauses 41–46). NEC4's *Managing Reality*[476] emphasises this importance by describing how the Scope should provide for testing and inspections in terms of:

○ The detail of tests and inspections to be carried out:
  – The test's or inspection's objective (including any standards to achieve),
  – Any specific methodology or guidelines to apply to the test or inspection,
  – Its location (e.g. whether it is within or outside of the working area),
  – When the test or inspection is due to be carried out.[477]
○ Who is responsible for the test or inspection:
  – The client (the supervisor or others engaged to inspect or test) or the contractor?
  – Who is obliged to provide testing and inspection facilities, along with any materials or samples?

---

472 FIDIC Yellow Book 2017, see clause 15.2.1(d).
473 FIDIC Yellow Book 2017, clause 7.6.
474 FIDIC Yellow Book 2017, clause 15.2.1(d).
475 NEC4 clauses 44.2, 44.3, and 44.4 relate to the correction of a Defect after Completion.
476 NEC4, *Managing Reality*, Book Three – *Managing the Contract*, 3rd ed., (2017), p. 49.
477 Ideally, the timing of tests and inspections should be shown on the contractor's 'programme' (see NEC4 clause 31.2).

Clause 40.2 obliges the contractor to provide a 'quality policy statement' and a 'quality plan' within the period set out in Contract Data Part One[478] and that meets the requirements set down in the Scope. For obvious reasons, it is essential that the Scope provides these requirements, as there is no fallback position for the contractor to provide a default policy statement or plan. The project manager accepts the quality policy statement and/or plan unless, somehow, either will prevent the contractor from fulfilling his obligation to provide the works.[479] In the latter case (and only simply described as 'If any changes are to be made…'), the contractor must change the quality plan and resubmit this to the project manager for acceptance. Clause 40.2 is silent regarding changes to the quality policy statement if the project manager does not accept it.[480] Clause 40.3 allows the project manager to instruct the contractor to comply with the quality plan if he fails to comply with it.[481]

Presumably, the quality policy statement and quality plan required by clause 40.2 comprise the 'quality management system' that the contactor is obliged to operate under clause 40.1. The NEC4 *User's Guide*[482] refers to an example of the quality management system being a form of 'corporate accreditation'. Accreditation comes in many forms, from BSI's ISO 9001 to covering various aspects of construction; from the more comprehensive (covering a company's profile, workload, finances, staff qualifications, and health and safety record) to those dedicated to a single aspect, such as health and safety. NEC4's *Managing Reality*[483] confirms the usual practice: that contractors provide details of their quality management systems with their tender documentation, and focus on the likely activities to be undertaken during the first six weeks. It also suggests that the Scope should set out if the project manager is to have any supervisory involvement of the contractor's quality management system and, if so, to what extent.

Clauses 41.1 and 42.1 deal with tests and inspections – which are intended to show the presence of any defect[484] in the work – and should one be present – how it will be corrected. A defect not only relates to the physical work not being in accordance with the Scope, but also to the contractor's design if it fails to conform with either the design that the project manager has accepted or any applicable law. NEC4 envisages a collaborative process, where both the contractor and those acting on the client's behalf (the supervisor – see below) notify each other[485] once they are aware of a defect.

Clause 42.1 concerns the testing or inspection of plant and materials[486] (identified in the Scope) before they can be brought to the working areas[487] (site), whereas clause 41 is more general and covers the tests and inspections required by either the Scope or the applicable law. The addition of 'applicable law' is important, as NEC4 is often used for projects for

---

478 NEC4, Contract Data Part One, Section 4 Quality Management. The period runs from the 'Contract Date' – see clause 11.2(4).

479 NEC4, 'Provide the Works' is defined in clause 11.2(15).

480 However, NEC4's *User's Guide: Managing an Engineering and Construction Contract*, vol. 4, p. 36, envisages that the contractor may change both the policy statement and the plan in the event of achieving new levels of corporate accreditation.

481 NEC4, clause 40.3, instructions are not compensation events.

482 NEC4, *Managing an Engineering and Construction Contract*, vol. 4, (June 2017), at p. 36.

483 NEC4, *Managing Reality*, Book Three – *Managing the Contract*, 3rd ed., (2017), p. 45.

484 NEC4, clause 11.2(6) defines a 'Defect'.

485 NEC4, clause 43.2.

486 NEC4, 'Plant and Materials' are defined in clause 11.2(14).

487 NEC4, 'Working Areas' is defined in clause 11.2(20).

industries that have their own regulations, such as highways, rail, water processing, power generation (including nuclear power), and aviation.

Instead of the project manager, it is the supervisor[488] who oversees the inspection and testing procedures under NEC4's clauses 41, 42, 43, and 44. The Scope provides that either the contractor or the supervisor may carry out inspections and testing, provided that each notifies the other beforehand.[489] It also sets out the facilities (including materials and samples)[490] that should be provided. In addition to specifying the requirements for each inspection or test in the Scope, it would be advisable for the programme[491] to identify the time or period for each test or inspection, and whether it will be undertaken by the contractor or the supervisor. Both the contractor and the supervisor must inspect and test at the appropriate time as follows:

o The contractor must notify the supervisor of the test or inspection and carry it out before any further work would obstruct the test or inspection.

o The supervisor may carry out an inspection as long as it does not cause 'unnecessary delay',[492] as otherwise, this entitles the contractor to a compensation event under clause 60.1(11). It appears that the supervisor may carry out any test (including those earmarked for the contractor) as long as he firstly notifies the contractor; and clause 43.2 provides that as soon as either the contractor or the supervisor becomes aware of a defect, they notify the other, and must do this until the 'defects date' (which is set out in Contract Data Part One, and expressed as a period after completion of all the works[493]).

Under clause 44.2, the contractor is obliged to correct all defects before the end of the 'defect correction period'. NEC4's use of the singular 'period' is a little unhelpful as initially, it may appear that a single 'defect correction period' will apply, after which, the contractor's obligation to rectify defects ceases, but this is not so. In practice, there may be many defect correction periods as described in the following scenarios:

o There may be more than one defect correction period set out in Contract Data Part One[494]:

  – Separate defects correction periods may apply for particular elements of work. Taking a construction project example, the defect correction periods for building services may be 12 months, but 6 months for all other elements of the work[495]; and

---

488 NEC4 – Other than for testing and defects under clauses 41–44, references to the supervisor and his duties are in clauses 11.2(7) – the supervisor issues the Defects Certificate (also see clauses 50.1 and 53.1), clause 14 (The Project Manager and the Supervisor), clause 27 (Other Responsibilities), 70 (The Client's Title to Plant and Materials) and 71 (Marking Equipment, Plant and Materials outside the Working Areas).
489 NEC4, clause 41.3.
490 NEC4, clause 41.2.
491 NEC4, refer to the last bullet point under clause 31.2 – 'other information which the Scope requires the Contractor to show…'.
492 NEC4, clause 41.5. Additionally, if a payment is dependent upon the supervisor's test, and the supervisor either fails to carry out the test or the relevant test is delayed through no fault of the contractor, then that payment becomes due after the later of the last *defect correction period* or the *defects date*.
493 NEC4, Contract Data Part One, Section '4 Quality Management' – the client inserts a period (e.g. 12 months) that applies between 'Completion' and the 'defects date'.
494 NEC4, clause 44.2 and the entries in Contract Data Part One – under '4 Quality Management'.
495 NEC4, *User's Guide* (2017 and 2020), p. 38.

- There may be separate defects correction periods for sections if section completion under secondary option X5 applies (see Section 8.3.3).
o Further (or individual) defect correction periods may arise in the following circumstances:
  - For defects notified before completion, the contractor must rectify them before the end of the defect correction period that commences at completion (i.e. as expressly stated in Contract Data Part One)[496]; and
  - For each defect notified after completion (but before the defects date), the contractor must rectify defects within the applicable defect correction period that commences when the defect is notified. Therefore, if a significant number of defects are notified after completion, the contractor will have to keep track of the defect correction period for each. If a defect is notified just before the defects date, the contractor's liability to correct defects may run for a significantly longer period than the defects correction period.

Unlike JCT DB 2016 or FIDIC Yellow Book 2017, NEC4 requires the contractor to search for defects if instructed to do so by the supervisor.[497] The supervisor's powers in this regard are relatively wide,[498] but he should be pragmatic when exercising them, as the contractor will be entitled to a compensation event under clause 60.1(10) if he is instructed to search for a defect but none is found.

If any notified defect[499] is found, then the following occurs:

o Under clause 44.1, the contractor must correct it in any event,[500] and the work is re-inspected and re-tested[501]; or
o The contractor and the project manager may agree that it may be more practicable to accept the defect (i.e. leave it uncorrected) in return for a reduction in the prices.[502] In the event that they do, the contractor issues a quotation, proposing the reduction in the prices and/or any earlier completion date, which may be accepted by the project manager, and importantly, the Scope is adjusted (commuted) accordingly, so the previously non-compliant work now does not contain a defect.

8.12.4 Summary of Testing and Defects Procedures

Table 8.18 summarises the testing and defects procedures in JCT DB 2016, FIDIC Yellow Book 2017, and NEC4 during RIBA Stage 5 (Manufacturing and Construction) – that is, before completion.

---

496 NEC4, Contract Data Part One, entry under '4 Quality Management' – setting out different 'defects correction periods' for different classes or types of defects.
497 NEC4, clause 43.1 provides that the supervisor may instruct the contractor to search for a defect until the *defects date.*
498 NEC4, clause 43.1 permits the supervisor to instruct tests and inspections that are *not* in the Scope, including opening up, taking down, and reinstating or re-assembling work, and instructing the contractor to provide for the supervisor to carry out inspections and tests (as per clause 41.2).
499 NEC4, i.e. 'notified' by either the contractor or supervisor under clause 43.2.
500 NEC4, see clauses 41.4 and 44.1. The contractor is obliged to rectify a defect, irrespective of whether it was notified by the supervisor (under clause 43.2), unless the project manager and the contractor agree otherwise under clause 45.2.
501 NEC4, clause 41.4.
502 NEC4, clause 45. Either the contractor or the project manager can propose to the other to reduce the prices in lieu of correcting a defect.

Table 8.18 Summary of testing and defects procedures.

Obligation or procedure	JCT DB 2016	FIDIC Yellow Book 2017	NEC4
Contractor to search for defects	No	No	Yes – clause 43.1
Contractor to notify when ready for inspection	No	Yes, clause 7.3	Yes, clause 41.3
Employer/Client instruction to test	Yes, clause 3.12	Yes, clause 7.4[a)]	Yes, clause 41.3, 43.1
Contractor entitled to be paid for testing if work complies with the contract	Yes, clause 3.12	No	No
Instructions to rectify defects	Clause 3.13	Yes, clause 7.6	No
Contractor obliged to rectify defects in any event	No	Yes, clause 7.5	Yes – clauses 44.1, 44.2
Employer may accept a defect and abate the contract price	Clause 2.35	No[b)]	Yes, clause 45
Employer/Client right to employ others if the contractor fails to rectify	Clause 3.11	Yes, clause 7.6	Yes, clause 46.1
Termination event	Clause 8.4.1.3	Yes, clause 15.2.1(d)	Yes, clause 91.2 (R11)

a) FIDIC Yellow Book 2017, clause 7.4 only enables the engineer to instruct 'additional testing'.

b) FIDIC Yellow Book 2017, the engineer may withhold amounts from Interim Payment Certificates (IPC's) if the contractor fails to construct the works in accordance with the contract under clause 14.6.2(a), but this only applies before the Taking Over Certificate is issued. However, the engineer's deductions under clause 14.6.2(a) are not thought to be an alternative to the engineer rejecting the contractor's work under clause 7.5 (Defects and Rejection).

9

Concluding the Contract

This chapter explores the ways in which the contract is concluded, and the events that lead up to its conclusion. Under normal circumstances, the contract is concluded after the design and physical works are completed, the contractor has rectified any defects present in the works, and the contractor's account is agreed upon with the employer. Concluding the contract under these normal circumstances will occur either during RIBA Stage 6 (Handover) or during the early part of Stage 7 (Use). There may also be a number of obligations for the contractor to perform before he completes the physical works such as testing and commissioning.

It is also important to note that the contract may be brought to an end prior to the works' completion (referred to as 'termination' or 'determination') – if either the employer or the contractor commits a breach of the contract that is sufficiently serious. Separate procedures for terminating or determining the contract will then apply. This chapter firstly looks at termination and then the procedures for concluding design and build contracts.

9.1 Termination

Termination (also referred to as 'determination' but not to be confused with the Engineer's 'Determination' under clause 3.7 of FIDIC Yellow Book 2017 – see Section 9.7.2.1) is the means for one party to bring the contract to a premature end. He can do this if the other party has become insolvent or commits a serious breach of that contract, or either party may terminate following some 'neutral' event resulting in it being impossible or impracticable to continue with the project. A party's right to terminate will normally arise after it has notified the other party, allowing both parties to begin to get their affairs in order before the termination takes effect, and then providing a procedure for concluding other aspects of the project (such as finishing any remaining work) once termination has occurred. Termination is usually a two-stage process; the first stage is a 'warning' to the other party, and the second stage usually concludes or confirms that termination has taken place. In the case of one party committing a serious breach, the 'first stage' notification period provides the innocent party with a 'self-help' remedy. It informs the other party to stop committing the breach (or warns them not to repeat it) and lets them know that if they fail to do so, the breach is sufficiently serious to allow the innocent party to terminate, usually by issuing a further, 'second stage' notice. Whilst the circumstances entitling a party to terminate can arise

Design and Build Contracts, First Edition. Guy Higginbottom.
© 2024 John Wiley & Sons Ltd. Published 2024 by John Wiley & Sons Ltd.

at any time after entering into the building contract, they usually arise when the parties have the most responsibilities to each other – during RIBA Stage 5 (Manufacturing and Construction).

As well as the 'contractual' termination provisions (summarised above, and dealt with in more detail below), a party may have a common-law right to terminate the contract. This is often referred to as repudiation or recission. If the other party commits a repudiatory breach (i.e. one that is sufficiently serious, often going right to the heart of his obligations under the contract – such as the contractor abandoning the works and leaving site without justification), then the other party can accept that repudiatory breach and bring the contract to an end.

Any party either considering or facing termination should promptly seek legal advice, as the consequences of failing to properly terminate a contract (under either the 'contractual' or 'common-law') routes are fraught with risk. If an (initially) innocent party fails to comply with the contractual termination procedure, or fails to act promptly or use clear wording, then that party may be at risk of committing a repudiatory breach himself. The following difficulties can arise:

o When attempting 'common law' termination:
  – The party intending to terminate may have failed to act with sufficient promptness after the other party committed the 'repudiatory' breach – effectively continuing with the contract (i.e. by continuing with the contract, his conduct 'affirms' that he intends to remain bound by the contract's terms).
  – If the contract between the parties contains a termination procedure, then the 'innocent' party's departure from that procedure may itself be a repudiatory breach in the circumstances, unless it clearly reserves its position in respect of the common-law termination.
  – The terminating party's actions may be unclear, either in terms of its conduct (e.g. if it effectively 'affirms' the contract – see above), or if it fails to use sufficiently clear words when accepting the 'repudiatory' breach.
o When attempting 'contractual' termination:
  – A party may not have committed a breach which entitles the 'innocent' party to terminate.
  – The 'innocent' party may have failed to issue the relevant notices (required by the contract) within the contractual timescales.
  – The purported notices of termination may be ineffective; they may not have been served correctly, or they may fail to contain wording that is either sufficiently clear or required by the contract.

Termination (under the terms of the contract) is often a procedure of last resort, and it is most commonly used when the other party is in default and clearly unable (or unwilling) to perform their obligations. It is rare to terminate for one of the 'neutral' reasons. Some common examples of a party terminating for a breach of the contract are:

o If the contractor fails to comply with an instruction from the employer within a particular period of time (and that instruction is usually to carry out a Variation or to rectify defective work), then the employer may engage and pay others to carry out the work that the contractor was instructed to carry out.

○ A simpler example is the contractor suspending performance of his obligations (e.g. in carrying out the work and its design) if the employer fails to pay a sum due under the contract.

The grounds for termination under JCT DB 2016, FIDIC Yellow Book 2017, and NEC4 are often similar, although there are specific grounds that apply to individual contracts, as summarised in Table 9.1.[1]

Once termination has taken place, the three design and build contracts contain procedures for concluding the contract, such as resolving the contractor's final account and dealing with any further payments from the employer to the contractor or from the contractor to the employer, as the case may be.

9.1.1 JCT DB 2016

The parties' rights for termination and the corresponding procedures in JCT DB 2016 are set out in Section 8 (Termination), and the parties may terminate the contractor's employment under the contract as follows:

○ The employer (following a contractor default) or either party (following a neutral event) may determine the contractor's employment before practical completion of the works[2]; or

○ The contractor (following the employer's default) or the employer (following the contractor's insolvency) may terminate the contractor's employment under the contract at any time.[3]

It appears that the principal reason for restricting the employer's termination rights before 'practical completion' is that once practical completion has been achieved, the contractor will have completed the majority of his obligations under the contract, save for making good any defects during the Rectification Period.[4] Even then, if the contractor fails to respond to an instruction to rectify defects[3] within seven days, the employer may engage others to carry out that instruction and deduct the costs from sums due to the contractor.[5] Conversely, the employer's fundamental obligation to pay the contractor continues after practical completion, so the contractor's right to terminate is preserved.

Under JCT DB 2016, the grounds for termination are split into the following three categories:

○ A party becoming 'insolvent'.

○ A party committing a particular breach or breaches of the contract (referred to as a 'specified default' or 'specified defaults'[6]- see Table 9.1).

○ 'Neutral' events permitting either party to terminate the contractor's employment.

---

1 Table 9.1 uses '(E)' for the Employer (having the same meaning as 'Client' for NEC4) and '(C)' for the Contractor where the respective party has termination rights.
2 JCT DB 2016, clause 8.4.1 (employer terminating following contractor's default), and clause 8.11.1 (either party terminating following a neutral event).
3 JCT DB 2016, contractor terminating for employer's default (clause 8.9.1) or insolvency of employer (clause 8.10.1), or employer terminating because the contractor is insolvent (clause 8.5.1).
4 JCT DB 2016, clause 2.35.2 (Defects – Schedules of Defects and Instructions).
5 JCT DB 2016, clause 3.6 (Non-compliance with Instructions).
6 JCT DB 2016, 'specified defaults' committed by the contractor (entitling the employer to begin the termination procedure) are set out in clause 8.4, and those committed by the employer (entitling the contractor to begin the termination procedure) are set out in clause 8.9.

Table 9.1 Grounds for termination.

Grounds for Termination[a]	JCT DB 2016 clauses	FIDIC Yellow Book 2017 clauses	NEC4 clauses[b]
Insolvency[c]:			
Of the Contractor	(E) 8.5.1	(E) 15.2.1(g)	(E&C)
Of the Employer	(C) 8.10.1	(C) 16.2.1(h)(i)/(ii)	91.1 (R1) to (R10)
Breach/Default by the Contractor:			
Fails to provide performance security	—	(E) 15.2.1(e)	(E) 91.2 (R12)
Wrongfully ceases or suspends work	(E) 8.4.1.1	(E) 15.2.1(b)	(E) 91.2 (R11)
Fails to comply with agreement or binding determination (inc by DAAB)	—	(E) 15.2.1(a)(ii) (E) 15.2.1(a)(iii)	—
Fails to proceed regularly and diligently	(E) 8.4.1.2	(E) 15.2.1(c)	(E) 91.3 (R14)
Fails to comply with a rectification instruction / Notice to Correct (FIDIC)	(E) 8.4.1.3	(E) 11.4(d)[d] (E) 15.2.1(a)(i) (E) 15.2.1(d)	(E) 91.6 (R18)
Fails to comply with clause 3.3 (subcontracting)	(E) 8.4.1.4	(E) 15.2.1(f)	(E) 91.2 (R13)
Fails to comply with clause 7.1 (assignment)	(E) 8.4.1.4	(E) 15.2.1(f)	—
Fails to comply with clause 3.16 (CDM)	(E) 8.4.1.5	—	(E) 91.3 (R15)
Engages in corrupt/fraudulent activity	(E) 8.6[e]	(E) 15.2.1(h)	(E) 91.8 (R22)
Termination for Employer's convenience[f]	—	(E) 15.5	—
Breach/Default by the Employer:			
Fails to sign the Contract Agreement	—	(C) 16.2.1(g)(i)	—
Fails to evidence financial arrangements	—	(C) 16.2.1(a)	—
Fails to provide Notice of Commencement Date	—	(C) 16.2.1(f)	—
Fails to provide a Payment Certificate	—	(C) 16.2.1 (b)	—
Fails to pay the amount due by the final date for payment	(C) 8.9.1.1	(C) 16.2.1(c)	(C) 91.4 (R16)
Fails to comply with an agreement or binding determination (inc by DAAB)	—	(C) 16.2.1(d)(i)/(ii)[g]	—
Fails to comply with clause 7.1 (assignment)	(C) 8.9.1.2	(C) 16.2.1(g)(ii)	—

Table 9.1 (Continued)

Grounds for Termination[a]	JCT DB 2016 clauses	FIDIC Yellow Book 2017 clauses	NEC4 clauses[b]
Fails to comply with clause 3.16 (CDM)	(C) 8.9.1.3	(C) 16.2.1(e)	—
Fails to substantially perform	—	(C) 16.2.1(h)[j]	—
Prolonged suspension by Employer	(C) 8.9.2[h]	(C) 16.2.1(h)(iii)	(C) 91.6 (R19)
Engaged in corrupt, fraudulent acts etc	—		—
Event not the fault of either party:			
If released from the contract by law	(E&C) 8.11.1.5	(E&C) 18.6	(E&C) 91.5 (R17)
Employer's representative does not instruct a re-start	(E&C) 8.8.1	—	(E&C) 91.6 (R20)
Prevents the contractor from completing the works[j]	(E&C) 6.14[k]/8.11.1[l]	(E&C) 18.5	(E) 91.7 (R21)
Grounds under s73 of the PC Regulations	(E&C) 8.11.3	—	—

a) Table 9.1 uses 'E' for the Employer (having the same meaning as 'Client' for NEC4) and 'C' for the Contractor where the respective party has termination rights.

b) NEC4 – the 'Reasons' for termination (R1) to (R22) are set out in clauses 91.1 to 91.8

c) JCT DB 2016 – clause 8.1, FIDIC Yellow Book 2017 – clause 16.2.1(h)(i), NEC4 – clause 91.1 reasons R1 to R10 define 'Insolvency'

d) FIDIC Yellow Book 2017 – clause 11.4(d) provides that clause 15.2 (Termination for Contractor's Default) does not apply if the contractor's failure substantially prevents the Employer from taking the benefit of the whole of the Works.

e) JCT DB 2016 – clause 8.6 (Corruption and regulation 83(1)(b) of the PC Regulations. Clause 1.1 – 'PC Regulations' means the Public Contracts Regulations 2015)

f) FIDIC Yellow Book 2017 – clause 15.5 (Termination for Employer's Convenience) – there is no breach of contract by the contractor.

g) FIDIC Yellow Book 2017, clause 16.2.1(d) provides that the Employer's failure under either sub-clauses (i) and / or (ii) must be a "material breach" to entitle the contractor to terminate.

h) JCT DB 2016 – the period of suspension under clause 8.9.2 is a 'specified suspension event', and must exceed the period of suspension set out in the Contract Particulars against the entry for clause 8.9.2 (the default period is 2 months)

i) FIDIC Yellow Book 2017, see clause 8.12 (Prolonged Suspension) sub-clauses (a) and (b)

j) NEC4, clause 91.7, the event also prevents the Contractor from completing the whole of the works within 13 weeks after planned Completion (as shown on the Accepted Programme),

k) JCT DB 2016, clause 6.14 (Loss or damage to Existing Structures – right of termination)

l) JCT DB 2016, clauses 8.11.1.1 (force majeure), 8.11.1.2 (Employer's instructions under clauses 2.13, 3.9 or 3.10 arising from Statutory Undertaker's default), 8.11.1.3 (loss or damages caused by an insurable risk, or an Excepted Risk – see clause 6.8), 8.11.1.4 (civil commotion, terrorism), 8.11.1.6 (delay in permission/approval of Development Control Requirements).

Insolvency generally has the effect that, from a financial perspective, a party cannot meet its obligations to the other, but its precise definition is set out in clause 8.1. This includes entering administration,[7] having orders for winding up[8] or bankruptcy[9] made, or a party entering into an arrangement with its creditors.[10] As it is unlikely that a party will escape from insolvency within a relatively short period of time[11] or at all, there seems little point in adopting a two-stage notification procedure for termination arising from insolvency. Therefore, termination for insolvency only requires a single-stage notification, and takes place upon the insolvent party's receipt[12] of a written notice (complying with clause 8.2) from the other party as follows:

o If the contractor is insolvent, the employer notifies the contractor under clause 8.5.1; and
o If the employer is insolvent, the contractor notifies the employer under clause 8.10.1.

Clause 8.2[13] of JCT DB 2016 provides that all notices under Section 8 (Termination) must comply with clause 1.7.4. Notices (and other communication) must be sent as follows:

o Addressed to either:
   – The recipient's address as set out against clause 1.7 in the Contract Particulars[14]; or
   – Such other address as has been notified by the recipient[2]; or failing that,
   – To the recipient's registered or principal office.[15]
o Delivered by:
   – Hand or
   – Recorded Signed For or Special Delivery Post.

Unless proof to the contrary can be provided (e.g. in the form of the Post Office's confirmation that the notice was received or proof that the notice was hand delivered), notices under clause 1.7.4 are deemed to be received on the second Business Day[16] after posting. It is essential to the operation of the termination procedures to establish when notices under clause 8 are received (or when deemed to have been received); in particular, to establish the date upon which termination takes place (i.e. the date that the relevant notice is received – see clause 8.2.2). It is also important that the sender obtains evidence of when the relevant notice was sent.

Where one party has committed a breach of contract that qualifies as a 'specified default',[1] JCT DB 2016 operates a two-stage termination process. The two stages of the termination process apply equally to the employer and the contractor.

---

7 JCT DB 2016, clauses 8.1.1.1 and 8.1.1.2.
8 JCT DB 2016, clauses 8.1.1.3, 8.1.1.4, and 8.1.2.1.
9 JCT DB 2016, clause 8.1.3.1.
10 JCT DB 2016, 8.1.4.1 except where the company is solvent, and the arrangement is for the purposes of reconstruction or amalgamation.
11 Such as the 14-day period specified in clauses 8.4.2 (for the employer) or 8.9.3 (for the contractor).
12 JCT DB 2016, clause 8.2.2.
13 JCT DB 2016, clause 8.2 (Notices under Section 8).
14 JCT DB 2016, clause 1.7.3.1.
15 JCT DB 2016, clause 1.7.3.2.
16 JCT DB 2016, clause 1.1 defines 'Business Day' as any day except Saturday, Sunday, or a Public Holiday.

The first stage is for the innocent party to send to the other a written notice[17] (complying with clauses 1.7.4 and 8.2) that identifies the relevant specified default or defaults[18] (the 'first-stage notice'). The first stage is intended to act as a warning to offending party, not to continue or repeat the 'specified default'. Whilst unlikely, it is possible that the offending party may not be aware that he has committed a 'specified default', so this 'first-stage notice' must clearly set out the 'specified default' that the innocent party relies upon. The second stage of the termination does not necessarily need to follow the first stage. The offending party may simply cease (or not repeat) the specified default so the first-stage notice has had its desired effect, so there is no need to terminate. Alternatively, the innocent party may simply choose not to pursue termination as a remedy.

If the recipient continues or repeats the specified default for a period of 14 days after receipt of the first-stage notice (hence the importance of establishing when the first-stage notice was received), then within a further period of 21 days, the innocent party may issue a second written notice (again, complying with clauses 1.7.4 and 8.2) ('the second-stage notice')[19] – hence the importance of the sender obtaining proof of when the second-stage notice was sent. It is thought that the recipient does not have to receive the second-stage clause 8 notice within the further period of 21 days, as this only limits the time in which the innocent party has to send the second-stage notice.

The importance of establishing when the recipient receives the second-stage clause 8 notice is paramount, as clause 8.2.2 provides that termination takes effect upon its receipt. If the innocent party does not issue the second-stage clause 8 notice within the 21-day period, then provided that the other party 'repeats'[20] the specified default, the innocent party may issue a notice[21] (complying with clauses 1.7.4 and 8.2) and terminate the contractor's employment, provided that notice is issued within a reasonable time after the other party has repeated the specified default.

Similar to the procedures for termination, JCT DB 2016 deals with the consequences of termination separately, depending upon whether termination was by the employer or by the contractor (albeit the procedures are the same where termination is by the contractor or by either party under clause 8.11). The consequences of termination allow the parties to conclude matters under the contract, including:

o Completion of the works (but see below), including the contractor leaving the site and/or providing any necessary documents to the employer; and
o Determining if any further payment is due to the contractor, and if so, how such a sum is calculated.

---

17 JCT DB 2016, the employer issues the 'first-stage clause 8 notice' under clause 8.4.1, and the contractor issues his equivalent under clause 8.9.1.
18 JCT DB 2016, for the contractor's specified defaults, refer to clause 8.4.1; and for the employer's specified defaults, refer to clause 8.9.1 as summarised in Table 9.1.
19 JCT DB 2016, the employer's second-stage clause 8 notice is issued under clause 8.4.2, and the Contractor's second stage clause 8 notice is issued under clause 8.9.
20 JCT DB 2016. For either party (employer or contractor), the specified default must be repeated (not continued), save that clause 8.9.4.2 permits the contractor to issue the 'further notice' (under clause 8.9.4) if the 'specified suspension' is repeated.
21 JCT DB 2016, the 'further notice' is issued by the employer under clause 8.4.3, or by the contractor under clause 8.9.4.

Table 9.2 summarises the actions for each party where termination is by the employer and contractor (or by either party).

The calculation of payments to the contractor will produce starkly differing results depending upon whether termination was by the employer or by the contractor, as the other (non-terminating) party will have committed a 'specified default'. The exception is that in the case of termination by either party under clause 8.11, the employer bears the risk of termination.

If the employer terminates, clause 8.7.4 permits him to recover the additional costs of completing the works, following the principle of common-law damages (see Section 8.11). It does so by requiring the employer to prepare a statement[22] within three months of completing the works (or the making good of defects if any are present), to calculate the difference between:

(1) The actual amount that the employer had paid for the works, including:
    - The amount already paid by the employer to the (original) contractor[23];
    - Amounts paid to a replacement contractor (or contractors) to complete the works[24]; and
    - Any further 'expenses' incurred by the employer from any other breaches of contract by the contractor (including those expenses arising from the termination)[25]; and
(2) The amount that the employer *would have* paid to the original contractor for completing the works.[26]

If the amount in (1) exceeds the amount in (2), then the difference shown on the employer's clause 8.7 statement is a debt due from the contractor to the employer.[27] This is likely to occur in the majority of cases as the employer will want to resume the works as quickly as possible in order to avoid the works suffering damage and to limit any additional financial effects (such as delayed sale or use of the project). Consequently, the employer may not have time to competitively obtain a price for completing the works, and a replacement contractor may not want to be held to a fixed price for completing another contractor's work.

However, in the (unlikely) event that that the sum in (ii) exceeds the sum in (i), and the employer would always have had to pay the original contractor the higher sum for completing the works,[28] then this produces a perverse result – as the difference is paid to the contractor. Whilst the employer may not have to pay any more for the works that he always would have done, it cannot be right that a contractor who commits a 'specified default' can be rewarded by being paid a sum greater than the value of work he has carried out. If the employer is able to use his commercial acumen to obtain a lower price for completing the work (i.e. mitigating his losses), then he should be able to gain its benefit in financial terms.

---

22 JCT DB 2016, clause 8.7.4. The employer's statement must be prepared within completion of the works or making good of defects.
23 JCT DB 2016, clause 8.7.4.2.
24 JCT DB 2016, clause 8.7.4.1.
25 See Table 9.2, note *a*. The 'expenses' incurred prior to the date of termination may include liquidated damages.
26 JCT DB 2016, clause 8.7.4.3.
27 JCT DB 2016, clause 8.7.5.
28 For example, if the economic conditions were such that it was a 'contractor's market' when the contract was concluded but subsequently and significantly changed to an 'employer's market' by the time the employer terminated the contractor's employment.

Table 9.2 Consequences of termination under JCT DB 2016.

Consequences post-termination	Termination by employer	Termination by contractor or by either party (under clause 8.11)
Completion of the Works		
– Cessation of the contractor's obligation and entitlement to continue with the Work[a]	8.5.3.2, 8.7.1	8.10.3, 8.12
– Employer may engage others to complete the Works	8.7.1	8.12
– Employer may use the contractor's facilities to complete the Work	8.7.1	—
– Contractor removes his facilities	Employer instructs – 8.7.2.1	8.12.2.1[b]
– Contractor assigns contracts for supply and sub-contracts to the Employer	8.7.2.3	—
Contractor to provide further documents to the Employer	Contractor's Design Documents in 14 days – 8.7.2.3	As-Built Drawings (under clause 2.37) – 8.12.2.2
Calculation of further sums to the contractor		
– Employer's statement, accounting for his costs (including completing the Works)	8.7.4[c]	—
– Contractor's 'termination account' (including costs of removing facilities and any loss and / or expense)	—	8.12.3, 8.12.4
Further sums due to the contractor	Not clear, but after 3 months after completing the Works, and making good defects – 8.7.4, 8.7.5[d]	Within 28 days after receiving the contractor's termination account – 8.12.1, 8.12.5

a) JCT DB 2016. Clause 8.5.3.2 (termination by Employer following contractor's insolvency) and 8.10.3 (termination by Contractor following Employer's insolvency) expressly state that the Contractor's obligations under Article 1 are suspended. Clauses 8.7.1 (termination by Employer) and 8.12 (Termination by Contractor or by either party under clause 8.11) imply that the Contractor's obligation and entitlement to complete the Works ceases.

b) JCT DB 2016, Clause 8.12.2.1 Contractor removes his facilities (or procures their removal) "*with all reasonable dispatch*"

c) JCT DB 2016, clause 8.7.4 provides that the Employer's "statement" must be provided not later than 3 months after the Works are completed.

d) JCT DB 2016, clause 8.7.5 provides that any further sum (either payable by the Employer to the Contractor or by the Contractor to the Employer) is a "debt due". Clause 8.7.5 does not specify any final date for payment

Clause 8.7 does not expressly state when the 'due date', or the 'final date for payment' occur for the 'difference' set out in the employer's clause 8.7 statement. Clause 8.7.5 merely states that the difference becomes a 'debt due'. It may be that the Scheme[29] applies, and the due date is likely to be either the date on which the employer sends the clause 8.7 statement or the date on which the contractor receives it. The final date for payment is likely to be 17 days after the due date.[30] This is important if the contractor disputes the 'difference' in the employer's clause 8.7 statement; then he can issue a 'pay less notice'. Clause 4.9.5.2 refers to the 'relevant statement' instead of the Final Payment Notice (which would be provided under normal circumstances where the contractor's employment is not terminated) and provides that a 'pay less notice' must be issued not later than five days before the final date for payment.

If the contractor has terminated (or either party has done so under clause 8.11), then the contractor's account under clause 8.12.3 is comparatively more straightforward. His account should be submitted either as soon as is 'reasonably practicable' after termination, or if either party has terminated under clause 8.11, then not later two months thereafter.[31] The account should show the following:

o The total of work properly carried out up to the date of termination (valued in accordance with the contract)[32] including goods or materials that the contractor has ordered and must pay for[33];

o Any loss and expense ascertained under clause 4.20, either occurring before[34] or after[35] termination;

o The costs of removing the contractor's facilities[36]; and

o Other amounts due under the contract.[37]

The contractor's clause 8.12 account does not need to show the amounts that the employer has paid to date, but the employer must account for such payments when making payment in respect of the contractor's account. The final date for the payment of the contractor's clause 8.12 account is 28 days after the employer receives it, and no retention can be deducted.[38] It appears that the 'due date' for payment of the contractor's clause 8.12 account is the date it is submitted to the employer, so again, it is advisable for the contractor to retain proof of the date upon which it was sent. In practical terms, and particularly where the clause 8.12 account is accompanied by significant documentation, it is unlikely that the employer has sufficient time to issue a 'payment notice'[39] within

---

29 The Scheme for Construction Contracts (England and Wales) Regulations 1998 SI 649, as amended – refer to Section 8.7.
30 The Scheme, Part II (Payment), paragraph 8(2).
31 JCT DB 2016, clause 8.12.3.
32 JCT DB 2016, clause 8.12.3.1.
33 JCT DB 2016, clause 8.12.3.4.
34 JCT DB 2016, clause 8.12.3.2. For 'loss and expense' see Section 8.11.2.
35 JCT DB 2016, clause 8.12.3.5, and in the case of termination by either party under clause 8.11.1.3 (where the works are damaged by an event that is an Excepted Risk, or covered by the works Insurance), only in respect of negligence or a default on the part of the employer (or any employer's person).
36 JCT DB 2016, clause 8.12.3.3.
37 JCT DB 2016, clause 8.12.3.1.
38 JCT DB 2016, clause 8.12.5.
39 HGCRA 1996, s110A(1)(a) and s110A(2) – see Section 8.7.

five days of the account's receipt. It is more likely that the employer will issue a 'pay less notice'[40] as he will have more time to do so. It appears that a pay less notice ought to be issued not later than five days before the final date for payment,[41] but as clause 8.12 does not stipulate such a period, a cautious employer may issue a pay less notice in respect of the contractor's clause 8.12 account not later than seven days before the final date for payment, so it complies with paragraph 10 of Part II of the Scheme.

The option of the employer deciding not to complete the works is considered in clause 8.8 and only seems to apply if the employer has terminated the contractor's employment under clause 8.4 (Default by Contractor) or 8.5 (Insolvency of Contractor). Clause 8.8 sets out a procedure whereby the employer must notify the contractor within six months (from the date on which the contractor's employment is terminated) that he does not intend to complete the works.[42] The employer must then provide a statement ('the clause 8.8 statement') to the contractor[43] showing:

o The total value of work that the contractor properly carried out at either the date of termination, or on the date when the contractor became insolvent, along with any other amounts due under the contract[44]; and
o The total (or 'aggregate') of expenses that the employer has properly incurred, including those arising from the contractor's defaults or breaches (either before termination or after).[45]

Somewhat imprecisely, clause 8.8.2 states that the amounts that the employer has already paid to the contractor must be taken into account, but doesn't say exactly how. However, firstly, taking the example of the employer's clause 8.7 statement, and secondly, the 'difference' (described in clause 8.8.2), it seems that the amounts previously paid to the contractor fall under the 'expenses' described in clause 8.8.1.2. The 'difference' (which seemingly means the total after subtracting the amount in clause 8.8.1.2 from the amount in clause 8.8.1.1) is paid by the employer to the contractor (if a positive sum) or by the contractor to the employer (if a negative sum).

9.1.2 FIDIC Yellow Book 2017

The procedures for termination under the FIDIC Yellow Book 2017 are similar to those under JCT DB 2016, albeit they are included in separate clauses and include an option for the employer to terminate for his convenience:

o Termination by employer:
   – Clause 15.2 – for the contractor's default
   – Clause 15.5 – for the employer's convenience

---

40 HGCRA 1996, s111(3).
41 JCT DB 2016, as is the case for interim payments or for the final payment under clause 4.9.5.
42 JCT DB 2016, clause 8.8.1.
43 JCT DB 2016, clause 8.8.1 provides that the employer provides the clause 8.8 statement either within a 'reasonable time' after notifying the contractor (in accordance with clause 8.8.1), or if no such notification is given, then within two months after the expiry of the six-month period in clause 8.8.1.
44 JCT DB 2016, clause 8.8.1.1.
45 JCT DB 2016, clause 8.8.1.2.

o Termination by the contractor:
   – Clause 16.2 – for the employer's default
o Termination by either party:
   – Clause 18.5 – Optional termination

Clause 15.1 of FIDIC Yellow Book 2017 contains an additional step for the employer in the event of the contractor's default, whereby he issues the contractor with a 'Notice to Correct' in the event he fails to carry out any obligation under the contract. The Employer's Notice to Correct is a precursor to the employer's entitlement to commence the contract's termination procedure under clause 15.2.1(a), and must:

o Set out a description of the contractor's failure, including the contract clause under which the contractor's obligation must be carried out – clauses 15.1(a) and 15.1(b); and
o Specify the period of time in which the contractor must 'correct' (i.e. carry out) the relevant obligation. This period of time must be reasonable – clause 15.1(c).

Given that the Notice to Correct can be issued following any failure by the contractor, it is likely that the employer will issue one in the majority of cases, unless the contractor is unable to take corrective measures, or it is not required by the contract.

If the contractor or the employer respectively commits one of the defaults set out in clause 15.2 (including the contractor's failure to act on a Notice to Correct) or 16.2 (see Table 9.1), then either the employer (not the engineer) or the contractor (in the case of the employer's default) may issue the first-stage notice under clause 15.2.1 or clause 16.2.1.[46] The defaults under clauses 15.2 and 16.2 (contractor's and employer's defaults, respectively) include 'insolvency'[47] (whereas this is dealt with separately under JCT DB 2016). If the recipient fails to comply with the sender's first-stage notice within 14 days,[48] then the sender may issue the second-stage notice, which immediately terminates the contract upon receipt.[49] Curiously, unlike the first-stage clause 15.2.1 notice, the second-stage clause 15.2.2 notice does not have to state that it is issued under clause 15.2.2, but for clarity, the employer should clearly state that it is.[50] Unlike the clause 8 notices under JCT DB 2016, FIDIC Yellow Book 2017 does not contain separate (or additional) requirements for sending termination notices, so they all (be they issued by the employer under clauses 15.2 or 15.5, by the contractor under clause 16.2, or by either party under clause 18.5) must simply comply with clause 1.3 (Notices and Other Communications).[51] Importantly (and this requirement remains from

---

46 FIDIC Yellow Book 2017. Clauses 15.2.1 and 16.2.1 stipulate that the first-stage notice must state that it is issued under either of those clauses.
47 FIDIC Yellow Book 2017, clause 15.2.1(g) – insolvency of the contractor and clause 16.2.1(h)(i) – insolvency of the employer.
48 FIDIC Yellow Book 2017, the contractor must comply with the employer's first-stage notice under clause 15.2.2, and the employer must comply with the contractor's first-stage notice under clause 16.2.2.
49 FIDIC Yellow Book 2017, the contract terminates when the contractor or employer receives the other's notice under clause 15.2.2 or 16.2.2, respectively.
50 Although, following the employer's clause 15.2.1 notice, it should be fairly obvious to the contractor that the subsequent notice is issued under clause 15.2.2.
51 FIDIC Yellow Book 2017, clause 1.3 (Notices and Other Communications) provides that the 'Notice' must (i) state it is a notice, and (ii) be either a paper original (signed by the parties' authorised representatives) sent by mail or delivered by hand (both with receipts) to the recipient's address as set out in the Contract Data (unless the recipient gives 'Notice' of another address), or (iii) if transmitted by

the 1999 Yellow Book[52]), clause 1.3 provides that a notice delivered by hand, by mail, or by courier must be done so 'against receipt' (i.e. a 'receipt' should be provided to confirm delivery), so the following can be established:

o The relevant periods of time for the recipient to comply with the first-stage notices[53];
o The dates for termination to take effect[54]; and
o The period that termination for the employer's convenience may take effect[55] (provided that the employer returns the Performance Security[56] to the contractor beforehand) can be accurately ascertained.

Clause 15.5 permits the employer to terminate the contract for his 'convenience'. In practice, such circumstances are relatively rare; relevant examples could be if the employer is unable to obtain sufficient funding to complete the project or, for some reason, the project is no longer required. If the employer opts to terminate under clause 15.5, then the contractor is compensated under clause 15.6 (Valuation after Termination for Employer's Convenience) for the project being brought to an end.

Firstly, the employer must notify the contractor that he will terminate for his convenience. As with the first-stage notices issued under clauses 15.2.1 and 16.2.1, the employer's notice under clause 15.5 must state it is issued under that clause. Secondly, the employer must return the Performance Security to the contractor.[57] Termination will take effect upon whichever of the following occurs last[58]:

o The expiry of 28 days after the contractor receives the employer's clause 15.5 notice, or
o When the Performance Security is returned to the contractor.

Clause 18.5 of FIDIC Yellow Book 2017 permits either party to terminate the contract if an Exceptional Event[59] occurs. Clause 18.1 defines an 'Exceptional Event' as an event which is:

o Neither within a party's control[60] nor substantially the fault of the other party[61]; and
o Could reasonably neither be avoided[62] nor overcome.[63]

---

electronic means, by the means either stated in the contract (e.g. to an email address for the intended recipient), or a means acceptable to the engineer.
52 FIDIC Yellow Book 1999, clause 1.3 (Communications).
53 FIDIC Yellow Book 2017, the recipient of the first-stage notice must comply with it within 14 days under clause 15.2.2 (if the recipient is the contractor), or clause 16.2.2 (if the recipient is the employer).
54 FIDIC Yellow Book 2017, termination takes place when the recipient receives the second-stage notice where the recipient is the contractor (under clause 15.2.2) or the employer (under clause 16.2.2).
55 FIDIC Yellow Book 2017, clause 15.5.
56 FIDIC Yellow Book 2017, clause 4.2 (Performance Security) provided by the contractor to the employer.
57 FIDIC Yellow Book 2017, clause 15.5(c).
58 FIDIC Yellow Book 2017, clause 15.5.
59 An 'Exceptional Event' under FIDIC Yellow Book 1999 was referred to as 'force majeure' (clause 19.1).
60 FIDIC Yellow Book 2017, clause 18.1(i).
61 FIDIC Yellow Book 2017, clause 18.1(iv).
62 FIDIC Yellow Book 2017, clause 18.1(iii) – a party's ability to avoid the Exceptional Event also includes that it would be unreasonable for a party to make some provision for it, prior to entering into the contract – clause 18.1(ii).
63 FIDIC Yellow Book 2017, clause 18.1(iii).

Clause 18.1 continues to give a non-exhaustive list[64] of events that may be considered to be an Exceptional Event, but only if *both* criteria (stated above) are satisfied. If an Exceptional Event occurs, then clause 18.2 (Notice of an Exceptional Event) provides that the 'affected party' must notify[65] the other party within 14 days of becoming aware of the Exceptional Event, whereupon he will be excused from performing all obligations[66] which the Exceptional Event prevents and from the date that such obligations were prevented. If a party fails to notify within 14 days, he will only be excused from performing his obligations from the date on which the other party receives his clause 18.2 notification.[66]

Returning to clause 18.5 (Optional Termination), if an Exceptional Event affects the works as follows:

o For a continuous period of 84 days; or
o The same Exceptional Event affects the works for more than one period, the aggregate of which is at least 140 days, then

the contract may be terminated by either party if they notify[66] the other accordingly, and the date of termination shall occur seven days after the clause 18.5 notice is received.

As with JCT DB 2016, the consequences of termination under the FIDIC Yellow Book 2017 depend upon whether the contract is terminated by:

o The employer following the contractor's default,[67]
o The employer for his convenience,[68]
o The contractor,[69] or
o Either party.

The relevant procedures in clauses 15 and 16 cover the contractor's exit from the site, use or return of the contractor's facilities (equipment and temporary works), completion of the works, valuation, and payment of the contractor or of the employer. These are summarised in Table 9.3.

---

64 FIDIC Yellow Book 2017, clauses 18.1(a) war and hostilities, 18.1(b) civil insurrection/war, terrorism, 18.1(c) civil commotion/riot, 18.1(d) industrial action, 18.1(e) encountering ammunition (included unexploded ordnance) or radioactive contamination, and 18.1(f) natural disasters – providing that the contractor is not responsible under subclause 18.1(c), (d), or (e).
65 FIDIC Yellow Book 2017, the notification under clause 18.2 must be a 'Notice' complying with clause 1.3 (Notices and Other Communications).
66 FIDIC Yellow Book 2017, clause 18.2 provides that a party's obligation to make payment to the other remains unaffected.
67 FIDIC Yellow Book 2017, clauses 15.2.3 (After Termination), 15.2.4 (Completion of the Works) and 15.3 (Valuation after Termination for Contractor's Default) apply if the employer terminated the contract.
68 FIDIC Yellow Book 2017, clauses 15.6 (Valuation after Termination for Employer's Convenience), 15.7 (Payment after Termination for Employer's Convenience), and 16.3 (Contractor's Obligations after Termination) apply if the employer terminates for his convenience.
69 FIDIC Yellow Book 2017, clauses 16.3 (Contractor's Obligations after Termination) and 16.4 (Payment after Termination by Contractor) apply if the contractor terminates the contract.

Table 9.3 Consequences of termination under FIDIC Yellow Book 2017.

Consequences post-termination:	Termination by employer	Termination for employer's convenience	Termination by contractor	Optional termination by either party (clause 18.5)
Completion of the works				
• Contractor to comply with instructions	15.2.3(a)	—	—	—
• Deliver goods to the employer	15.2.3(b)(i)	—	—[d]	—[d]
• Contractor to cease work	—	15.5 (16.3(c))	16.3(a)	16.3(a)
• Leave site[a]	—[b]	15.5 (16.3(c))	16.3(c)[e]	16.3(c)
• Employer to complete the works using (i) the contractor's equipment and temporary works and (ii) documents	15.2.3(c)	—	—	—
• Employer gives notice to contractor (post-completion) to remove its equipment and temporary works	15.2.4	15.5 (16.3(c))	—	—
• Employer has no right to use contractor's facilities	15.2.4	15.5(b)[c]	—[f]	—[f]
• Return the Performance Security to the contractor	—	15.5(c)	—	—
Contractor's documents				
• Contractor to provide further documents to employer	15.2.3(b)(ii), 15.2.3(b)(iii)	—	16.3(b)	16.3(b)
• Employer has no right to use the contractor's documents	—	15.5(a)	—	—
Calculation of further sums to contractor				
• Engineer to determine (or agree) the values of the permanent works, goods, and contractor's documents *without* the contractor's involvement	15.3	—	—	—
• Contractor submits details[g] of the value of work done, additions/deductions and loss and expense to the engineer, who then issues a payment certificate	—	—	—	—
• Contractor submits details of the amount payable for work done[h]; plant and materials, other contractor's costs or liabilities, removing temporary works and equipment, and any staff repatriation costs.	—	15.6(a)(i) and (ii)	16.4(a) (18.5)	18.5
• Employer pays the contractor loss of profit, other losses or damages resulting from the termination.	—	15.6(b)	16.4(b)	18.5(a) to (e)
				—Ω–

(Continued)

Table 9.3 (Continued)

Consequences post-termination:	Termination by employer	Termination for employer's convenience	Termination by contractor	Optional termination by either party (clause 18.5)
Further payment to contractor				
• Only due when all the employer's losses or damages are established	15.4	—	—	—
• 112 days after the engineer receives the contractor's submission under clause 15.6		15.7	—	—
• 'Promptly'			—	—
• In accordance with clause 14.6 (Issue of IPC), 56 days later (clause 14.7)			16.4(a) (18.5) (18.5)	18.5 (14.6)
Payment to employer (from contractor)[j]				
• Additional costs of clearing site after termination	15.4(a)	—	—	—
• Costs of completing the works	15.4(b)	—	—	—
• Delay damages	15.4(c)	—	—	—
• Employer may sell contractor's equipment and temporary works[j]	15.2.4	—	—	—

a) FIDIC Yellow Book 2017, clause 15.2.3(c) permits the employer to expel the contractor from the site if he refuses to leave.
b) Implied by clause 15.2.3(c).
c) FIDIC Yellow Book 2017, clause 15.5(b) applies if clause 4.6 (Co-operation) applies.
d) FIDIC Yellow Book 2017, clause 16.3(b) obliges the contractor to deliver 'other works' for which the contractor has been paid which may include 'Goods'.
e) FIDIC Yellow Book 2017, clause 16.3(c) permits the removal of goods on the proviso that it is safe to do so.
f) Implied by clause 16.3(a).
g) FIDIC Yellow Book 2017, clause 15.6 provides that the contractor only provides 'details' to the engineer and not a 'Statement', as is otherwise required under clause 14.3 (Application for Interim Payment) or clause 14.10 (Statement on Completion).
h) FIDIC Yellow Book 2017, clause 18.5(a) – the price for work done which is set out in the contract (i.e. within the 'Schedules' in 2(f) of the Contract Agreement – see Section 3.6).
i) FIDIC Yellow Book 2017, clause 15.4 (Payment after Termination for Contractor's Default) is subject to clause 20.2 (Claims for Payment and/or EoT) – refer to Section 8.4.10.1.
j) FIDIC Yellow Book 2017, clause 15.2.4 permits the employer to sell the contractor's temporary works and equipment if the contractor fails to make a payment due to the employer.

9.1.3 NEC4

NEC4 adopts a different procedure for termination to that found in JCT DB 2016 or FIDIC Yellow Book 2017. JCT and FIDIC both include a two-stage procedure whereby the aggrieved party firstly notifies the other of the default, and if the default is not rectified within a certain period, then by a further notice, the innocent party may effect termination.

NEC4's[70] termination procedure is set out in clauses 90–93 and is fundamentally different. Instead of either the contractor or the employer notifying each other, clause 90.1 provides that the innocent party notifies[71] the project manager of the 'reason' for wanting to terminate. The 'reason' for terminating available to both the client[72] and to the contractor[73] (see Table 9.1 – Grounds for Termination) are set out in the 'Termination Table' in clause 90.2, and are 'fully described'[74] in greater detail in clause 91.1 (Reasons for Termination – R1 to R22). If Option X11 (Termination by the Client) is selected, then the client may terminate for any reason[75] (i.e. a reason not identified in the clause 90.2 Termination Table). The termination table (in clause 90.2) and NEC4's *User's Guide*[76] suggest that the clause 90.1 notice can only contain a single 'reason'; however, in practice, a party intent on terminating may want to rely upon more than one reason, particularly if the other party has committed more than one default. The notifying party may either issue a clause 90.1 notice where the second (or any subsequent) reason is stated to be 'without prejudice' to the first (or preceding) reason, or he may simply issue a separate clause 90.1 notice for each reason available to him, on the basis that the more reasons (or notices) that the project manager has to consider, the greater the chances of persuading him that he must issue a termination certificate.

Upon receipt of either party's clause 90.1 notice, the project manager must 'promptly' issue a 'termination certificate' if (i) the reason given in the notice accords with the contract,[77] and (ii) the notice gives 'details' of that reason. No further guidance[78] is given in terms of what 'details' should be provided, but it is advisable for the notifying party to provide sufficient details to allow the project manager to verify the reason given, and issue the termination certificate promptly, if the reason complies with the contract.

Clause 90.1 does not define 'promptly', but it is suggested that it should be a sufficient period of time for the project manager to not only check that the reason given in the clause 90.1 notice is one of the reasons in the Termination Table but also that the reason

---

70 NEC3 contains a similar procedure, but unlike its predecessor, NEC4's core clauses do not permit the Client (referred to as the 'Employer' in NEC3) to terminate for 'any reason' (NEC3, clause 90.2). Secondary Option Clause X11 (Termination by Client) must be selected to permit the client to terminate for a reason not set out in the Termination Table (in clause 90.2).
71 NEC4, clause 13 (Communications), and clause 13.7 govern 'notices'.
72 NEC4, the table in clause 90.2 shows that the client may notify the project manager (under clause 90.1) of its intent to terminate for (i) reasons R1–R15, R18, or R22, (ii) R7 or R20, or (iii) R21.
73 NEC4, the table in clause 90.2 shows that the contractor notifies the project manager (under clause 90.1) of its intent to terminate for (i) R1–R10, R16, or R19, and (ii) R17 or R20.
74 NEC4, *User's Guide* (2020), p. 78.
75 NEC4, *Managing Reality,* Book Two, 3rd ed., (2020), p. 91, recommending that Option X11 is selected if a client wants to terminate the contract for any reason. *Managing Reality,* Book Five, 3rd ed., p. 62.
76 NEC4 see (i) clause 90.2 termination table only appears to permit a single reason, and (ii) *Managing an Engineering and Construction Contract* – vol. 4, (2020), p. 78.
77 NEC4, clause 90.1.
78 For example, in NEC4's *User's Guide,* or the *Managing Reality* books.

is valid. NEC4's *Managing Reality* suggests that the project manager should check for the following[79]:

o If the clause 90.1 notice is issued by the client;
  – the four-week period for the contractor to remedy the defaults in clauses 91.2[80] and 91.3[81] has expired; or
  – the 13-week period for instructing the contractor to restart or start removing work under clause 91.6[82] has expired.
o If the clause 90.1 notice is issued by the contractor:
  – The 13-week period applying to both non-payment[83] or the instruction to re-start or start removing work has expired.[84]

In practice, the project manager should exercise extreme caution before issuing a termination certificate, as he may be liable to the client if he does so wrongfully or carelessly. In addition to simply checking the expiry periods (as detailed above), it is suggested that the project manager should also make reasonable investigations into the facts behind any of the stated reasons for termination (R1–R22 in clause 91), and then seek legal advice – as not only must he act with skill and care,[85] but also his decision making must be fair, unbiased, and holding the balance between the employer and the contractor.[86] The client is likely to incur significant losses if the contract is terminated incorrectly, irrespective of whether the clause 90.1 notice was issued by the client or by the contractor.

Once the project manager issues the termination certificate, clause 90.3 provides that the termination procedures in clause 92 (procedures P1–P4) commence immediately. Unlike JCT DB 2016 or FIDIC Yellow Book 2017,[87] NEC4 does not expressly state when termination occurs, but once the clause 92 termination procedures commence, it should be clear that termination has taken place.

The procedures in clause 92 are as follows:

o Procedure P1: Whatever the reason for either party to give notice of termination (reasons R1–R22), the client can complete the works and in doing so, he may use any plant and materials from which title has passed[88] to him from the contractor.
o Procedures P2 and P3: These are available to the client only, when terminating for reasons R1–R15, R18, or R22, and are relevant to the client completing the works.

---

79 NEC4, *Managing Reality*, Book Five, 3rd ed., (2020), pp. 62–63.
80 NEC4, clause 91.2, client terminating for reasons R11, R12, or R13.
81 NEC4, clause 91.3, client terminating for reasons R14 or R15.
82 NEC4, clause 91.6, client terminating for reason R18 or R20.
83 NEC4, clause 91.4, contractor terminating for reason R16.
84 NEC4, clause 91.6, contractor terminating for reason R19.
85 Assuming that the project manager is appointed under NEC4's Professional Services Contract ('NEC4 PSC'). Refer to NEC4 PSC – clause 20.2 (and Section 4.2.1).
86 See Section 4.2.8 and *Costain v Bechtel*, EWHC 1018, [2005].
87 Clauses stipulating when termination occurs are JCT DB 2016, clause 8.2.2 (for either party), or FIDIC Yellow Book 2017, clause 15.2.2 (if the employer terminates) or 16.2.2 (if the contractor terminates).
88 NEC4. Title of Plant and Materials passes from the contractor to the client if (i) they are within the Working Area (clause 70.1), or (ii) if not, if the supervisor has 'marked' them (clause 70.2). Plant and materials are defined in clause 11.2(14).

- Procedure P2 – The client can instruct the contractor to assign the benefit of any contract or subcontract to him, along with instructing the contractor to leave the site, and remove any equipment,[89] plant, and materials; and
- Procedure P3 – When completing the works, the client may use any of the contractor's equipment (to which the contractor has title), which the contractor must then remove from the site when the project manager 'informs' him that it is no longer needed.
- Procedure P4: This mainly applies to the contractor when he terminates for reasons R1–R10, R16, or R19, but also both parties for 'neutral' termination events that are not the fault of either party (reasons R17, R20, or R21). Procedure P4 simply states that the contractor leaves (and removes the equipment from) the working areas.

Clause 93 (Payment on Termination) sets out four stages of calculating the amount due (A1[90] and A2–A4[91]) on termination which are available to each party as follows:

- The amount due under method A1 is due for any reason for terminating,[92] and includes[93]:

Amount assessed for 'normal payments'	£A
Excluding amounts to be permanently paid by the contractor:	
– Failing to meet a Key Date under clause 25.3	
– Client's cost of rectifying an uncorrected defect under clause 46.1	
Defined Cost for Plant and Materials:	£B
– In the Working Areas; or	
– Where title has passed to the client and the contractor will accept delivery	
Any other Defined Cost that the contractor 'reasonably incurred'[a]	£C
Amounts that the client has retained[b]	£D
Less any unpaid instalment or balance of an advanced payment[c]	(£E)
Total amount due under A1:	**£A + £B + £C + £D-£E**

a) For example, in NEC4 – Option A (Priced Contract with Activity Schedule) – including the value of uncompleted activities as at the date of termination.
b) For example, retention under Option X16, or amount retained under clause 50.5 if the contractor has not issued its first programme (under clause 31.1).
c) NEC4, Option X14 (Advanced Payment to the Contractor), clause X14.3.

- A1 is used by the parties as follows:
  - By either party in the event of the other's insolvency (reasons R1–R10), if the law releases either of them from further obligations (R17), or if an instruction to stop work arises from no fault of either party (R20);
  - By the client only for reasons R11–R15 or R22 (all contractor default) or if an unforeseeable event (which is neither party's fault) prevents completion (R22); or

89 NEC4, clause 11.2(9) defines 'Equipment'.
90 NEC4, Clause 93.1 – A1.
91 NEC3, Clause 93.2 – A2 to A4.
92 NEC4, *User's Guide*, (2020), p. 80.
93 NEC4, clause 93.1.

- By the contractor only in the event of the client's failure to pay (R16) or from a client default (R19).
  o A2 comprises the forecast Defined Cost[94] of removing equipment and is used by the parties as follows:
    - By the contractor when the client is insolvent (reasons R1–R10), fails to pay (R16), or following an instruction to stop work arising from the client's default (R19);
    - By either party for reasons R17 (released from obligations by law) or R20 (instruction to stop work not arising from any party's default); or
    - By the client in the event an unforeseeable event prevents completion (R20).
  o A3 is used for the additional cost to the client (as forecast) of completing the works. A3 is solely available to the client.
  o A4 is the application of the 'fee percentage,[95] which is solely available to the contractor when terminating for the client's insolvency (R1–R10), when the client fails to pay (R16), and there is an instruction to stop work arising from the client's default (R19). Under A4, the fee percentage is applied depending upon which Main Option clause applies – as shown in Table 9.4.

There are additional subclauses applying to Option A (Priced Contract with Activity Schedule),[96] Option C (Target Cost Contract with Activity Schedule),[97] and Option D (Target Cost Contract with Bill of Quantities)[98] concerning the calculation of the amount due on termination.

NEC4's reasons for termination, the applicable procedures, and methods for calculating the amount due on termination are shown in Table 9.5.

Table 9.4 NEC4 – Clause 93 (A4): application of 'fee percentage' on termination.

Main contract options	Fee percentage applied to:
Option A: priced contract with activity schedule Option B: priced contract with bill of quantities Option C: target cost contract with activity schedule Option D: target cost contract with bill of quantities	'fee percentage' is applied to £X Where £X = (total of the prices at the contract date) – (price for work done to date)
Option E: cost reimbursable contract	'fee percentage' is applied to £Y Where £Y = [first forecast of defined cost for the works] – [(price for work done to date) – (the fee)]

94 See Section 8.9.3.2 and Table 8.13.
95 NEC4, clause 11.2(10) defines the 'Fee'.
96 NEC4, Option A, clause 93.3 – Grouping of activities is not taken into account when assessing the amount due on termination.
97 NEC4, Option C, clause 93.4 – The project manager assesses the contractor's share after issuing the termination certificate. Clause 93.6 – If the project manager's assessment of the contractor's share is positive (in the event of a saving) it is added to the amount due to the contractor on termination. Conversely, if the assessment is negative (an excess) it is deducted from the amount due to the contractor on termination.
98 NEC4, Option D, clauses 93.5 and 93.6 (having similar effect as clauses 93.4 and 93.6 under Option C).

Table 9.5 Summary of NEC4's termination reasons, procedures, and methods of calculating the amount due on termination.

	Reason (clause 91.1) "E" = Client, "C" = contractor	Procedure (clause 92)				Calculation of Amount Due (clause 93)			
		P1	P2	P3	P4	A1	A2	A3	A4
Individual Insolvency	(R1) Other party presents bankruptcy application	E / C	E	E	C	E / C	C	E	C
	(R2) Other party has bankruptcy order against it	E / C	E	E	C	E / C	C	E	C
	(R3) Receiver appointed (over assets)	E / C	E	E	C	E / C	C	E	C
	(R4) Made arrangement with creditors	E / C	E	E	C	E / C	C	E	C
Company Insolvency	(R5) Winding up order made against other party	E / C	E	E	C	E / C	C	E	C
	(R6) Provisional liquidator appointed	E / C	E	E	C	E / C	C	E	C
	(R7) Resolution for winding up passed[a)]	E / C	E	E	C	E / C	C	E	C
	(R8) Administrator appointed / administration order	E / C	E	E	C	E / C	C	E	C
	(R9) Receiver appointed over substantial assets	E / C	E	E	C	E / C	C	E	C
	(R10) Made arrangement with creditors	E / C	E	E	C	E / C	C	E	C

Table 9.5 (Continued)

Reason (clause 91.1) "E" = Client, "C" = contractor	Procedure (clause 92)				Calculation of Amount Due (clause 93)			
	P1	P2	P3	P4	A1	A2	A3	A4
(R11) Contractor substantially fails to satisfy his obligations	E	E	E	—	E	—	E	—
(R12) Contractor fails to provide bond or guarantee	E	E	E	—	E	—	E	—
(R13) Contractor sub—contracts a substantial part of the Works without Project Manager's approval	E	E	E	—	E	—	E	—
(R14) Contractor substantially hinders the Client	E	E	E	—	E	—	E	—
(R15) Contractor commits a serious health and safety breach	E	E	E	—	E	—	E	—
(R16) Client fails to pay contractor within 13 weeks	C	—	—	C	C	C	—	C
(R17) Law releases either party from further obligations	E C	—	—	E C	E C	E C	—	—
(R18) Instruction arises from contractor's default	E	E	E	—	E	—	E	—
(R19) Instruction arises from Client's default	C	—	—	C	C	C	—	C
(R20) Instruction does not arise from a party's default	E C	—	—	E C	E C	E C	—	—
(R21) Unforeseeable, non-fault[b] event prevents completion	E C	—	—	E C	E	E	—	—
(R22) Contractor commits a "corrupt act"	E	E	E	—	E	—	E	—

Instruction to stop work refers to rows (R18)–(R20).

a) NEC4, clause 91.1 (R7) – resolution for winding up passed exception for the purposes of amalgamation or reconstruction
b) NEC4, clause 91.7

9.2   Obligations Prior to Completing the Physical Works

Under most forms of building contract, it is common for the contractor to perform some further obligations prior to – and often to facilitate – completion of the physical works. These obligations may include providing particular notifications to the employer (or his representative), carrying out administrative tasks, along with taking steps to demonstrate that the works are nearing completion. The meaning of 'completion' under each of the three design and build contracts is considered in Section 9.3.

The extent of the contractor's obligations depends upon the nature of the project being constructed and the contents of the employer's requirements (or Scope). As covered in Section 8.12, an engineering project often requires far more testing than for a construction project, because erecting the physical works (for example, in the case of a chemical plant, gas or oil refinery) is only part of the overall project. Testing, commissioning, and operating the plant to achieve the specified performance criteria may also take up a significant part of the project's overall programme.

To a large extent, this is reflected in the design and build contracts. JCT DB 2016 does not contain any specific testing and commissioning procedures prior to completion, other than satisfying the general requirement of achieving 'practical completion' (which is covered in Section 9.3.1). At the other end of the spectrum, FIDIC Yellow Book 2017 contains detailed testing procedures that the contractor must satisfy to achieve completion, which FIDIC defines as 'Taking Over' (see Section 9.3.2). Given that NEC4 seems to be adopted for an increasing proportion of infrastructure projects in the UK,[99] it is perhaps surprising that NEC4 does not provide an intermediate position between JCT and FIDIC for testing and commissioning before completion (see Section 9.3.3).

9.2.1   JCT DB 2016

Prior to completion (or 'practical completion' in this case[100]), JCT DB 2016 does not contain many obligations for the contractor, other than his principal obligation (under clause 2.1) to carry out and complete the works' design and the works in order to achieve 'practical completion' (see Section 9.3.1). Clause 2.37 obliges the contractor to carry out the following activities prior to practical completion of either the works, or of a section[101]:

o Obligations under the CDM Regulations[102] in relation to the health and safety file[103]: and
o Provide the Contractor's Design Documents,[104] which:
  – Give the 'as-built' information (either by description or shown on a drawing); and
  – Provide information relating to the works' (or section's) operation and maintenance.

---

99 National Construction Contracts & Law Report (2018) RIBA (NBS) indicates that for project values in excess of £25 M, circa 65% are let under NEC4 (or its predecessor NEC3), compared to circa 35% let under JCT contracts.
100 JCT DB 2016, clause 2.27 (Practical Completion).
101 See Section 8.3.1.
102 Construction (Design and Management) Regulations 2015.
103 Health & Safety Executive, Managing Health and Safety in Construction L153 (2015) – see 'Guidance 12' paragraphs 114 and 115.
104 JCT DB 2016, 'Contractor's Design Documents' are defined in clause 1.1, and clause 2.37 also obliges the contractor to provide the same, as set out in the Contract Documents, or as the employer reasonably requires.

JCT DB 2016s conditions do not include any specific testing or commissioning procedures that are necessary to achieve practical completion, although the employer is obviously at liberty to introduce such procedures via the Employer's Requirements (see Part 2, Section 12.4).

9.2.2 FIDIC Yellow Book 2017

Given that the purpose of FIDIC Yellow Book 2017 is for the design and building of 'plant', it is unsurprising that it contains a detailed and structured regime of testing to be carried out prior to completion (or 'Taking Over'). This procedure is contained in clause 9 (Tests on Completion) and is summarised in Figure 9.1.

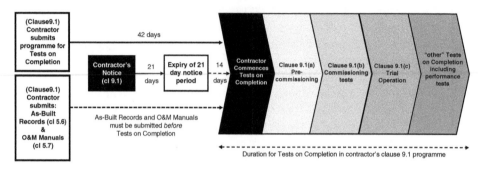

Figure 9.1 FIDIC Yellow Book 2017 (tests on completion).

Clause 9.1 obliges the contractor to carry out a number of important administrative and operational functions before he may begin the 'Tests on Completion' as follows:

o He must provide his programme[105] showing the 'Tests on Completion' activities not later than 42 days before intending to commence them. The engineer reviews the contractor's clause 9.1 programme, and either:
  – Issues a Notice of No-Objection, whereupon the contractor may begin the Tests on Completion; or
  – If the clause 9.1 programme does not comply with the contract, he notifies the contractor within 14 days. The contractor shall then rectify the programme. If no notice is given by the engineer within 14 days, he is deemed to have given a Notice of No-Objection.
o Prior to or within the 42-day period for providing the clause 9.1 testing programme, the contractor must:
  – Submit all O&M Manuals[106] and As-Built Records[107]; and
  – Notify the engineer that, within 21 days, he is ready to commence testing under clause 9.1, and should do so within 14 days after receiving the engineer's instruction in this regard; and

---

105 FIDIC Yellow Book 2017, clause 8.3(c) requires the contractor's programme to show the dates on which the contractor intends to submit the 'Contractor's Documents' which include the O&M Manuals and the as-built records.
106 FIDIC Yellow Book 2017, clause 5.7 (O&M Manuals).
107 FIDIC Yellow Book 2017, clause 5.6 (as-built records).

o Until a 'Notice of No-Objection' is issued by the engineer (or is deemed to have been so issued), the contractor shall not begin the Tests on Completion.[108]

Clause 8.3(c) provides that the contractor's programme (and any revised programme) should already show the 'anticipated timing' of 'testing, commissioning and trial operation' so submitting a further programme under clause 9.1 may only be necessary if the contractor has subsequently developed or firmed up the sequence and timing for the Tests on Completion (see Table 8.5 and Section 8.6.2).

Once the contractor has received the engineer's instruction, he can commence up to four stages of Tests on Completion as follows:

o 'Pre-commissioning' under clause 9.1(a),
o 'Commissioning' under clause 9.1(b),
o 'Trial Operation' under clause 9.1(c), and
o 'Other' tests (during 'Trail Operation') including performance tests.

When the contractor considers that the relevant works have passed one of the four stages (described above), he provides the test results to the engineer in the form of a 'certified report', which the engineer then reviews. Within 14 days the engineer may either (i) notify the contractor of the degree that the test results fall short of the contract's requirements, or (ii) accept the test results. Curiously, clause 9.1 does not provide any express means for the engineer to accept the test results. Instead, if the engineer doesn't notify the contractor within 14 days of receiving the 'certified report', he is deemed to have issued a Notice of No-Objection.[109]

Until the relevant tests are passed, the contractor is prohibited from proceeding with the subsequent stages of testing, so for reasons of clarity, the engineer should be encouraged to notify the contractor that the test results are in accordance with the contract. If the engineer notifies the contractor that the test results are rejected, then he has the following options under clause 9.4 (Failure to Pass Tests on Completion):

o Instruct the contractor to 'retest' the work under clause 9.3 (Retesting).[110] This imposes the same procedure as in clause 7.5 (Defects and Rejection) covered in Section 8.12.2;
o If the condition of a section, or of the works is such that it substantially deprives the employer of the respective benefit of a section,[111] or all of the works,[112] then the employer may terminate the contract, applying the procedure under clause 11.4 (Failure to Remedy Defects)[113]; or
o If the employer requests, issue the Taking Over Certificate to the contractor, and the employer shall be entitled to an appropriate reduction in the Contract Price, to reflect the reduced performance.[114]

---

108 FIDIC Yellow Book 2017, clause 9.1.
109 FIDIC Yellow Book 2017, 'Notice' is defined in clause 1.1.56 and 'No-Objection' is defined in clause 1.1.55.
110 FIDIC Yellow Book 2017, clause 9.4(a).
111 FIDIC Yellow Book 2017, clause 9.4(c).
112 FIDIC Yellow Book 2017, clause 9.4(b).
113 FIDIC Yellow Book 2017, clause 11.4(d) – also see Table 9.1 (in Section 9.1) and 9.1.2.
114 FIDIC Yellow Book 2017, clause 94(d) – the reduction in the Contract Price, to be paid by the contractor to the employer is treated as a 'Claim' under clause 20.2 – refer to section 8.10.4.1. The reduction in the Contract Price is described in clauses 11.4(b)(i) and 11.4(b)(ii).

Clause 9.2 (Delayed Tests) provides that the contractor and the employer are liable to the other in the event that one of them unduly delays the Tests on Completion after the contractor has given his clause 21-day notice to the engineer under clause 9.1. The following procedures apply in each case:

o If employer has caused a delay of either 14 continuous days or delays totalling 14 days in the aggregate,[115] then clause 10.3 (Interference with Tests on Completion)[116] applies:
  – The contractor describes the preventing act in a notice to the engineer[117];
  – The works are treated as having been taken over by the employer when, but for the employer's delay, the Tests on Completion would have been completed[118]; and
  – A Taking Over Certificate (for the works or section) is immediately issued by the engineer,[119] following which, the engineer notifies the contractor to begin the Tests on Completion within 14 days, which must be carried out as soon as is practicable.[120]
  – The 'Claims' procedure (clause 20.2) will apply if the contractor suffers delay or incurs additional costs.[121]
o If the contractor delays Tests on Completion, then clause 9.2 (Delayed Tests) stipulates:
  – The engineer notifies the contractor to carry out the relevant tests within a period of 21 days. The contractor gives the engineer not less than 7 days' notice of when (within the 21-day period) he will carry them out; or
  – If the contractor fails to carry out the tests within the 21-day period, the Employer's Personnel[122] may carry out the testing instead.[123] The tests will be deemed to have been carried out irrespective of whether the contractor witnesses them, but he has the right to do so.[124]
  – The engineer provides the test results to the contractor within 28 days of completing the tests[125]; and
  – The employer may seek payment from the contractor (under clause 20.2) of any additional costs incurred from carrying out testing.[126]

9.2.3 NEC4

NEC4 shares a similar approach with JCT Design & Build 2016 insofar as the contractor's principal obligation in terms of completion is to provide the works in accordance with the

---

115 FIDIC Yellow Book 2017, clause 10.3 (Interference with Tests on Completion) – the delay must prevent the contractor from carrying out the Tests on Completion.

116 Also See Section 9.3.2.

117 FIDIC Yellow Book 2017, clause 10.3(a).

118 FIDIC Yellow Book 2017 clause 10.3(b). Good evidence of when Testing on Completion would have been completed would be set out in the contractor's clause 9.1 programme (if not already set out in his clause 8.3 programme).

119 FIDIC Yellow Book 2017, clause 10.3(c).

120 FIDIC Yellow Book 2017, clause 10.3. If the Taking Over Certificate is issued under clause 10.3(c), the contractor must complete the Tests on Completion before the Defects Notification Period (DNP) expires.

121 FIDIC Yellow Book 2017, clause 10.3, the contractor may be entitled to 'Costs Plus Profit' (as defined in clause 1.1.20).

122 FIDIC Yellow Book 2017, clause 1.1.32 defines 'Employer's Personnel'.

123 FIDIC Yellow Book 2017, clause 9.2(a).

124 FIDIC Yellow Book 2017, clause 9.2(b) and clause 9.2 generally.

125 FIDIC Yellow Book 2017, clause 9.2(c).

126 FIDIC Yellow Book 2017, clause 9.2(d).

Scope[127] so that he completes the work (including its design[128]) on or before the Completion Date[129] (i.e. so there are no defects preventing the client from using the works[130]). NEC4's conditions do not include any specific procedures that the contractor must carry out in the same way that clause 9 of FIDIC Yellow Book 2017 provides with the 'Tests on Completion'.

As is the case with many aspects of NEC4, any specific procedures must be set out in the Scope, which could include:

o Tests and inspections under clause 41.1, including those for any plant or materials before they are delivered (clause 42.1); or
o If a 'Defect' includes substandard performance under Option X17 (Low Performance Damages), the contractor may have to test the works and correct that defect.

9.3 Meaning of 'Completion'

In general terms, 'completion' signifies the end of RIBA Stage 5 (Manufacturing and Construction), so the contractor's obligations to complete the physical works are at an end, and it is the start of its obligations to rectify defects or non-compliant work or (depending upon the particular form of contract) carry out further testing.

Section 8.12 (Testing and Defects) briefly described the different emphasis that each design and build contract places upon the contractor in order to 'complete' the physical works. In short, construction projects (such as those let under JCT DB 2016) provide a higher obligation, so the contractor must design and construct the works so that they are not only fully functional but also aesthetically acceptable; whereas engineering projects place greater emphasis upon the works achieving a functional capacity, or a specific level of performance, so rectifying aesthetic defects is relatively less important. The meaning of 'completion' can therefore differ; not just under the conditions of the three design and build contracts, but also because of the varying nature of the projects for which they are used. It may also include administrative functions, without which 'completion' cannot occur.

In addition to completing the design and the physical works, further requirements to achieve 'completion' may be set out in the contract's conditions (refer to Section 9.2), or in the Employer's Requirements (JCT and FIDIC) or the Scope (NEC4).

9.3.1 JCT DB 2016 – 'Practical Completion'

JCT DB 2016 expresses 'completion' as 'practical completion', which rather unhelpfully, remains undefined in the 2016 editions of all of the JCT building contracts. The term 'practical completion' has been used since the 1963 (then RIBA) edition of the standard building contract, and its meaning has been considered by the Courts as follows:

o 'Practical completion' means completion of all the work that has to be done[131];

---

127 NEC4, clause 20.1.
128 NEC4, clause 21.1.
129 NEC4, clause 30.1.
130 NEC4, clause 11.2(2) definition of 'Completion' – which also provides that no defects must be present to prevent 'Others' (see clause 11.2(12)) from working.
131 *J Jarvis & Sons v Westminster Corp*, 1 WLR 637 at 646D-F, (1970).

o 'Practical completion' can occur, notwithstanding the presence of latent defects[131] however, the presence of patent defects ordinarily prevents the certificate (or 'statement' in the case of JCT design and build contracts) of practical completion being issued[132]; and

o The supervising officer (in this case, the architect) has discretion to certify practical completion where there are minor items of incomplete or defective work (the de minimis principle).[133]

Ordinarily, all of the physical work, save for latent or very minor (de minimis) defects or items of incomplete works, must be complete, along with the design, for which the 'as-built' information must have been provided (see Section 9.2.1). The Courts' interpretation of 'practical completion' accommodates the 'construction' nature of JCT DB 2016, allowing more emphasis to be placed upon the aesthetic element of completion, rather than focusing more on the building's 'practical' or 'functional' aspects. For example, a dwelling house may fulfil the function of being fit for habitation,[134] but its aesthetic standard of finish may be such to prevent a sale or lease at the desired price.

Clause 2.27 provides that the employer issues a 'Practical Completion Statement' (or in the case of Sectional Completion,[135] a 'Section Completion Statement') when the works as a whole (or in the relevant section) have reached a state of practical completion. This has the following effects:

o The contractor's obligations to carry out and complete the physical works and its design are at an end, save for rectifying any latent defects (i.e. those which could not reasonably be discoverable), during the Rectification Period.

o If retention has been deducted from amounts due to the contractor in interim payments, then in accordance with clause 4.18.2.1, half of the retention may be deducted from subsequent interim payments until the Notice of Completion of Making Good Defects is issued (under clause 2.36) – see Section 8.7.1.

o The Rectification Period commences. This will be as per the entry in the Contract Particulars against clause 2.35, or if no period is entered, a default period of six months applies (or different Rectification Periods may apply if the works are being completed by sections).

o The contractor's obligation to insure the works under Insurance Option A (All Risks Insurance of the Works by the Contractor) shall cease.[136]

o The employer is no longer entitled to instruct Changes (see Section 8.8.3) as the works are complete.

---

132 *Kaye Ltd v Hosier & Dickinson,* 1 WLR 146, (1972).

133 *HW Nevill (Sunblest) v William Press* 20 BLR 78, (1981), at 87; see also *Emoson Eastern (in receivership) v EME Developments Ltd,* 55 BLR 114, (1991) and *Walter Lilly & Co Ltd v Mackay & Another (No. 2),* EWHC 1773 (TCC), [2012], at 372.

134 As per s1(1) of the Defective Premises Act 1972.

135 See Section 8.3.1.

136 JCT DB 2016, clauses 2.33 and 6.7.2 (Insurance Options and Period) – also see Schedule 3 (Insurance Options), Insurance Option A, paragraph A.1 (Contractor to effect and maintain a Joint Names Policy).

It is often the case that the Practical Completion Statement (or Section Completion Statement) is accompanied by a list of work that is either incomplete or defective (usually called a 'snagging list'). Snagging lists are slightly controversial for the following reasons:

o Unless the items of work described in the snagging list (both individually and collectively) can be considered as minor (or de minimis), then they shouldn't be included because (on a strict interpretation) they would prevent 'practical completion'. If the employer wants to occupy and use the works before 'practical completion' then he effectively has to take 'partial possession' (see below) of the works (or a section).

o Ordinarily, the Employer's Agent has no obligation to produce a snagging list; particularly if he is an architect. The *RIBA Good Practice Guide: Inspecting Works* gives the following guidance[137]:

  – At an initial (or early) project team meeting, the Employer's Agent should point out '*...that under the contract, the architect has not duty to prepare "snagging" lists towards the end of the job, and that any such lists compiled by anyone involved in the project should not be considered by the contractor to be exhaustive or definitive*'[138]

  – At completion: '*Under the standard forms of contract the contractor has no right to demand a snagging list from the architect. If the contractor asks the architect to prepare a snagging list, the architect should resist doing so. The architect is not normally paid by the client to prepare such lists*[139] *for the contractor, and to do so can waste a great deal of the architect's time*'.[140]

  – Furthermore, '*...it is the contractor's job to ensure that the work is properly completed and ready to hand over, and that the architect will not be providing snagging lists*'.[141]

When strictly considering the contractor's overriding obligation to complete the works in accordance with the contract under clause 2.1, a refusal to provide a snagging list does not seem particularly controversial. However, contractors, employers, and their consultants should adopt a collaborative approach where appropriate, so it is often easier for both parties for the employer to provide a snagging list, rather than simply telling the contractor that the works haven't reached 'practical completion' and keeping him guessing as to the reasons why.

Clauses 2.30–2.34 provide for the employer taking 'partial possession' of parts of the works or of a section before the Practical Completion Statement (or relevant Section Completion Statement) is issued. In the event that the employer intends to take partial possession, the contractor shall provide the following under clause 2.30:

o His consent to the employer taking partial possession; and

---

137 This guidance is intended where the Employer's Representative is an architect or contract administrator (e.g. under JCT SBC 2016) but is also applicable under JCT DB 2016 where the architect is acting as the Employer's Agent.
138 *RIBA Good Practice Guide: Inspecting the Works,* (2009), p. 69.
139 The RIBA's Standard Professional Services Contract (2020) does not specify any 'snagging lists' under Stage 6 – Handover, in the Schedule of Services, although the 'Lead Designer' may be obliged to '...advise on the resolution of defects' (see also Sections 4.2.3 and 4.2.7).
140 *RIBA Good Practice Guide: Inspecting the Works,* (2009), p. 96.
141 *RIBA Good Practice Guide: Inspecting the Works,* (2009), p. 97.

o A notification to the employer of the following that identifies:
  - The part or parts of which the employer has taken possession (referred to as the 'Relevant Part'); and
  - The date on which the employer took partial possession (referred to as the 'Relevant Date').

Clause 2.31 confirms that when partial possession is taken, 'practical completion' is deemed to have occurred on the Relevant Date for the Relevant Part, and this has the following effects:

o The Rectification Period commences from the Relevant Date for the Relevant Part,[142] and under clause 2.32, the employer must issue a separate notice that all defects have been rectified for each Relevant Part; and
o For the Relevant Part, half of the retention is released[143];
o Clause 2.33 provides that insurance under Options A, B, or C2 shall cease for the Relevant Part; and
o Clause 2.34 provides that any liquidated damages are reduced for the works (or for a section) proportionally to the value of the Relevant Part (against either the Contract Sum, or relevant Section Sum[144]).

The employer should carefully consider his options before seeking to take partial possession. He may be taking possession of a part of the works which, under normal circumstances, would not have met the criteria for practical completion (e.g. the Relevant Part may be uncompleted, or have patent defects present). As soon as the employer takes partial possession, that Relevant Part is deemed to be practically complete, so the contractor has no obligation to complete any outstanding work or rectify any patent defects.

It is entirely possible for the employer to progressively take 'partial possession' of all parts of the works. Whilst this presents the contractor with the advantages of having a shorter Rectification Period for the works overall, progressive and earlier release of the first half of retention, a similar reduction in his insurance liability and reduced liquidated damages, as the employer may never issue either a Practical Completion Statement (or Section Completion Statement), this may affect the contractor's release from any bonds. The contractor will have to ensure that the wording of any bond for which the surety's liability expires at 'practical completion' also provides for its liability to expire in relation to the Relevant Part from the Relevant Date.

9.3.2 FIDIC Yellow Book 2017 – 'Taking Over'

FIDIC Yellow Book 2017 describes 'completion' (in the context of this Section 9.3) in terms of the works reaching a state of readiness, so that they may be 'taken over' by the employer.

---

142 JCT DB 2016, the entry in the Contract Particulars against clause 2.35 states that the Rectification Period is a period of months (if uncompleted, the default period is six months for the works or for a section).
143 JCT DB 2016, see clause 4.18.2 (Retention – amounts and periods).
144 JCT DB 2016, the entry in the Contract Particulars against clause 2.34 sets out the Section Sums.

Clause 10.1 (Taking Over the Works and Sections) is far more specific than JCT DB 2016's description of 'practical completion', as it sets out the various steps in terms of the contractor's obligations for completing the physical work and administrative functions as follows:

o Contractor's obligations for completing the physical works:
 – The works have been completed to the extent that the works[145] can be used safely with minor defects or outstanding work remaining,[146] and comply with the Contract[147];
 – Tests on Completion[148] have been passed (see Section 9.2.2 and Figure 9.1); and
 – The contractor does not have to complete all the work specified in the Employer's Requirements, as (i) the Taking Over Certificate may exclude minor examples of uncompleted work or defects, and (ii) the reinstatement of surfaces or ground.[149]
o Administrative obligations:
 – The engineer has issued (or is deemed to have issued) Notices of No-Objection to the 'as-built' records,[150] and the O&M Manuals[151] (see Section 9.2.2 and Figure 9.1);
 – The contractor has carried out any required training[152];
 – The engineer has issued (or is deemed to have issued) the Taking Over Certificate[153]; and
 – The contractor has notified the engineer of the date when he thinks the works (or a section) are ready for Taking Over (clause 10.1).

The issue (or deemed issue) of the Taking Over Certificate by the engineer formally signifies that the following shall be taken over by the employer:

o The works,
o A section (see Section 8.3.2), or
o A part[154] of the works.

The notification procedure to facilitate Taking Over (or the Engineer's Notice of Rejection) is shown in Figure 9.2.

---

145 FIDIC Yellow Book 2017 defines the 'Works' as the 'Permanent Works' *and* the 'Temporary Works' (clause 1.1.89). 'Permanent Works' (defined in clause 1.1.65) is self-explanatory as works of a 'permanent nature' which the contractor must construct, whereas 'Temporary Works' (defined in clause 1.1.82) are to facilitate construction of the Permanent Works (but exclude 'Contractor's Equipment' – defined in clause 1.1.15).
146 FIDIC Yellow Book 2017, clause 10.1(i).
147 FIDIC Yellow Book 2017, clause 10.1(a).
148 FIDIC Yellow Book 2017, clause 9 (Tests on Completion).
149 FIDIC Yellow Book 2017, clause 10.4 (Surfaces Requiring Reinstatement).
150 FIDIC Yellow Book 2017, clause 10.1(b), see also clause 5.6 (As-Built Records).
151 FIDIC Yellow Book 2017, clause 10.1(c), see also clause 5.7 (Operation and Maintenance Manuals).
152 FIDIC Yellow Book 2017, clause 10.1(d), also see clause 5.5 (Training).
153 FIDIC Yellow Book 2017, clause 10.1(e), 'Taking Over Certificate' is defined in clause 1.1.81.
154 FIDIC Yellow Book 2017, clause 1.1.58 defines a 'Part', also see clause 10.2 (Taking Over Parts).

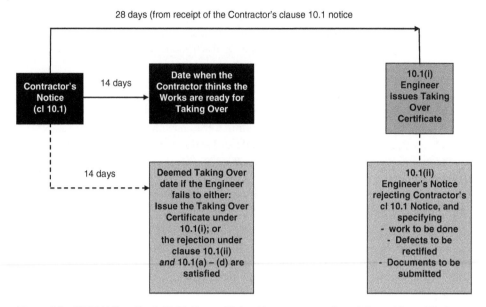

Figure 9.2 FIDIC Yellow Book 2017, Clause 10.1, taking over procedure. Adapted from FIDIC Yellow Book 2017, Clause 10.1.

Clause 10.1 provides the default position that the works (or a section) are 'deemed' to have been taken over if the engineer fails to issue either the Taking Over Certificate,[155] or his rejection[156] of the contractor's clause 10.1 notice within 28 days; in each case, on the proviso that the requirements of clauses 10.1(a) to 10.1(d) have been met.[157]

Similar to 'partial possession' under clauses 2.30 to 2.34 of JCT DB 2016, clause 10.2 of the FIDIC Yellow Book 2017 permits the employer to take over 'Parts' of the works. 'Parts' of the works means part of the works or a section that is being used by the employer or has been deemed to have been Taken Over.[158] Clause 10.2 provides that the employer may take over 'part' of the Permanent Works only[159] and in the following circumstances:

○ Generally, if the engineer has issued a Taking Over certificate for that 'part', then[160]:
  – The employer shall ensure that the contactor may attend to any Tests on Completion (see Section 9.2.2) and any remedial or outstanding work set out in the Taking Over Certificate at the earliest opportunity;
  – The contractor carries out such work before the Defects Notification Period (DNP) expires, and in any event, as soon as is practicable.

---

155 FIDIC Yellow Book 2017, clause 10.1(i).
156 FIDIC Yellow Book 2017, clause 10.1.(ii).
157 See footnotes 147, 150, 151, and 152.
158 FIDIC Yellow Book 2017, clause 1.1.58 defines 'Part'.
159 FIDIC Yellow Book 2017, clause 10.2. This should be contrasted with Taking Over the Works (or a Section) under clause 10.1 which may also include taking over the 'Temporary Works' (i.e. within the definition of 'Works' – see footnote 145).
160 FIDIC Yellow Book 2017, clause 10.2.

- The daily rate of any Delay Damages[161] applicable to the remaining parts (i.e. those not taken over or deemed to have been taken over) is reduced (in proportion to the part taken over).
o If the engineer has not issued a Taking Over Certificate for the relevant 'part' but the employer uses that 'part', then:
  - The contractor notifies the engineer, identifying each 'part' that the employer is using, and describing the nature of its use.
  - The 'part' is deemed to have been 'taken over' as soon as the employer uses it,[162] and the engineer shall immediately issue a Taking Over Certificate[163] for that 'part'.[164]
  - Responsibility for caring for that 'part' passes from the contractor to the employer.[165]

In general, the employer should not hinder the contractor in bringing the works (or a section, or a 'part') to the state when Taking Over can occur. If the employer does hinder the contractor, then the contractor may be entitled to Cost Plus Profit[166] under clause 20.2 (Claims for Payment and/or EoT[167]) in the following circumstances:

o If the employer taking over a 'part' causes the contractor to incur a cost[168]; or
o If he interferes with Tests on Completion – clause 10.3 (see Section 9.2.2).

The final aspect under clause 10 (Taking Over), clarifies that any Taking Over Certificate, whether issued for the works (as a whole), a section, or a 'part', does not certify that the contractor has completed the reinstatement of any surfaces (or ground).[169]

After the Taking Over Certificate has been issued, the contractor should issue his Statement[170] (under clause 14.3) to include the first half of the Retention Money[171] (see Section 8.7.2).

9.3.3 NEC4 'Completion' and 'Taking Over'

Owing to NEC4's ability to be used as a 'hybrid' contract for both 'construction' and 'engineering' projects, it is unsurprising that its procedures for completion contain similar steps to those found in JCT DB 2016 and FIDIC Yellow Book 2017.

Firstly, 'Completion' is a defined in clause 11.2(2) as follows:

o Where the work required to achieve Completion is set out in the Scope;
  - It must all be completed by the Completion Date[172]; and

---

161 FIDIC Yellow Book 2017, clause 10.2 – the maximum amount of delay damages remains unaffected by Taking Over 'parts'.
162 FIDIC Yellow Book 2017, clause 10.2(a).
163 FIDIC Yellow Book 2017, clause 10.2(c) – the Taking Over Certificate for that 'part' shall set out any outstanding work to be completed, and/or defects to be rectified.
164 In the event that the engineer does not issue a Taking Over Certificate for the relevant Part – as provided in clause 10.2(c) – then it appears that the employer's use of a 'part' of the works (or section) is sufficient proof for Taking Over to be deemed to apply – as per clause 10.2(a). .
165 FIDIC Yellow Book 2017, clause 10.2(b).
166 FIDIC Yellow Book 2017, 'Cost Plus Profit' is defined in clause 1.1.20.
167 See 8.10.4.1 and 8.11.3.
168 FIDIC Yellow Book 2017, clause 10.2.
169 FIDIC Yellow Book 2017, clause 10.4 (Surfaces Requiring Reinstatement).
170 FIDIC Yellow Book 2017, clauses 14.9(a) or 14.9(b).
171 FIDIC Yellow Book 2017, clause 14.3(iii).
172 Similar to the standard required to achieve 'practical completion' in JCT DB 2016.

 – No notified defects remain which would prevent the client (or others[173]) from using the works.[174]
 o However, if the Scope does not describe the work that the contractor must complete by the Completion Date, then the contractor must carry out all the work 'necessary' for the client (or others) to use the works.

Clause 30.2 provides that the project manager decides when the 'Date of Completion' occurs, and that he must certify it within one week.

Similar to the provisions in JCT DB 2016 for 'partial possession' and in FIDIC Yellow Book 2017 (Taking Over Parts), the client may 'take over' the works or parts of them, as provided by clause 35 (Take Over) in the following circumstances:

 o Before 'Completion' the client may take over the works, or 'part' of it except[175]:
   – If stated in the Contract Data[176];
   – When using the works benefits the contractor's method of working; or
   – The reasons for using that 'part' of the works is set out in the Scope; or
 o The client takes over the works not later than two weeks after 'Completion'.[177]

Within one week of the client doing so, the project manager must certify[178]:

 o The date when the client took over the works, or a 'part'; and
 o The extent (or 'part') of the works taken over by the client.

Following Completion, the project manager must provide the following:

 o If Option X16 (Retention) is adopted, then half of the retention must be included in the subsequent assessment of the amount due to the contractor (under clause 50.1)[179] – see Section 8.7.3; and/or
 o If either of the target cost options (C or D[180]) is adopted, the project manager must provide the 'preliminary assessment' of the contractor's share[181] – see Section 9.6.3.

9.3.4 Summary of Procedures and Obligations on 'Completion'

Table 9.6 summarises the procedures and obligations under JCT DB 2016, FIDIC Yellow Book 2017, and NEC4 which occur on 'completion'.

---

173 NEC4, clause 11.2(12) defines 'Others'.
174 Similar to the standard required to achieve 'Taking Over' in FIDIC Yellow Book 2017 (save for the contractor's administrative obligations). It appears that defects that have not been notified and which affect the use of the works would not prevent 'Completion' – however, given the client's (via the supervisor) and contractor's obligations under clause 43.2 (Searching for and notifying defects), it is unlikely that any significant defects would go unnoticed (see Section 8.12.3).
175 NEC4, clause 35.2.
176 NEC4, clause 35.1 – see Section 3 (Time) in Contract Data Part One.
177 NEC4, clause 35.1.
178 NEC4, clause 35.3.
179 NEC4, Option X16 (Retention), clause X16.2.
180 NEC4, Option C (Target Contract with Activity Schedule), or Option D (Target Contract with Bill of Quantities).
181 NEC4, Option C – clause 54.3, or Option D – clause 54.8.

Table 9.6 Summary of obligations/procedures on 'Completion'.

Procedure/Obligation	JCT DB 2016	FIDIC Yellow Book 2017	NEC4
Administrative			
– 'Completion' is recorded in writing	2.27[a], 2.31[b]	10.1(e), 10.1(i)	30.2[c]
– Contractor's Documents do not have to comply with laws/standards enacted *after* Taking Over	—	5.4	—
– Period of retaining contractor's documents is known	—	—	X15.4[d]
Employer's rights and obligations			
– Employer takes over the works	—	—	35.1 (2 weeks after)
– Employer's right to vary the works ceases	—	13.1	—
– Employer (or Engineer) can no longer instruct certain remedial works	2.35.1	7.6	—
Contractor's obligations			
– Contractor to complete 'outstanding work' by date in Taking Over Certificate	—	11.1(a)	—
– Period for rectifying defects begins	2.35	11.1(b)	44.1, 44.2[e]
– Period for Tests on Completion begins	—	12.1	—
Payment			
– Half the retention is released	4.18.2	14.9	X16.2
– Unpaid balance of advanced payment due	—	14.2.3	—
– Engineer's right to withhold IPC ceases	—	14.6.2	—
– Early completion bonus calculated (Option X6)	—	—	X6.1
– Preliminary assessment of contractor's 'share' (Options C and D)	—	—	Option C – 54.1 Option D – 54.7
Delay damages cease to apply[f]	2.29.2	8.8	X7.1
Responsibility for the Works / insurance			
– Contractor's obligation to insure ceases	2.33, 6.7.2	17.1, 19.2	80.1
– Contractor ceases having to provide access, facilities etc, guarding/watching the Works	—	4.8(f)(i)	—
'End of liability' date is known (Option X18)	—	—	X18.6[g]

(Continued)

Table 9.6 (Continued)

Procedure/Obligation	JCT DB 2016	FIDIC Yellow Book 2017	NEC4
Reporting / monitoring / forecasting			
– No further programmes are provided	—	—	32.2
– Monthly progress reports cease	—	4.20	15.1

– Delay notification / early warnings cease:
  (i) EWN delaying completion
  (ii) EWN meetings
	—	—	15.2

– Contractor's forecasts for Defined Costs cease (Options C, D and E)	—	—	20.4
– Contractor's obligation to submit 'Revised Estimates' ceases	—	14.3	—

– Early contractor involvement (X22):
  (i) Contractor forecasts cease
  (ii) Budget incentive prepared
	—
	—
		—
		—
			X22.2(5)
			X22.7(2)

a) JCT DB 2016, clause 2.27: the Statement of Practical Completion is issued for the Works, or a Sectional Completion Statement for each Section of the Works
b) JCT DB 2016, clause 2.31 provides that "practical completion" of the Relevant Part is deemed to have occurred from the Relevant Date
c) NEC4, clause 30.2 the Project Manager certifies completion
d) NEC4, Contract Data Part One, entry for Option X15 (The Contractor's Design)
e) NEC4, clause 44.2 the *"defect correction period"* begins at Completion
f) Refer to section 9.4 of this book.
g) NEC4, Contract Data Part One, entry for Option X18 (Limitation of Liability)

9.4   Damages for Late Completion

It is common for construction contracts to include a provision for the employer to recover monies (delay damages) from the contractor as compensation for him failing to complete the works by the completion date because of delays for which the contractor is culpable. The employer may incur the following types of losses or additional costs on the project:

o Loss of revenue:
   – Loss of rent for commercial property,
   – Additional costs of financing (e.g. if sales are delayed for a residential project),
   – Loss of operating revenue (e.g. for a power station, chemical processing plant, high-ways or rail project, etc.).
o Loss of amenity for public projects
o Additional, time-related costs for the period of delay:
   – Temporary accommodation costs
   – Additional consultant's fees
   – Additional insurance premiums

The employer usually selects one of the following methods of recovering delay damages when drafting the building contract:

o Unliquidated damages. The employer will have to follow the usual rule applicable to 'general damages' and prove the amount of any losses caused by the contractor's failure to complete the works on time. This can be a laborious process, as the employer may not incur the full extent of his financial liabilities as soon as the contractor fails to complete – the costs may be incurred over time. This can put the employer at a disadvantage, as towards the end of a project, there may not be sufficient money owing to the contractor from which the employer can cover his losses. In such cases, he will have to seek repayment of the shortfall from the contractor.

o Liquidated damages (or liquidated and ascertained damages – LADs). This is the most common method. The employer inserts a sum (often expressed[182] per as £/week, or £/per day) into the contract, so the contractor knows the amount he will have to pay for each day or week that the works remain uncompleted. Of course, prior to entering into the building contract, the contractor may not agree to the initial liquidated damages sum imposed by the employer, so the final sum will be agreed upon following negotiation. The employer cannot just select any sum per week or per day; it has to be a genuine pre-estimate[183] of his likely losses. In addition, when considering the liquidated damages clause and calculating the rate, the employer (and his advisors) should consider '*whether any (and if so what) legitimate business interest is served and protected by the clause, and second, whether, assuming such an interest to exist, the provision made for the interest is nevertheless in the circumstances extravagant, exorbitant or unconscionable*'.[184]

The main advantage to the employer for choosing liquidated damages is that he can immediately deduct the sum for each period (week, day, etc.) once the contractor has failed to complete the works by the completion date, provided he follows any contractual or statutory[185] notification procedures. The main disadvantages of liquidated damages are firstly, that they are an 'exclusive remedy'.[186] This means that the employer cannot claim any additional costs arising from the contractor's failure to complete on time. The second disadvantage (and closely related to the first) is that the employer is stuck with the rate of delay damages set out in the contract and cannot increase it, even if it proves to be insufficient.

Ordinarily, the employer's ability to deduct liquidated damages depends upon the following factors:

o The absence of any culpable delay caused by the employer or, in the event of employer delay, the existence of a contractual mechanism to extend the completion date, so as to preserve the employer's right to claim liquidate damages; and

o The liquidated damages clauses being simple and clear to operate.

---

182 However, for engineering projects that will generate a significant amount of revenue for the employer (such as a chemical engineering plant, an oil refinery, a power station, an oil or gas extraction platform, etc.), the delay damages may be expressed as an amount per hour.

183 *Dunlop Pneumatic Tyre Co Ltd v New Garage & Motor Co Ltd,* AC 79 HL, [1915], at 88.

184 *Cavendish Square Holding BV v Talal El Makdessi*, UKSC 67, [2015].

185 In the UK, complying with the Housing Grants Construction and Regeneration Act 1996 (as amended) s110A(1)(a), s110A(2), and/or s111(3) – see Section 8.7.

186 *Temloc Ltd v Errill Properties Ltd*, 39 BLR 30, (1987), and *Piggott Foundations Ltd v Shepherd Construction Ltd,* 67 BLR 48, (1993).

If the employer causes delay and there is no contractual right to extend the time for completion (or postpone the completion date), then he loses the right to claim liquidated damages.[187] Therefore, it is essential that the building contract contains effective extension of time provisions to preserve the employer's right to liquidated damages, which enable employer delays and contractor delays to be distinguished from one another.

Another key point is that liquidated damages clauses must be simple and clear to operate, as is discussed below:

o Owing to their relatively serious, and potentially draconian nature, liquidated damages clauses are interpreted contra proferentum,[188] which means:

> *'against the profferer', that is, against the person who drafted or tendered the document. If there is an ambiguity in a document to which all other methods of construction have failed to resolve, so that there are two alternative meanings to certain words, the court may construe the words against the party who put forward the document and give effect to the meaning more favourable to the other party'.*[189]

o When considering a JCT building contract with contractor's design,[190] Dyson J illustrated the relative simplicity of calculating liquidated damages:

> *'The reality is that there are only 3 elements to a claim for [liquidated damages]. The first is the date when the Works or Section of the Works ought to have been completed. For the purposes of a deduction of [liquidated damages], this is fixed, for the time being at least, by the giving of a notice under clause 24.1. The second is the date of practical completion of the Works or the relevant Section. The third is the application of the amount or amounts included in the contract for LADs to the period between the two dates just mentioned, a matter of simple arithmetic'.*

o The calculation of liquidated damages can be complicated by matters such as:
  – The employer taking possession of parts of the works or a section[191]; or
  – Inconsistent, incomplete, or ambiguous entries in the contract for (i) the rates of liquidated damages in the contract,[189] or (ii) the sections of the works to which the relevant rates of liquidated damages are to be applied.[192]

Therefore, it is essential that the particular building contract contains provisions for adjusting the calculations for liquidated damages when the employer takes over a part, and that the relevant entries in the Contract Particulars (JCT DB 2016) or Contract Data (FIDIC and NEC4) are fully completed.

---

187 *Percy Bilton v Greater London Council,* 20 BLR 1, [1982], at 13.

188 *Peak Construction (Liverpool) Ltd v McKinney Foundations Ltd,* 1 BLR 111, (1970).

189 *Keating on Construction Contracts,* 10th ed., (2016), at 3-044.

190 *Bovis Lend Lease Ltd v Braehead Glasgow Ltd,* EWHC Technology 118, [2000], at paragraph 23. Whilst the judgement states that the contract was for 'the design and construction of a shopping centre' it is not clear which JCT contract.

191 For the employer taking 'partial possession' under JCT DB 2016 – see 9.3.1, and for the employer taking over 'parts', see Sections 9.3.2 (FIDIC Yellow Book 2017) and 9.3.3 (NEC4). See *Bramall & Ogden Ltd v Sheffield City Council,* 29 BLR 73, (1983).

192 *Vinci Construction UK Ltd v Beumer Group UK Ltd,* EWHC 2196 (TCC), [2017].

9.4.1 JCT DB 2016

The procedures for the employer to deduct or withhold liquidated damages under JCT DB 2016 are more extensive than the '3 elements' discussed in the case of *Bovis v Braehead*,[190] including the employer's notifications intended to verify the contractor's failure to complete on time, and to notify him of that the employer may seek payment or deduction of liquidated damages. In terms of the 'three elements', the relevant Completion Date for the Works (or a Section) will either be as set out in the Contract Particulars,[193] or will be at such later date (or dates) as are fixed by the employer, following the operation of the extension of time procedures (see Section 8.10.3). The second and third elements (practical completion and the rate of liquidated damages in the contract) are self-explanatory.

Where the works are divided into sections, it is essential that the Contract Particulars[194] accurately identify each section. In the unfortunate event that the sections are not accurately described or an element of the works falls outside the sections' descriptions, then the employer may not be able to deduct liquidated damages because of the uncertainty surrounding the parts of the relevant sections that remain uncompleted.[195]

Assuming that the extension of time provisions have been operated correctly and that firstly, there has either been no employer delay, or any employer delay has already been properly accounted for when fixing a later date as the Completion Date(s) and, secondly, the works (or a section) have not reached practical completion[196] by the relevant Completion Date(s), then under clause 2.27, the employer issues a Non-Completion Notice to the contractor.

If the employer intends to deduct liquidated damages from sums that are otherwise due to the contractor, then he must carry out the procedure in clause 2.29, as shown in Figure 9.3.

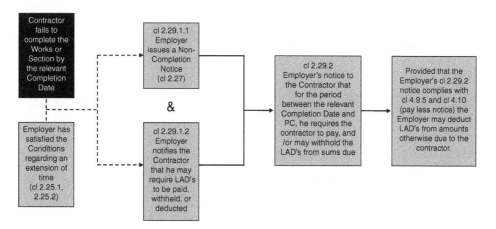

Figure 9.3 JCT DB 2016, procedure for the employer to deduct liquidated damages.

---

193 JCT DB 2016, the entry (Date for Completion of the Works) or entries (Dates for Completion of Sections) against clause 1.1 in the Contract Particulars.
194 JCT DB 2016, the entry (Descriptions of Sections) against the Fifth Recital.
195 *Taylor Woodrow Holdings Ltd v Barnes & Elliot Ltd*, EWHC 3319 (TCC), [2014].
196 See Section 9.3.1.

The clause 2.29 procedure requires the employer to issue the following three essential documents to the contractor:

o *Both* of the following:
  - (Document 1) The Non-Completion Notice (under clause 2.27); and
  - (Document 2) A notification (under clause 2.29.1.2)[197] that the employer may require the contractor to pay liquidated damages to the employer, or the employer may withhold or deduct liquidated damages;

then followed by:

o (Document 3) A notification to the contractor under clause 2.29.2, stating that for the period between the relevant Completion Date(s), and the applicable date of practical completion (for the works or a section(s)), the employer will:
  - Require the contractor to pay liquidated damages[198] (either at the rate in the Contract Particulars, or a lesser rate[199]) for the period described in the clause 2.29.2 notice; and/or[200]
  - Withhold or deduct liquidated damages (at the rates and for the period as mentioned above) from sums due to the contractor.[201]

Provided that the employer's notice either complies with the requirements of a pay less notice (clauses 4.9.5 and 4.10 refer) or any separate payment notice[202] or pay less notice refers to the employer's clause 2.29.2 notice, then the employer may deduct liquidated damages from sums due to the contractor.

In the event that the employer has taken partial possession (see Section 9.3.1) of parts of the works or a section(s), then the rate of liquidated damages applicable to works or a section, other than the 'Relevant Part', shall be reduced in proportion (clause 2.34). The relevant proportion of liquidated damages is calculated as the value of the Relevant Part, compared to either

o For the works as a whole, the Contract Sum[203]; or
o For a section, the Section Sums[204] in the Contract Particulars.

It is therefore essential that the Contract Sum and Section Sums are properly set out in the Contract Particulars, otherwise the employer may not be able to deduct liquidated damages in the event of partial possession.

---

197 JCT DB 2016, the employer's notice under clause 2.29.1.2. must be issued before the due date for the final payment (under clause 4.24.4.5).
198 JCT DB 2016, clause 2.29.2.1.
199 JCT DB 2016, a lesser rate of liquidated damages may be payable under clause 2.34 if the employer has taken 'partial possession' (see 9.3.1).
200 JCT DB 2016. Clause 2.29.2 refers to the employer requiring the contractor to pay liquidated damages 'and/or' the employer withholding or deducting liquidated damages from sums due to the contractor, whereas the employer's clause 29.1.2 notice must state one or other (i.e. payment *or* withholding/deduction).
201 JCT DB 2016, clause 2.29.2.2.
202 JCT DB 2016, clause 4.7.5 refers to a payment notice for interim payments, and clause 4.8 refers to a payment notice for the final payment (under clause 4.24.5).
203 JCT DB 2016, the Contract Sum is set out in Article 2 (clause 1.1).
204 JCT DB 2016, the entry (for Section Sums) is against clause 2.34 in the Contract Particulars.

9.4.2 FIDIC Yellow Book 2017

Clause 8.8 (Delay Damages) of FIDIC Yellow Book 2017 sets out the procedure for the employer claiming delay damages from the contractor. Similar to JCT DB 2016, the rate of 'delay damages' is intended to be a liquidated sum (i.e. a pre-agreed estimate set out in the Contract Data[205]) for the works or each section, payable for 'each day of delay', and is the employer's exclusive remedy for the contractor's failure to complete by the relevant Date for Completion.[206] Clause 8.8 and the Contract Data also contain a provision for a 'cap' on the amount of delay damages that the employer can recover,[207] provided that such a limit is set out therein.

Clause 8.8's mechanism for the employer's deduction of delay damages is far less procedural than its equivalent under JCT DB 2016. The employer (or the engineer) does not have to notify the contractor of any failure to complete the works,[208] so it appears that the employer is entitled to deduct delay damages if the relevant Date for Completion passes *before*:

o The date when the contractor thinks the works (or a section) are ready for Taking Over, as notified to the engineer under clause 10.1 notice (Taking Over);
o The contractor issues his clause 10.1 notice, but the engineer issues his notice of rejection under clause 10.1(ii); or
o The contractor issues his clause 10.1 notice.

At first glance, it appears that he employer is entitled to payment of the delay damages after the first day that the relevant Date for Completion is delayed[209] by the contractor. However, clause 8.8 is very clear that payment of delay damages is subject to clause 20.2 (Claims for Payment and/or EoT – see Section 8.10.4.1). In order not to fall foul of the 28-day period for issuing the 'Notice of Claim' under clause 20.2.1, a prudent employer may issue his Notice of Claim at the following stages before the relevant Date for Completion:

o Within 28 days after receiving:
  – The contractor's revised programme[210] if it shows the 'actual progress'[211] being insufficient to complete the works (or section) by the relevant Date for Completion; or

---

205 FIDIC Yellow Book 2017, the first entry against clause 8.8 in the Contract Data.
206 FIDIC Yellow Book 2017, clause 8.2 (Time for Completion). Clause 8.8 reinforces the importance of the contractor's obligation under clause 8.2, as it clearly confirms that the deduction (or payment) of delay damages does not relieve the contractor of his obligation to complete the works or from carrying out any other obligations, duties, or responsibilities (either under or in connection with the contract).
207 FIDIC Yellow Book 2017, the second entry against clause 8.8 in the Contract Data. This may be expressed as a sum of money, or as a percentage of the Contract Price. However, clause 8.8 confirms that the 'cap' on delay damages does not apply in the event that the contractor commits any fraud, gross negligence, deliberate default, or reckless misconduct.
208 Although the employer may be required to provide an 'advance warning' to the contractor and the engineer under clause 8.4(d).
209 FIDIC Yellow Book 2017, clause 8.8: '…which shall be paid for every day which shall elapse between the relevant Time for Completion and the relevant Date of Completion of the Works or Section'.
210 It would be unlikely that the contractor's first clause 8.3 programme would show completion of the works or section being achieved later than the relevant Date for Completion.
211 FIDIC Yellow Book 2017, clause 8.3(j) refers.

- An 'Advance Warning' from the contractor, advising that there may be a delay in completing the works or a section[212]; or
o 28 days before the relevant Date for Completion of the works or section.

In practice, it should be fairly straightforward for the employer to issue his 'Fully Detailed Claim' (or an Interim Fully Detailed Claim[213] if the delay continues beyond one month) for delay damages under clause 20.2.4 as is shown below:

o The detailed description of the event or circumstance would be the contractor's failure to complete the works (or a section) by the relevant Date for Completion.[214] It would be advisable for the employer (or the engineer on his behalf) to record the contractor's failure to do so in writing[215];
o The employer's statement of the contractual (or legal) basis of the claim would be as provided in clause 8.8 (Delay Damages)[216];
o The 'detailed supporting particulars' should simply consist of a description, identifying the works, section, 'part', or 'parts' that the contractor has failed to complete, and a calculation showing:
  - The rate of the Delay Damages[217] applying to each day of delay, up to the maximum allowable (if applicable)[217]; and
  - Any reduction in Delay Damages if the employer has taken over a part (or parts) of the Permanent Works.[218]

If the project is subject to UK law, the employer should also ensure that any claims for Delay Damages are properly notified to the contractor by either a payment notice[219] or a pay less notice[220] (refer to Section 8.7.2).

9.4.3   NEC4

Unlike JCT DB 2016 or FIDIC Yellow Book 2017, NEC4 does not provide an automatic entitlement for the client to deduct liquidated damages if the contractor fails to complete by the Completion Date or the relevant Completion Date for a Section if Option X5 (Sectional Completion) is selected. The client's right to deduct liquidated sums as delay damages is provided by selecting Secondary Option X7 (Delay Damages).[221] In the absence of selecting

---

212 FIDIC Yellow Book 2017, clause 8.4(d).
213 FIDIC Yellow Book 2017, clause 20.2.6 (Claims of continuing effect).
214 FIDIC Yellow Book 2017, clause 20.2.4(a).
215 FIDIC Yellow Book 2017, clause 20.2.4(c) provides that the 'Fully detailed Claim' shall include all contemporary records on which the employer (in the case of a clause 8.8 Delay Damages claim) relies. As FIDIC Yellow Book 2017 does not require any formal notification of the contractor's failure to complete the works or a section (in the same way as clause 2.27 of JCT DB 2016 requires the employer to issue a Non-Completion Notice – see Section 9.4.1), it would be advisable for the employer to record the contractor's failure in order satisfy clause 20.2.4(c).
216 FIDIC Yellow Book 2017, clause 20.2.4(b).
217 As set out in the Contract Data, against the entry for clause 8.8 (for either the works or for a section).
218 FIDIC Yellow Book 2017, clause 10.2 (Taking Over Parts), reduction of Delay Damages to reflect the 'parts' not taken over.
219 HGCRA 1996 (as amended), s110(1)(a).
220 HGCRA 1996 (as amended), s111(3).
221 NEC4, Contract Data Part One, entry for 'Secondary Options' in section '1 General'.

Option X7, it is thought that the client would be able to deduct unliquidated damages if the contractor was in breach of his completion obligations. Similar to FIDIC Yellow Book 2017, under NEC4 the project manager does not have to notify the contractor of a failure to complete the works by the Completion Date.

Clause X7.1 provides that delay damages are:

o Calculated on a daily basis;
o At the rate (or rates if Option X5 is selected) stated in the Contract Data[222];
o Between the Completion Date[223] and the *earlier* of:
  – Completion[224]; or
  – When the client takes over[225] the works (or section); and
o Are paid by the contractor.[226]

As the project manager assesses[227] the 'amount due' for payment to the contractor under clause 50, it appears that unlike JCT DB 2016 or FIDIC Yellow Book 2017, under NEC4 it is the employer's representative (the project manager), and not the client who calculates the delay damages that the contractor must pay.

Clause X7.2 deals with repayment of delay damages to the contractor in the event of a change in the Completion Date (extension of time) being awarded (refer to Sections 8.9.3.2 and 8.10.5). In addition to repaying the delay damages to the contractor, the client must also pay interest,[228] calculated from the date that delay damages were either deducted from (or paid by) the contractor to the date when they are repaid by the client.

Clause X7.3 reduces the delay damages if the client takes over part of the works (or a section) before Completion. The damages are reduced in proportion to the benefit (as assessed by the project manager) that the client gains for taking over a part, compared to the benefit the client would obtain if he took over the whole of the works. The 'benefit' may be subjective and unclear. For example, if the client took over all of the works with exception of the power supply, it could be argued that he has obtained no benefit at all, because without power, the works cannot be operated. NEC4's *User's Guide* gives what is seemingly meant to be a simple assessment of the proportion of benefit the client may obtain and how this is applied to the daily rate of delay damages.[229] However, this example considers many variables for a manufacturing client, including:

o The client's forecast profit for each of two sections of the works (assuming Option X5 applies); and
o The actual profit that the client would obtain for using each section as at the Completion Date.

---

222 NEC4, Contract Data Part One – the relevant entries for 'delay damages' are found under 'X7: Delay Damages', including when Option X5 (Sectional Completion) is also selected.
223 NEC4, clause 11.2(3).
224 NEC4, clause 11.2(2).
225 NEC4, clause 35 (Take over).
226 NEC4, clauses 50.3 and 50.4 provide that the 'amount due' must account for '…amounts to be paid by or retained from the Contractor'.
227 NEC4, clause 50.2.
228 NEC4, clause 51.4 and the relevant entry for the 'interest rate' is set out in Contract Data Part One, in '5 Payment'.
229 NEC4 *User's Guide*, (2020), p. 88.

Given that liquidated sums for delay damages are intended to simplify matters by providing a pre-agreed sum for each period of delay (per day in the case of NEC4) that, for convenience, is accepted by the contractor as the amount he must pay, it seems peculiar to apply a subjective and significantly more complicated calculation of what may be an 'intangible' benefit to the client, in order to proportionally reduce the delay damages to reflect the work taken over by the client. Considering the variable elements of NEC4's example (above), there is simply more for the parties to argue over in terms of disputing the project manager's assessment under clause X7.3. It would be far more sensible to adopt a similar and far less subjective approach based upon the value of the works taken over (as provided by clause 2.34 of JCT DB 2016 and clause 10.2 of FIDIC Yellow Book 2017 – see Section 9.4.1).

9.4.4 Damages for Late Completion Summary

The delay (or liquidated) damages provisions for the three design and build contracts are summarised in Table 9.7.

Table 9.7 Summary of damages for late completion.

Delay (liquidated) damages provision	JCT DB 2016	FIDIC Yellow Book 2017	NEC4
Liquidated (delay) damages apply	Clause 2.29	Clause 8.8	Option X7
Unliquidated (common law) damages apply	No	No	Not selecting Option X7
Employer has to notify failure to complete	Yes, clause 2.38	No	No
Employer must notify contractor (in addition to a payment/pay less notice)	Clause 2.29.1.2 Clause 2.29.2	No	No
Liquidated damages are an exclusive remedy	Yes	Yes, clause 8.8	Yes
Separate rates of damages for sections	Yes	Yes, clause 8.8	Yes, Options X5 and X7
Unit of time applying to delay damages	Per week	Per day	Per day
Damages reduced for 'partial possession' or 'taking over parts'	Clause 2.34	Yes, clause 10.2	Yes – X7.3
Limitation of delay (liquidated) damages	No	Yes, clause 8.8	No – X18.5

9.5 Rectification or Completing Outstanding Works After Completion

Section 9.3 set out the requirements under each of the three design and build contracts for achieving completion. The completion 'bar' is set highest for JCT DB 2016, which provides that the contractor must effectively complete all of the works, save for only minor (de minimis) items. A more practical approach is taken by FIDIC Yellow Book 2017 and

NEC4, as completion depends upon whether the employer (or client in the case of NEC4) can safely use the works, despite some defects and uncompleted work remaining.

Given the nature of construction and process engineering work, the three design and build contracts envisage that remedial work will be required after the project has reached 'completion'. This is perfectly sensible, and is more a reflection of the works 'settling' over the next 6 or 12 months (for example, being subject to seasonal changes in weather and temperature) or being operated for a similar period of time, rather than an assumption that to some degree the contractor's works will always be defective.

9.5.1 JCT DB 2016

JCT DB 2016 does not strictly permit any uncompleted work to remain after practical completion; it only allows minor (de minimis) defects to remain (see Section 9.3.1). The contractor's obligations to correct defects after completion will include making good:

o Any minor (or de minimis) defects that would not otherwise prevent practical completion; and
o Any 'defects, shrinkages or other faults' found in the works or a section (collectively referred to herein as 'defects') at his own cost.

Clause 2.35 of JCT DB 2016 sets out the procedure for the contractor to 'make good' (or rectify) any defects after practical completion during the Rectification Period. The Rectification Period[230] is set out in the Contract Particulars against clause 2.35 as follows:

o As a single period of months commencing from the date of practical completion of the works; or
o As individual periods of months from the date of practical completion for each section. The individual Rectification Periods could all be different (e.g. if a section consisted of the mechanical and electrical works, this may have a longer Rectification Period than the period or periods for the construction work); and
o If the parties fail to state a period (or periods for each section, where the work is to be completed in sections), then Contract Particulars provides a default period of six months.

These are quite wide categories of defect,[231] which may include work that is not in compliance with the specification (a defect), or initially compliant work that has subsequently been affected (e.g. by external conditions [such as shrinkages] or operating errors in equipment – other faults).

At any time during the Rectification Period (or Periods), the employer may instruct the contractor to make good any defects in the works.[232] The contractor's final obligation

---

230 JCT DB 2016, clause 1.1 defined the 'Rectification Period'. It was previously referred to as the 'Defects Liability Period' in previous editions up to JCT Standard Form of Building Contract 1998 Edition with Contractor Design.
231 'Defect' is defined as 'shortcoming, imperfection or lack', *Oxford Dictionary of English*, (2010),and 'Something that renders an object less than the specification, not as required, and not perfect. Defects include scratches to surfaces, poor workmanship, error, or damage to the work', *Oxford Dictionary of Construction, Surveying & Civil Engineering,* 2nd ed., (2020).
232 JCT DB 2016, clause 2.35.2.

to rectify defects under the terms of the JCT DB 2106 contract comes in the form of the employer's schedule (which acts as an instruction for the contractor to rectify), which must be provided by the employer not later than 14 days after the expiry of the relevant Rectification Period.[233] The contractor is obliged to make good all defects within 'a reasonable time after receipt of such schedule or instruction'.[234] What is a 'reasonable time' will depend upon the facts and the circumstances of each defect, so the contractor may be obliged to rectify an urgent defect (such as a significant roof leak) far sooner than for a defect that is far less urgent (such as a loose external fence panel). It is common for an employer to amend this provision to either oblige the contractor to rectify a defect within the time specified in the employer's clause 2.35 instruction or schedule or to specify the period in which a contractor has to rectify a particular category of defect (see Section 11.2).

In the alternative, clause 2.35 permits the employer to accept a defect but abate (or reduce) the contract price by instructing the contractor accordingly, and making an appropriate deduction from the Contract Sum. As it is the employer who is making the decision to accept a defect (it is the employer who issues the instruction), there is less risk for any consultants engaged by the employer, who could otherwise be liable for any consequences of the defect's continued presence.[235] However, it would be advisable for the employer and any consultants to carefully consider the effect a remaining defect may have before issuing such an instruction (for example, there should not be any detrimental effects to the building itself or to its use or maintenance under the CDM Regulations 2015). The contractor should also insist that the employer's instruction reduces the specification accordingly, so the contractor will not be in breach – so the instruction has a similar effect to clause 45 of NEC4, where the Scope is changed so the defect does not have to be corrected.[236] Additionally, the contractor may insist upon an indemnity, absolving him of any responsibility from the employer for accepting the defect.

Clause 2.36 provides that the employer issues a 'Notice of Completion of Making Good' stating the date upon which the contractor has completed making good any defects that the employer has required him to rectify. As the employer may issue the Notice of Completion of Making Good at the time when he is satisfied that the contractor has carried out the remedial work he requires, it could be issued at any time, either before or after the Rectification Period has expired. It is usually the case that the employer will hold the contractor to his obligations for the entirety of the Rectification Period, so it is unlikely that the Notice of Completion of Making Good will be issued beforehand. The employer is not entitled to retain retention from any interim payments after the Notice of Completion of Making Good has been issued.[237] Issuing the Notice of Completion of Making Good is also one of the three events that triggers the due date for the final payment (refer to Section 9.8.1).

---

233 JCT DB 2016, clause 2.35.1.
234 JCT DB 2015, clause 2.35.
235 Contrast with clause 2.38 of the JCT SBC 2016, where the architect/contract administrator instructs the contractor not to make good a defect in return for a reduction in the Contract Sum (albeit, with the employer's consent).
236 NEC4, clause 45. In return for the project manager (and obviously the client) agreeing to reduce the Scope in order to accept a defect, the contractor submits a quotation to reduce the 'Prices' and/or achieve an earlier Completion Date for the project manager to accept.
237 JCT DB 2016, clause 4.18.2.1.

9.5.2 FIDIC Yellow Book 2017

FIDIC Yellow Book 2017 contains the following extensive provisions that apply after 'Taking Over':

o Clause 11 – Defects after Taking Over, and
o Clause 12 – Tests after Completion.

Clause 11 sets out mandatory obligations for the contractor to remedy defects after completion, whereas any further testing under clause 12 is optional and is carried out by the employer (with limited involvement from the contractor).

The contractor's obligations under clause 11 and (if applicable) clause 12 last for the duration of the Defects Notification Period[238] (or DNP), until the Performance Certificate[239] is issued.

Clause 11.1[240] obliges the contractor to carry out the following work[241]:

o Complete any uncompleted work[242] that was set out in the Taking Over Certificate, either within:
   – The time specified in the Taking Over Certificate, or
   – A reasonable period as instructed by the engineer.
o Rectify any defect or damage for which notice is given[243] to the contractor:
   – That the contractor must complete before the expiry of the DNP. It is implied that such defects or damage are discovered and notified to the contractor before the Taking Over Certificate (and listed therein[244]); or
   – Any defect or damage, appearing or occurring during the DNP. Following an inspection[245] by the Employer's Personnel[246] and the contractor, the contractor submits a proposal[247] for the remedial work. The procedure in clause 7.5 (Defects and Rejection) then applies[248] (see Section 8.12.2).
   – All work carried out by the contractor to rectify any defect or damage is carried out at the contractor's cost and risk,[249] as per clause 11.2 (Cost of Remedying Defects) unless

---

238 FIDIC Yellow Book 2017, 'Defects Notification Period' or 'DNP' is defined in clause 1.1.27. However, see clause 11.3 (Extension of Defects Notification Procedure). The Defects Notification Period (or DNP) is entered in the Contract Particulars against clause 1.1.27 and is expressed as a period of 'days'.
239 FIDIC Yellow Book 2017, 'Performance Certificate' is defined in clause 1.1.62.
240 FIDIC Yellow Book 2017, clause 11.1 (Completion of Outstanding Work and Remedying Defects).
241 FIDIC Yellow Book 2017, clause 11.1, excluding any 'fair wear and tear'.
242 FIDIC Yellow Book 2017, clause 11.1(a).
243 FIDIC Yellow Book 2017, clause 11.1(b), and clause 11.1 (fourth paragraph) – the notice to the contractor is given by the employer (or on the employer's behalf).
244 FIDIC Yellow Book 2017, clause 10.1(i).
245 FIDIC Yellow Book 2017, clause 11.1(i).
246 FIDIC Yellow Book 2017, 'Employer's Personnel' is defined in clause 1.1.32.
247 FIDIC Yellow Book 2017, clause 11.1(ii).
248 FIDIC Yellow book 2017, clause 11.1(iii) – Clause 7.5 (Defects and Rejection) applies, save for paragraph 1 of that clause.
249 FIDIC Yellow Book 2017, if any such remedial work under clause 11.1(b) is attributable to the contractor's design (clause 11.2(a)), workmanship, materials, or plant not in accordance with the contract (clause 11.2(b), incorrect maintenance or operation for which the contractor is responsible under clauses 5.5 (Training), 5.6 (As-Built Records), and 5.7 (Operation and Maintenance Manuals) (clause 11.2(c)), or any other failure by the contractor to comply with obligations under the contract (clause 11.2(d)).

it is attributable to any design for which the employer was responsible.[250] If the contractor considers that the remedial work (under clause 11.1(b)) arose from any other cause than listed in clause 11.2, he can notify the engineer to proceed with a clause 3.7 determination (see Section 9.7.2.1). If the engineer's determination concurs that the remedial work arose from any other cause, the remedial work is treated as if it was instructed under clause 13.3.1 (Variation by Instruction);

- With the employer's consent, the contractor may remove items of plant to rectify any defect or damage off site, if the remedial works can be carried out more 'expeditiously' than if the repairs are carried out in situ.[251] In order to do so, the contractor must firstly notify the engineer, providing the details required by clauses 11.5(a) to 11.5(f),[252] along with any further details the engineer may reasonably require. The employer may require the Performance Security[253] (see Section 8.1.2) to be increased to cover the costs of replacing the relevant plant; and

- Within seven days of completing the relevant remedial works, the contractor shall notify the engineer, firstly giving details of the remedied works,[254] and secondly proposing the particular tests to comply with clause 9 (Tests on Completion – see Section 9.2.2) or, if applicable, clause 12 (Tests after Completion). The engineer should then notify the contractor if he accepts the proposed testing[255] or instructs the repeated tests[256] he feels are necessary to ensure the remedied works comply with the Contract.[257] He must do this within seven days after receiving the contractor's notice. If the contractor fails to provide his clause 11.6 notice within seven days, the engineer notifies the contractor, instructing him to carry out the necessary repeated tests.[258]

It should be noted that clause 10.1 only requires 'defects' and not 'damage' to be listed in the Taking Over Certificate. It would therefore be advisable for the engineer to notify[259] the contractor of any damage (that would not prevent taking over) either on or before issuing the Taking Over Certificate.

---

250 FIDIC Yellow Book 2017, clause 11.2(a).
251 FIDIC Yellow Book 2017, clause 11.5 (Remedying Defective Work Off Site).
252 FIDIC Yellow Book 2017, clause 11.5 requires the contractor's notice to identify (i) the damage or defect to be repaired, (ii) the proposed repair's location, (iii) method of transport and associated insurance cover, (iv) off-site inspections and testing, (v) the duration until the plant is returned to site, and (f) the planned duration for re-installation and any applicable re-testing complying with clauses 7.4 (Testing by the Contractor) and clause 9 (Tests on Completion).
253 FIDIC Yellow Book 2017, 'Performance Security' is defined in clause 1.1.64.
254 FIDIC Yellow Book 2017, clause 11.6 (Further Tests after Remedying Defects) including a 'Section', 'Part' and/or 'Plant'.
255 FIDIC Yellow Book 2017, clause 11.6(a).
256 FIDIC Yellow Book 2017, clause 11.6 provides that any 'repeated' tests must be carried out in the same way as the previous tests, at the risk and cost of the party who is liable under clause 11.2 (Cost of Remedying Defects).
257 FIDIC Yellow Book 2017, clause 11.6(b).
258 FIDIC Yellow Book 2017, clause 11.6 provides that in the event of the contractor's failure to issue his clause 11.6 notice within seven days, the engineer's notice (and instruction) may be issued 14 days after the contractor has completed the relevant remedial work.
259 FIDIC Yellow Book 2017, the engineer must issue a 'Notice' (as defined in clause 1.1.56).

Clauses 11.7 (Right of Access after Taking Over) and 11.8 (Contractor to Search) facilitate the identification, resolution, and discovery of defects, along with their responsibility as follows:

o Clause 11.7 entitles the contractor to access to the works in order to observe how the employer is using the work, along with inspecting any records of their operation, performance, and maintenance. This is an important right, as any remedial work due to 'improper operation or maintenance'[260] by the employer would mean that the rectification work would not be at the contractor's risk and cost (under clause 11.2), and he would be entitled to a variation instruction (under clause 13.3.1). The contractor's right of access is as follows:

  – It lasts until 28 days after the engineer issues the Performance Certificate[261];
  – It is subject to the contractor's notice to the employer,[262] which must set out the contractor's reasons for requiring access, and giving reasonable notice of the date(s) on which access is preferred;
  – It is given by either (i) the employer's Notice,[263] consenting[264] to the contractor's request which must be provided within seven days, or (ii) failing that, the employer is deemed to have provided consent[265]; and
  – It is subject to the employer's security restrictions (e.g. if the works are a secure installation such as a nuclear power station).

  Subject to the contractor making a successful 'Claim' under clause 20.2, the employer must pay Cost Plus Profit to the contractor if he unreasonably delays providing access to the contractor.

o Clause 11.8 permits the engineer to instruct the contractor to search for 'the cause of any defect', which he must do under the engineer's direction and on the dates set out in the engineer's instruction. Provided that the 'defect' is not caused by one of the matters in clause 11.2, for which remedial works are not carried out at the contractor's risk and cost, then the contractor may submit a claim under clause 20.2 (see Section 8.10.4.1) for payment of Cost Plus Profit[266] for carrying out the search. If the contractor fails to search for a defect's cause in accordance with the engineer's instruction, then the employer may submit a claim under clause 20.2 for the reasonable costs incurred by the Employer's Personnel for carrying out the search.[267]

Clause 11.3 (Extension of the DNP) entitles the employer to extend the DNP by up to a further two years if the damage or defect[268] prevents the works (including a section or a

---

260 FIDIC Yellow Book 2017, clause 11.2(c).
261 FIDIC Yellow Book 2017, clause 11.9 (Performance Certificate).
262 FIDIC Yellow Book 2017, clause 11.7(a).
263 FIDIC Yellow Book 2017, clause 11.7(b).
264 FIDIC Yellow Book 2017, clause 11.7(b)(i).
265 FIDIC Yellow Book 2017, clause 11.7(b)(ii).
266 FIDIC Yellow Book 2017, 'Cost Plus Profit' is defined in clause 1.1.20.
267 FIDIC Yellow book 2017, clause 11.8 requires that the employer issues a notice to the contractor, stating the date(s) when the Employer's Personnel will be carrying out the search or searches.
268 FIDIC Yellow Book 2017, clause 11.3, provided that the 'defect or damage' is attributable to any of the matters in clause 11.2 for which the contractor is responsible.

part) or a 'major item of Plant'[269] being used for its intended purpose, provided he complies with the 'Claim' procedure in clause 20.2 (see Section 8.10.4.1). The extension will not apply where the works have been suspended under clause 8.9 (Employer's Suspension) unless the contractor caused the reason for suspension, or clause 16.1 (Suspension by Contractor).

Clause 11.4 (Failure to Remedy Defects), provides the employer with wide-ranging remedies if the contractor either delays or fails to remedy any damage or defects (for which the contractor is responsible under clause 11.2). Firstly, and in the event of such failure or delay by the contractor, the employer must notify the contractor of a date upon which he must rectify the damage or defects (allowing the contractor a reasonable time in which to do so). Secondly, and in the event of the contractor's failure to rectify the damaged or defective work in accordance with the employer's clause 11.4 Notice, the employer has the following options:

o Undertake the remedial works, either by himself of by engaging others.[270] The employer should carefully consider this option as the contractor will no longer be responsible for the rectified work, which may have consequences if it could invalidate any warranties or guarantees provided by the contractor.
o Accept the damaged or defective work in return for financial compensation, such as:
  – The contractor paying Performance Damages,[271] or
  – A reduction in the Contract Price.[272]
o Immediately terminate the Contract with the contractor, and elect to dismantle and clear the site and return all plant and materials to the contractor.[273] This is a very extreme measure and is only likely if the employer has no realistic prospect of using the works (or a significant element of them) for their intended purpose.
o Treat the unremedied work as an omission as if instructed under clause 13.3.1 (Variation by Instruction), if the works (or any part of it) cannot be used for its intended purpose.[274]

In the first three options, the employer's claim for reasonable costs of completing the remedial works,[270] claiming Performance Damages,[271] a reduction in the Contract Price,[272] or recovery of sums arising from termination (and any ancillary measures[275]), is subject to the 'Claims' procedure in clause 20.2. The employer's rights when effectively 'omitting' the works or terminating the Contract are without prejudice to any other rights (under the Contract or otherwise) that he may have.

---

269 FIDIC Yellow Book 2017, whilst 'Plant' is defined in clause 1.1.66, clause 11.3 does not give any guidance on what could be a 'major' item of plant. What constitutes a 'major item of plant' could therefore be subjective and a source of disputes.
270 FIDIC Yellow Book 2017, clause 11.4(a).
271 FIDIC Yellow Book 2017, clause 11.4(b)(i). 'Performance Damages' are defined in clause 1.1.63.
272 FIDIC Yellow Book 2017, clause 11.4(b)(ii).
273 FIDIC Yellow Book 2017, clause 11.4(d) – clause 15.2 (Termination for Contractor's Default) does not apply.
274 FIDIC Yellow Book 2017, clause 11.4(c).
275 FIDIC Yellow Book 2017, clause 11.4(d) – costs include all sums paid for the works, and financing charges.

The final testing procedures for the works are optional, and if used are contained in clause 12 (Tests after Completion). However:

o Testing under clause 12 is optional, as it must be specified in the Employer's Requirements[276] (unlike testing under clause 7.4 [Testing by the Contractor], clause 9 [Tests on Completion], or clause 11 [Defects after Taking Over]).
o The testing is carried out by the employer[277] (and not the contractor), albeit with some limited involvement from the contractor who may attend.[278] Clause 12 therefore envisages that the employer has been using the works (or a section or part) during the DNP (i.e. after the engineer has issued the Taking Over Certificate[279]), and the contractor is either in the process of or has:
  – Finished any items of uncompleted work, and has rectified any defects and damage under clause 11.1 (Completion of Outstanding Work and Remedying Defects) whether on, or off site[280]; and
  – Completed any further tests under clause 11.6 (Further Tests after Remedying Defects).

The employer therefore must:

o Include the following information in the Employer's Requirements:
  – Specify each test to be carried out under clause 12;
  – Identify the timing of the clause 12 tests[281] (as the contractor's clause 8.3 programme[282] will not normally show any activities after the Time for Completion[283] has expired, and his clause 9.1 programme will only show details of Tests on Completion).
o Provide sufficient facilities and resources to undertake the tests.[284]
o (Via the engineer) give 21 days' prior notice to the contractor of the following[285]:
  – The date and location of the clause 12 tests; and
  – A programme showing the clause 12 testing's predicted duration.
o When carrying out the tests, ensure they comply with:
  – The Employer's Requirements,[286]
  – The O&M Manuals,[287] and
  – Any guidance that the contractor is required to give.[288]

---

276 FIDIC Yellow Book 2017, clause 12.1 (Procedure for Tests after Completion).
277 FIDIC Yellow Book 2017, clause 12.1(a).
278 FIDIC Yellow Book 2017, clause 12.1. Both the employer and the contractor may request the contractor attends the clause 12 testing.
279 FIDIC Yellow Book 2017, clause 10.1.
280 FIDIC Yellow Book 2017, clause 11.5 (Remedying of Defective Work Off Site).
281 FIDIC Yellow Book 2017, clause 12.1 provides that if the Employer's Requirements do not specify the 'timing' of any clause 12 tests, they are to be carried out 'as soon as is reasonably practicable' after the works (or a relevant part) have been Taken Over.
282 See Section 8.6.2.
283 FIDIC Yellow Book 2017, 'Time for Completion' is defined in clause 1.1.86.
284 FIDIC Yellow Book 2017, clause 12.1(a).
285 FIDIC Yellow Book 2017, clause 12.1.
286 FIDIC Yellow Book 2017, clause 12.1(b)(i).
287 FIDIC Yellow Book 2017, clause 12.1(b)(ii), refers to the O&M Manuals (see clause 5.7) for which the engineer has given (or is deemed to have given) a Notice of No-Objection (see clause 5.2.2).
288 FIDIC Yellow Book 2017, clause 12.1(b)(iii).

o Carry out the tests during the DNP (or within any other period agreed with the contractor), otherwise the works (or section) will be deemed to have passed the clause 12 testing.[289]

The contractor's attendance is important, not just for practical reasons (as it gives him the opportunity to facilitate the employer's compliance with clause 12.1(b) above), but it may affect how the test results are treated. In any event, the tests are deemed to be carried out in the contractor's presence, and proper allowance should be made for any prior use by the employer. Whilst the clause 12 test results are 'compiled and evaluated' by both the employer and the contractor, if the contractor does not attend the testing, he is deemed to accept that the test results are accurate.[290]

If the works (or a section) do not pass the Tests after Completion, it appears that the employer has three options:

o Clause 12.3 (Retesting) provides that:
  – The contractor rectifies all damage and defects in accordance with clause 11.1(b); and
  – Further testing under clause 11.6 shall apply;
o Clause 12.4 (Failure to pass Tests after Completion) allows the employer to deduct Performance Damages if they are set out in the Schedule of Performance Guarantees[291]; or
o Additionally under clause 12.4, the contractor may notify the employer, proposing alterations to the works (or a section), whereupon:
  – The employer may notify the contractor of a reasonable time when he may have a right of access to undertake the alterations in the contractor's clause 12.4 notice,[292] and the contractor carries out the same in order to satisfy the relevant test; or
  – If the employer fails to provide his clause 12.4 notice within the DNP, the relevant work is then deemed to have passed the test and the contractor is no longer liable to carry out his proposed alterations.[293]

The parties' rights to claim additional costs are subject to complying with clause 20.2 (Claims) as follows:

o The employer may claim costs for:
  – Retesting under clause 12.3; and
  – Performance Damages under clause 12.4.
o The contractor may claim additional costs if the employer:
  – Delays tests under clause 12.2; and/or
  – Unreasonably delays the contractor carrying out his proposed alterations under clause 12.4.

The contractor's obligation to remedy defects and damages, along with attending any testing after Completion, ends when the engineer issues the Performance Certificate under

---

289 FIDIC Yellow Book 2017, clause 12.2 (Delayed Tests).
290 FIDIC Yellow Book 2017 clause 12.1.
291 FIDIC Yellow Book 2017, 'Schedule of Performance Guarantees' is defined in clause 1.1.74, and is included in the 'Schedules' (attached to the Letter of Tender and included in the Contract).
292 FIDIC Yellow Book 2017, clause 12.4(ii).
293 FIDIC Yellow Book 2017, clause 12.4(iii).

clause 11.9 (Performance Certificate). The Performance Certificate is the only document under the Contract that signifies that the works are accepted,[294] and it must be issued within 28 days of the following:

o The expiry of the DNP, or the last DNP to expire (where sections or parts have been taken over); or
o After the contractor has:
 – Provided all the Contractor's Documents,[295] including the updated as-built records[296] to which the engineer has given a Notice of No-Objection[297]; and
 – Completed all testing[298] and remedying defects required by and in accordance with the Contract.[299]

The contractor should also issue a Statement (under clause 14.3) 'promptly' after the latest of the DNPs, that includes the second half of the Retention Money.[300]

If the engineer fails to issue the Performance Certificate within the 28-day period, clause 11.9 provides that it is deemed to have been issued (by default), 28 days after the date he should have issued it.

The contractor's final obligation is to clear the site as provided by clause 11.11 (Clearance of the Site), which he must do 'promptly' after the engineer issues the Performance Certificate. The contractor must remove and reinstate the items set out in subclauses 11.11(a) to 11.11(c). If the contractor fails to do so within 28 days after the Performance Certificate is issued, the employer may sell or dispose of any of the contractor's items or equipment after that period. Further, or in the alternative, the employer may clear the site himself, at the contractor's cost (subject to clause 20.2). Clause 11.10 (Unfulfilled Obligations) ensures that each party remains liable to the other for any unfulfilled obligations after Performance Certificate is issued. Following the Performance Certificate's issue, both the contractor and employer remain liable for any obligations that were not performed, and the contract remains in force.[301] The contractor's liability for defects or damage to plant is limited to two years after the expiry of the DNP (for plant), unless the law in that particular jurisdiction provides otherwise.[304] In the UK, the Limitation Act 1980[302] ordinarily imposes periods of liability of six years (if the contract is executed under hand), or 12 years (if the contract is executed as a deed), and these periods usually run from when the physical works are completed (or taken over in the case of FIDIC Yellow Book 2017).

---

294 FIDIC Yellow Book 2017, clause 11.9 final paragraph.
295 FIDIC Yellow Book 2017, 'Contractor's Documents' is defined in clause 1.1.14.
296 FIDIC Yellow Book 2017, the 'updated as-built records' that the contractor provides are described in clauses 5.6(b)(i) and 5.6(b)(ii).
297 FIDIC Yellow Book 2017, clause 11.9(a).
298 FIDIC Yellow Book 2017, including any testing under clause 12 (Tests after Completion) specified in the Employer's Requirements.
299 FIDIC Yellow Book 2017, clause 11.9(b).
300 FIDIC Yellow Book 2017, clause 14.9 (Release of Retention Money).
301 FIDIC Yellow Book 2017, clause 11.10 (Unfulfilled Obligations).
302 Limitation Act 1980 c.58.

9.5.3 NEC4

NEC4's procedures that deal with defects after Completion[303] and during 'defects correction period' (or periods) are similar to those for inspecting, searching, testing, and rectifying defects before completion (see Section 8.12.3); the main difference is that after completion, there is a procedure to conclude matters under NEC4's 'Quality Management' provisions.[304]

When the client is using the works (i.e. after 'completion'), he can elect if he wants the contractor to return to correct defects. If the client does want the contractor to correct the remaining notified defects, then the project manager arranges access to the relevant part of the works for the contractor to do so.[305] The contractor must then rectify the defect within the defect correction period, failing which the project manager assesses the cost of the client engaging others to correct the defect, and the Scope is changed to accept the defect.[306]

If the client does not want the contractor to correct the defect, he has the following options:

o He can simply deny the contractor access to correct the defect, in which case, the project manager assesses the amount it would cost the contractor to correct the defect (which the contractor must pay to the client)[307]; or

o The project manager proposes to the contractor that the defect is accepted without it being corrected, whereupon the contractor submits a quotation to reduce the prices.[308]

For each of the above client's options, the Scope is changed so the defect does not have to be corrected.[309]

Clause 44.3 provides that the supervisor issues the Defects Certificate[310] under the following circumstances:

o If there are no notified defects – at the 'defects date'[311]; or

o When the earliest of the following occur:
  – When the last 'defect correction period' has expired, or
  – When the contractor has corrected all notified defects.

The Defects Certificate contains the following information[312]:

o A list of defects (notified by the supervisor) that the contractor has not corrected before the defects date; or

o A statement that no defects were either:
  – Notified by the supervisor, or
  – Remain to be corrected by the contractor; and

---

303 NEC4, clause 11.2(2) defines 'Completion'.
304 NEC4, section '4 Quality Management' containing clauses 40–45.
305 NEC4, clause 44.4 – expresses the contractor's access to correct defects as 'use' of the relevant part of the works that have been taken over.
306 NEC4, clause 46.1.
307 NEC4, clause 46.2.
308 NEC4, clauses 45.1 and 45.2.
309 NEC4, clauses 45.1, 45.2, and 46.2.
310 NEC4, clause 11.2(7) defines the 'Defects Certificate'.
311 NEC4, the 'defects date' is set out in Contract Data Part One, under the entry against section '4 Quality Management' and is expressed as a period of time after Completion.
312 NEC4, clauses 11.2(7) and 44.3.

Table 9.8 Summary of rectification provisions after 'completion'.

Defects rectification provision	JCT DB 2016	FIDIC Yellow Book 2017	NEC4
Contractor rectifies defects after 'completion'	Clause 2.35	Clause 11.1	Clause 44.2
Contractor can rectify work off site	No	Clause 11.5	No
Contractor to search for defects	No	Clause 11.8	Clause 43.1
Contractor completes outstanding work after 'completion'	No	Clause 11.1	No[a]
Further testing of rectified work	No	Clause 11.6 Clause 12	No
Employer (or employer's representative) certifies completion of rectification work	'Notice of Completion of Making Good' Clause 2.36	Performance certificate Clause 11.9	Defects certificate Clause 44
Contractor's entitlement to the second half of retention after completion of defects: – Employer pays – Contractor must claim	Yes, clause 4.18.2 No	No Yes, clause 14.9	Yes, X16.2 No
Contractor rectifies work at his own cost	Yes, clause 2.35	Clause 11.2	Clause 44.2
Contractor has access to the works, and records of operation and maintenance after completion	Yes[b]	Clause 11.7	Clause 44.4
Contractor pays 'damages'	No.	Clause 11.4(b)(i)	X17.1
Employer may accept defect and abate the contract price	Yes, clause 2.35	Clause 11.4(b)(ii)	Clauses 45.1, 45.2, and 46.2
Employer can extend the period for rectifying defects	No	Yes Clause 11.3	No[c]
Employer can remedy defects himself if the contractor fails to	Yes, clause 3.6	Yes Clause 11.4. (a)	Clause 46.1
Employer can terminate the contract if the contractor fails to rectify defects[d]	Yes, clause 8.4.1.3	Clause 11.4(d)	Clause 91.2(R11)

a) NEC4, unless the uncompleted work falls outside what the Scope requires to be done by the Completion Date as per clause 11.2(2).

b) JCT DB 2016, the contractor's right of access is implied.

c) NEC4, whilst client (employer in NEC4) has no power to extend the defect correction period, fresh defects correction periods can apply if a defect is notified after Completion (clause 44.2).

d) Refer to Table 9.1.

o If Option X17 (Low Performance Damages) applies, in respect of each relevant defect, it shall state the achieved level of performance (compared with the performance level stated in Contract Data Part One[313]), and where the achieved performance level is lower, the contractor shall pay the performance damages set out in the Contract Data.

If Option X16 (Retention) applies, then following the Defects Certificate, the project manager assesses the amount due[314] to include the remaining amount of retention.[315]

9.5.4 Summary of Rectification Procedures After 'Completion'

A summary of the three contracts' requirements for the contractor to rectify defects (and other works) after completion is summarised in Table 9.8.

9.6 Final Account

The term 'final account' is most commonly used to describe the process to arrive at the final sum that the contractor is entitled to be paid in respect of carrying out and completing his obligations under the building contract. It should be contrasted against the contractor's 'interim' account for the works during their construction, and perhaps for a period after 'completion' (see Section 9.3). The majority of the contractor's entitlement to payment will comprise the contract price, adjusted for the following:

o Additions to the contract price:
  - Variations (see Sections 8.8 and 8.9);
  - Additional costs, such as loss and expense (see Section 8.11);
  - Price fluctuations (see Section 8.9);
  - 'Contractor's share' (gain share) paid by the client to the contractor (see Section 9.6.3).
o Reductions of the contract price:
  - Omitted works;
  - Any deductions from or abatement of the contract price for defective work;
  - 'Contractor's share' (pain share) paid by the contractor to the client (see Section 9.6.3);
  - Damages to be paid by the contractor – such as delay damages (see Section 9.4) or low-performance damages (see Sections 9.5.2 and 9.5.3).

Quite obviously, the contractor's final account should be assessed when as many of the matters that occur after completion have been resolved. For each party, these are typically as follows:

o The contractor:
  - Provides all documents necessary for finally adjusting the contract price;
  - Completes all the rectification works (including and testing and re-testing) that he is obliged to undertake after completion.

---

313 NEC4, Contract Data Part One, entries against 'X17: Low Performance Damages' specifying the amount of damages, and the applicable performance level.
314 NEC4, clause 50.1 (Assessing the amount due).
315 NEC4, Option X16 (Retention), clause X16.2. See Section 8.7.3.

o The employer:
  - Issues the formal statement of 'completion';
  - Provides an assessment of the contractor's final account documents, or in the event that the contractor has failed to provide them, a notification that the employer will assess the contractor's final account in the continued absence of the contractor's documents;
  - Provides a formal statement that any testing, re-testing, and rectification works have been completed; and
  - Produces a formal, final document (such as certificate, notice, or statement) that concludes the various aspects (including the financial aspects) of the contract.

In order for both the contractor and the employer to assess the final account's value, the ascertainment process ought to take place at some point after the works are completed. This will allow the contractor to receive and collate the relevant information from his subcontractors and suppliers, and to present this to the employer. It also allows the employer to assess the works and any defects that may remain – so he can decide whether to accept them in return for reducing the contract price by an appropriate amount.

9.6.1 JCT DB 2016

The final account procedure under JCT DB 2016 commences after practical completion either of the works or, if there is completion by sections, of the last section. It is essentially a two-stage process involving:

o The submission of the Final Statement:
  - By the contractor (in which case it must include all supporting documents that the employer may reasonably require) within three months after practical completion[316]; or
  - By the employer, if (i) the contractor fails to submit the Final Statement within three months after practical completion, and (ii) if the contractor's failure continues for two months after the employer notifies him; and
o The final payment becoming conclusive evidence of the following financial matters comprising the 'sum due' under clause 4.24.2:
  - The Contract Sum as adjusted[317]; and
  - Any direct loss and/or expense (see Section 8.11.2).[318]

Figure 9.4 summarises the procedure for submitting the Final Statement.

The procedure for submitting the Final Statement (by the contractor) or, in the event of the contractor's continued default, the employer's Final Statement (either of which is referred to as the 'relevant Final Statement') is set out in clause 4.24, and commences after practical completion – of either the works or, if applicable, the last section. The three-month period for the contractor to submit the Final Statement is shorter when compared to JCT

---

316 JCT DB 2016, clauses 4.24.1 and 4.24.3.
317 JCT DB 2016, clause 4.24.2.
318 JCT DB 2016, clause 1.8.1.3.

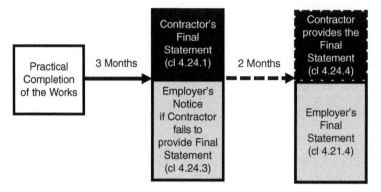

Figure 9.4 JCT DB 2016, procedure for submitting the Final Statement.

SBC 2016, where it is six months.[319] This is based upon the premise that under JCT DB 2016, there are fewer opportunities for variations that require the Contract Sum to be adjusted.

Under clause 4.24.3, if the contractor fails to provide the Final Statement within three months, then the employer must notify him of this failure. It is advisable for the employer to issue this notice immediately upon the expiry of the three-month period, as the two further months for the contractor to provide the Final Statement, run from the date of the employer's notice, and not the expiry of the period of three months following practical completion. It appears that it is only the contractor's failure to submit the Final Statement in three months that triggers the employer's entitlement to issue the clause 4.24.3 notice, and not any failure to provide 'supporting documents' (see below).

Clause 4.24.2 sets out that the Final Statement (or the employer's Final Statement) must show the 'balance' that is either due to the contractor (from the employer), or from the employer (to the contractor) as the case may be,[320] calculated as the difference between:

o The Contract Sum as finally adjusted under clause 4.2. This comprises all adjustments as set out below:
  – Amounts for Changes, either agreed between the employer and the contractor, or valued in accordance with the Valuation Rules in clause 5.2 (see Section 8.9.1)[321] (referred to as a 'Valuation'[322]);
  – Amounts agreed under a Confirmed Acceptance[323] of an Acceleration Quotation[324] (see Section 8.5.1)[325];
  – If Fluctuations apply, any price adjustments arising therefrom (see Section 8.9.1)[326];

---

319 JCT SBC 2016, clause 4.24.1.
320 JCT DB 2016, clause 4.24.2. It is expected that the 'balance' will be due from the employer to the contractor in the Final Statement, with a greater likelihood of being due from the contractor to the employer in the employer's Final Statement.
321 JCT DB 2016, clause 4.2.1.
322 JCT DB 2016, clause 1.1 defines 'Valuation' as 'a valuation in accordance with the Valuation Rules, pursuant to clause 5.2'.
323 JCT DB 2016, clause 1.1 defines 'Confirmed Acceptance' as 'the Employer's Instruction under Supplemental Provision 4 confirming Acceptance of an Acceleration Quotation'.
324 JCT DB 2016, clause 1.1. defines 'Acceleration Quotation' as 'a quotation by the Contractor'.
325 JCT DB 2016, clause 4.2.2.
326 JCT DB 2016, clause 4.2.3.

- Direct loss and/or expense (see Section 8.11.2) unless already included in the accepted Acceleration Quotation[327];
- Any deductions[328] for (i) reduction of the Contract Sum for accepting defects (under clause 2.35), (ii) employer's costs of employing others to give effect to instructions under clauses 3.5 and 3.9 where the contractor has failed to rectify defective work (clause 3.6), (iii) change in the rate of Pool Re[329] cover (clause 6.10.2), (iv) the costs of insurance if the contractor fails to insure (clause 6.12.2), and (v) the contractor's failure to carry out Remedial Measures under the Joint Fire Code (clause 6.19.2);
- The deduction of any Provisional Sums that have not been expended[330]; and
- Any other amount which is to be added to the Contract Sum[331]; and
o The total amount that the employer has already paid to the contractor.

Clause 4.24.1 provides that the Final Statement shall also state the basis on which the 'balance' (expressed in clause 4.24.1 as the 'amount') has been calculated, and details of all adjustments (under clause 4.2) must also be provided. This is the equivalent to the requirements for providing payment notices and pay less notices (see Section 8.7.1). Whilst clause 4.24.1 also obliges the contractor to provide any 'supporting documents' that the employer may reasonably require, it appears that the Final Statement is intended to be virtually 'self-contained', and not simply a summary to which the details are attached. However, clause 4.24.1 clearly anticipates that the employer may then require some additional information once he has received the Final Statement.

Clause 4.24.6 provides that the Final Statement (or the employer's Final Statement) is intended to conclude the financial aspects of the contract. It achieves this aim by stating that unless the party receiving the relevant Final Statement[332] gives notice[333] to the other, disputing anything in the relevant Final Statement before the 'due date for the final payment', then it shall be conclusive as to the 'balance' in clause 4.24.2,[334] subject only to any formal dispute (which must be either adjudication, arbitration, or litigation) already underway[335] concerning any matter covered by the relevant Final Statement (clause 1.8.2 – see Section 9.8.1). The effect of the relevant Final Statement becoming 'conclusive' in this regard is explored in more detail in Section 9.8.1, but it essentially means that neither party would be able to

---

327 JCT DB 2016, clause 4.2.4.

328 JCT DB 2016, clause 4.2.4 refers to deductions under clause 4.12.3 (if Alternative A applies) or 4.13.3 (if Alternative B applies) refer to Section 8.7.1.

329 The public financial guarantee provided by HM Government via the Government, Pool Reinsurance Company Ltd. ('Pool Re') – a reinsurer for terrorism risk.

330 JCT DB 2016, clause 4.2.5.

331 JCT DB 2016, clause 4.2.6 (e.g. interest for late payments).

332 JCT DB 2016, clause 4.24.6, the party receiving the relevant Final Statement will be (i) the contractor receiving the employer's Final Statement or (ii) the employer receiving the Final Statement from the contractor.

333 JCT DB 2017, the 'notice' would seemingly have to comply with clause 1.7 (Notices and other communications).

334 JCT DB 2016, clause 4.24.6 expresses the 'balance' as the 'sum due under clause 4.24.2'.

335 JCT DB 2016, clause 1.8.2.1, the adjudication, arbitration, or litigation must have commenced not later than 28 days after the date that the relevant Final Statement is issued, or if a dispute is referred to adjudication, which commences in accordance with clause 1.8.2.1 and a party wants to finally determine that dispute, it must commence arbitration or litigation not later than 28 days after the date of the adjudicator's decision (clause 1.8.2.2).

contest the 'balance' due (under clause 4.24.2) in the relevant Final Statement issued by the other. The 'balance' will then become final and binding on the parties.

It is therefore critical for both parties to know when the 'due date for the final payment' occurs. Unfortunately, this may not be straightforward, as firstly, it depends upon whichever of the following matters in clause 4.24.5 occurs last:

o The expiry of the Rectification Period (or the last Rectification Period, if there are sections);
o The date set out in the Notice of Completion of Making Good; or
o The provision on the relevant Final Statement;

following which, the 'due date for the final payment' will occur one month thereafter.[336] These events are shown in Figure 9.5.

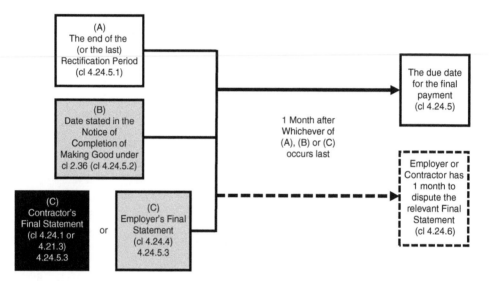

Figure 9.5 JCT DB 2016, Final Statement and Final Payment procedure.

In most cases, the last of the three matters in clause 4.24.5 to occur will be the date stated in the Notice of Completion of Making Good,[337] so this should alert either the employer or the contractor that they have one month to dispute anything (including the clause 4.24.2 'balance') in the relevant Final Statement.

9.6.2 FIDIC Yellow Book 2017

FIDIC Yellow Book 2017 concludes the 'final account' after the contractor has completed the work (see Section 9.3.2 – 'Taking Over'), and the following stages have been completed:

o The contractor has completed:
  – Outstanding Works and Remedying Defects (clause 11.1);

---

336 JCT DB 2016, clause 4.24.5.
337 JCT DB 2016, clause 4.24.5.2.

- Further Tests after Remedying Defects (clause 11.6);
- Any Tests after Completion (clause 12); and
o The engineer has issued the Performance Certificate (clause 11.9).

Similar to JCT DB 2016, the contractor must compile the 'final account' within a particular period, otherwise (and following the engineer's notice of his failure to do so), it can be compiled by the engineer. FIDIC differs from JCT in that the contractor's 'final account' is submitted in two stages; the first is in draft so the engineer has the opportunity to request further information if he either disagrees or cannot verify the contractor's draft. The second stage requires the contractor to submit the final documents, and this triggers the engineer's obligation to issue the concluding certificate.

Figure 9.6[338] summarises this procedure.

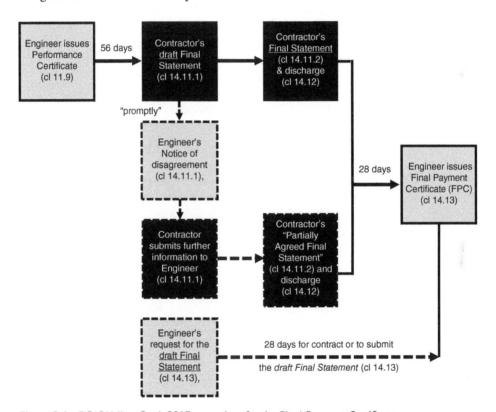

Figure 9.6 FIDIC Yellow Book 2017, procedure for the Final Payment Certificate.

Clause 14.11 (Final Statement) of the FIDIC Yellow Book 2017 sets out the procedure for concluding the final account. Clause 14.11.1 provides that not later than 56 days after the engineer has issued the Performance Certificate (under clause 11.9) the contractor submits the 'draft final Statement' to the engineer. The 'draft final Statement' must contain the

---

338 Figure 9.6 only denotes the durations between stages where such periods are specified in clause 14.11 (Final Statement).

information required by – and be in the form set out in – clauses 14.11.1(a) to 14.11.1(c) as follows:

o The same format as the contractor's Statements for interim payments under clause 14.3[339] (see Section 8.7.2), and in hard copy (paper) and in electronic format[340]; and
o Include supporting documents.[341]

The 'draft final Statement' contains one or both of the following categories of financial information, shown in detail[342]:

o Firm or established amounts that the contractor either knows or considers are due:
  – In respect of the value of all the work he has done[343]; and
  – Any further sums (either under the contract or otherwise) that the contractor considers are due as at the date when the Performance Certificate is issued[344]; and separately
o The contractor's 'estimate' of amounts (again, arising under the contractor or otherwise) that he considers may or will become due after the Performance Certificate is issued, including the following matters for which an outcome may be pending[345]:
  – Claims under clause 20.2 (Claims for Payment and/or EoT)[346];
  – Any matter being considered by the Dispute Avoidance/Adjudication Board[347] (DAAB) under clause 21.4 (Obtaining DAAB's Decision)[348];
  – Any matter for which a Notice of Dissatisfaction (NOD)[349] is pending.[350]

If the contractor fails to provide the draft final Statement within the 56-day period in clause 14.11.1, then clause 14.13 provides that the engineer 'requests' that he does so.[351] If the contractor's failure to provide the draft final Statement continues for a further 28 days, the engineer must issue the Final Payment Certificate (FPC) for 'such an amount as the Engineer fairly considers to be due'.[352]

If the engineer either cannot verify or disagrees with any of the amounts the contractor claims are due up to and including the date of the Performance Certificate, he shall 'promptly' notify the contractor.[353] The engineer's notice specifies the information he

---

339 FIDIC Yellow Book 2017, clause 14.11.1 (a).
340 FIDIC Yellow Book 2017, clause 14.11.1(b), or as many hard copies as are set out in the corresponding entry in the Contract Data.
341 FIDIC Yellow Book 2017, clause 14.11.1(c). Contrast with JCT DB 2016 (see Section 9.6.1), where further substantiation may be provided after the Final Statement is submitted, is as per the employer's reasonable requirement.
342 FIDIC Yellow Book 2017, clause 14.11.1(c).
343 FIDIC Yellow Book 2017, clause 14.11.1(c)(i).
344 FIDIC Yellow Book 2017, clause 14.11.1. (c)(ii).
345 FIDIC Yellow Book 2017, clause 14.11.1. (c)(iii) referring to the matters in clause 14.10(c)(i) to 14.10(c)(iii) (Statement at Completion).
346 FIDIC Yellow Book 2017, clause 14.10(c)(i).
347 FIDIC Yellow Book 2017, clause 1.1.22 defines 'DAAB' or 'Dispute Avoidance/Adjudication Board'.
348 FIDIC Yellow Book 2017, clause 14.10(c)(ii).
349 FIDIC Yellow Book 2017, clause 1.1.57 defines 'Notice of Dissatisfaction' or 'NOD'.
350 FIDIC Yellow Book 2017, clause 14.10(c)(iii).
351 FIDIC Yellow Book 2017, clause 14.13 refers to a 'request' from the engineer, not a Notice (as defined in clause 1.1.56).
352 FIDIC Yellow Book 2017, clause 14.13 (Issue of FPC).
353 FIDIC Yellow Book 2017, clause 14.11.1, the engineer's Notice must be issued 'promptly', although not definite period of time for its issue is specified.

needs from the contractor to enable such verification or agreement, along with the period of time within which the contractor must provide it. It appears that once the contractor has provided the engineer with the substantiation he requires, the contractor makes such changes to the draft final Statement as are agreed upon between them.

Clause 14.11.2 provides that if the draft final Statement does not contain any amounts that may arise after the Performance Certificate is issued,[354] then the contractor may proceed with issuing the Final Statement[355] to the engineer. If any elements of the draft final Statement cannot be agreed upon,[356] or any amounts estimated to arise after the Performance Certificate is issued,[357] then the contractor submits a 'Partially Agreed Final Statement' to the engineer, that clearly identifies:

o Amounts agreed,
o Estimated amounts, and
o Amounts not agreed.

It does not appear that the contractor issuing a 'Partially Agreed Final Statement' is essential. Clause 14.13 (Issue of FPC) enables the engineer to deem a 'Partially Agreed Final Statement' has been issued if the contractor has not issued a Partially Agreed Final Statement, but his draft final Statement contains amounts that may not be considered to be 'final' (such as sums denoted as 'on account' or 'to be agreed'), and to which the discharge (under clause 14.12) could not apply. In such cases[358] (or if the contractor submits a Partially Agreed Final Statement[359]), clause 14.13 provides that the engineer proceeds in accordance with clause 14.6 (Issue of IPC) and issues an Interim Payment Certificate.

Along with whichever of the Final Statement or Partially Agreed Final Statement (hereafter referred to as 'either Statement') that the contractor is to submit to the engineer, clause 14.12 stipulates that either of these documents must be accompanied by what is called a 'discharge' – which is a written statement that, save for the outcome of any Dispute,[360] either Statement shows the total sum in 'full and final settlement' of all amounts due to the contractor, both under and in connection with the contract, and becomes effective after:

o The amount in the FPC[361] has been paid in full; and
o The contractor has received the Performance Security.[362]

If the contractor fails to provide the clause 14.12 'discharge' with either Statement, it is deemed to have been provided and takes effect when full payment of the FPC amount and Performance Security are received.[363]

---

354 FIDIC Yellow Book 2017; that is, such amounts included under clause 14.11(c)(iii).
355 FIDIC Yellow Book 2017, clause 1.1.41 defines the 'Final Statement'.
356 FIDIC Yellow Book 2017, clause 14.12(b).
357 FIDIC Yellow Book 2017, clause 14.12(a).
358 FIDIC Yellow Book 2017, clause 14.12(ii).
359 FIDIC Yellow Book 2017, clause 14.12(i).
360 FIDIC Yellow Book 2017, clause 14.12 provides that the 'discharge' neither affects the employer's nor the contractor's liability for a Dispute (defined in clause 1.1.29) for which an arbitration or a DAAB proceeding is underway – under clause 21 (Disputes and Arbitration).
361 FIDIC Yellow Book 2017, clause 14.12(a). Clause 1.1.40 defines 'Final Payment Certificate' or 'FPC'. The FPC is issued under clause 14.13 (Issue of FPC).
362 FIDIC Yellow Book 2017, clause 14.12(b). Refer to clause 4.2 (Performance Security).
363 FIDIC Yellow Book 2017, clause 14.12.

Clause 14.13 (Issue of FPC) provides that the engineer must issue the FPC[364] to the employer (with a copy to the contractor) not later than 28 days after the contractor issues either Statement[364] and the clause 14.12 discharge. The FPC must show the following:

o The amount that the engineer considers is fairly due, including any additions or deductions[365]:
  – Under clause 3.7 (Agreement or Determination); or
  – Under or in connection with[366] the Contract.
o The balance due,[367] either from the employer to the contractor, or from the contractor to the employer (as the case may be), after giving credit for:
  – All amounts paid previously by the employer;
  – All sums which the employer is entitled to receive; and
  – Amounts under the Performance Security (paid by the contractor or received by the employer).

The conclusion of FIDIC Yellow Book 2017's final account procedure is set out in clause 14.14 (Cessation of Employer's Liability). This limits the employer's liability to paying the amount due under the FPC along returning the Payment Security, effectively excluding any further liability of the employer to the contractor 'for any matter or thing under or in connection with the Contract or execution of the Works' with the following exceptions:

o If the contractor makes a claim under clause 20.2 (Claims for Payment and/or EoT) within 56 days after the engineer issues the FPC[368] – refer to Section 8.10.4.1;
o That the contractor has included a specific amount for the particular 'matter or thing' referred to in clause 14.14 in either[360]:
  – The Final Statement or Partially Agreed Final Statement.[369] In addition (and not mentioned in clause 14.14), the contractor will either have issued or will be deemed to have issued the 'discharge' under clause 14.12;
  – The 'Statement on Completion' (under clause 14.10), except if that 'matter or thing' (in clause 14.14) has arisen after the Taking Over Certificate[370]; or
  – In the case of any gross negligence, reckless misconduct, fraud, or deliberate default by the employer.

In the event that the contractor disputes anything in the FPC, it is therefore vital that he raises a claim under clause 20.2 within the 56-day period after it is issued by the engineer.

9.6.3 NEC4

NEC4 contains the 'Final Assessment' procedure in clause 53. This is a procedure that was not included in NEC3. Clause 53's intention is to provide a 'final account' procedure

---

364 'either Statement' means the 'Final Statement' (in clause 14.11.2) or the 'Partially Agreed Final Statement' (in clause 14.13).
365 FIDIC Yellow Book 2017, clause 14.13(a).
366 FIDIC Yellow Book 2017, clause 14.13(a) refers to 'under the Contract or otherwise'.
367 FIDIC Yellow Book 2017, clause 14.3(b).
368 FIDIC Yellow Book 2017, clause 14.14.
369 FIDIC Yellow Book 2017, clause 14.14(a).
370 FIDIC Yellow Book 2017, clause 14.14(b).

whereby the client can conclude parts of the contract's financial entitlements (in similar but fewer ways than are available to the employer in JCT DB 2016 and FIDIC Yellow Book 2017). Clause 53.1 provides that the project manager assesses the 'final amount due' and certifies a final payment (if indeed, there is one) not later than:

- o Four weeks after the Defects Certificate is issued[371]; or
- o In the event of termination (see Section 9.1.3), 13 weeks after the 'termination certificate' is issued by the project manager.

It appears that the scope of any assessment of the 'final amount due' is particularly limited for the following reasons:

- o The only way that the prices can be changed under NEC4 is via the compensation event procedure (see Section 8.8.5).
- o Once implemented, clause 66.3 provides that a compensation event is 'not revised' (see Section 8.9.3) – except as stated in the contract, and the project manager's ability to do so seems limited correcting an assumption upon which his previous assessment was based. This, in itself, is a compensation event under clause 60.1(17) – see Section 8.8.5.1.
- o Clause 61.7 provides that a compensation event cannot be notified after the supervisor issues the Defects Certificate (see Section 8.8.5.2). Therefore, the only compensation events that could be included in the project manager's assessment of the final amount due are:
  - – Those for which the contractor has issued a clause 61.3 notification not later than four weeks before the supervisor issued the Defects Certificate; or
  - – Those for which the project manager or supervisor notifies prior to the Defects Certificate being issued;

and in both cases, would include the project manager's correction of an assumption under clause 60.1(17).

The final payment is then made within either:

- o Three weeks of the project manager's assessment, or
- o A different period which must be set out in the Contract Data.[372]

In cases of compensation events being notified as described above, there appears to be a two-week discrepancy between (i) clause 53.1's period of four weeks for 'assessing the final amount due' and (ii) the overall six-week period from receiving the contractor's clause 61.3 notification, to the project manager implementing[373] the compensation event by accepting a quotation[374] (refer to Figure 8.3 in Section 8.8.5.2).

This discrepancy may be resolved by clause 53.2, which provides that the contractor issues his assessment[375] of the final amount due to the client if the project manager fails to provide his assessment within the four-week period in clause 53.1. Despite clause 53.2 permitting

---

371 NEC4, clause 44.3.

372 NEC4, Contract Data Part One. Entry in section 5 (Payment) against 'The period within which payments are made is…'.

373 NEC4, clause 66.1.

374 NEC4, clause 62.3.

375 NEC4, clause 53.2. The contractor must provide details of how he has assessed the 'final amount due'.

the client to agree with the contractor's assessment directly, it is doubtful that he would bypass the project manager, as the client is likely to require the project manager's views of the contractor's assessment of the final amount due. Whilst clause 53.2 provides that the final payment is made within either three weeks or the alternative period set out in the Contract Data, it is not immediately clear at which time the applicable period begins. Clause 53.2 states the final payment is made 'within three weeks of the assessment', so given that it contains the intermediate step of the 'client's agreement' (in between the contractor's assessment and the final payment), it is firstly difficult to see how the relevant period for payment runs from the date when the contractor provides his assessment. However, the NEC4 *User's Guide* gives a clear indication that the period of making the final payment runs from the date when the contractor provides his assessment, because under clause 53.2, the final payment due is determined by the contractor.[376]

The three other financial matters that NEC4 expressly includes in the 'final amount due' in clause 53 are:

o The contractor's share under Option C (Target Contract with Activity Schedule) in clauses 54.1 to 54.4;
o The contractor's share under Option D (Target Contract with Bill of Quantities) in clauses 54.5 to 54.8; and
o The Incentive payment under Option X22 (Early Contractor Involvement), clause X22.7(3) – see Section 5.4.3.

The clauses governing the contractor's share under Options C and D are identical. The contractor's share is assessed by the project manager on the following two occasions, albeit on a slightly differing basis:

o A 'preliminary assessment' is assessed at Completion[377] using:
  – A forecast of the Price for Work Done to date; and
  – The final total of the Prices; whereas
o The 'final assessment' is made and included in the 'final amount due'[378]; using:
  – The final Price for Work Done to Date, and
  – The final total of the Prices.

The calculation for the 'Contractor's share' firstly uses the difference[379] between the following elements, as set out in Table 9.9.

The calculation of the 'difference' is as follows:

'Total of the Prices'	£A
Less: the 'Price for Work Done to Date'	(£B)
Difference	**£C**

376 *NEC4: A User's Guide – Managing an Engineering and Construction Contract*, vol. 4, (2020). Note in relation to clause 53.3: 'This clause confirms the status of the final amount due, determined through either 53.1 by the Project Manager, or <u>53.2 by the Contractor</u>' [underlining added for emphasis], p. 50.
377 NEC4, Option C, clause 54.3; or Option D, clause 54.7.
378 NEC4, Option C, clause 54.4; or Option D, clause 54.8.
379 NEC4, Option C, clause 54.1; or Option D, clause 54.5.

Table 9.9 NEC4: Elements of the 'contractor's share' calculation.

Element of the 'difference' for the 'contractor's share' calculation	Description
The 'total of the Prices'; and	The 'Prices' are essentially the 'value' of the works, or the 'target' – i.e. the total of the prices from the relevant pricing document, as adjusted (increased or decreased) by compensation events.
	o Option C, clause 11.2(32)[a] the total of the activities in the Activity Schedule[b] as adjusted in accordance with the contract;
	o Option D, clause 11.2(35)[a] means the total of:
	– The quantity of work completed multiplied by the rate in the Bill of Quantities[c] and
	– A portion of each lump sum of work in the Bill of Quantities that the contractor has completed.
The 'Price for Work Done to Date'	The 'Price for Work Done to Date' is calculated as the 'Defined Cost' (i.e. the costs[d] used for interim payments) plus the Fee.
	o Option C, clause 11.2(32); and
	o Option D, clause 11.2(35) means the total Defined Cost plus the Fee.

a) NEC4, Option D, 'the total of the Prices', is not defined in Option C but is defined in clause 11.2(35) of Option D.
b) NEC4, Option C, 'Activity Schedule' is defined in clause 11.2(21).
c) NEC4, Option D, 'Bill of Quantities' is defined in clause 11.2(22).
d) NEC4, see clause 52.1 and refer to Section 8.9.3.1.

The 'Difference' is then expressed as follows[380]:

o A 'saving' if the Price for Work Done to Date is less than the total of the Prices, in which case the client pays the 'contractor's share' to the contractor; or
o An 'excess' if the Price for Work Done to Date is greater than the total of the Prices, in which the contractor pays the 'contractor's share' to the client.

The final stage of calculating the contractor's share is to apply the 'contractor's share percentages' to the 'share ranges' set out in Contract Data Part One.[381] The 'share ranges' are expressed as percentages applying to the 'difference' (i.e. both the 'saving' and the 'excess'), and are divided into four 'increments' or bands. A separate 'contractor's share percentage' is then set against each of these increments (or bands). The following two examples show how the contractor's share range is calculated against a 'saving' and against an 'excess':

Saving example

o The 'difference' between the 'total of the prices' (which is £2 M) and the 'Price for Work Done to Date' (which is £1.8 M) is £200 000.

---

380 NEC4, Option C, clause 54.2; or Option D, clause 54.6.
381 NEC4, Option C, clause 54.1; or Option D, clause 54.5.

o The 'share range' and the corresponding 'contractor's share percentage' are as follows:

Share range:		Contractor's share percentage:
Less than	1%	10%
From	1% to 5%	20%
From	6% to 10%	40%
Greater than	10%	50%

o The calculation of the 'contractor's share' is as follows:

Share range:		£
Less than 1%	Up to 1% (£2000), the contractor is paid 10%:	£200
From 1% to 5%	For the next 1–5% (£8000), the contractor is paid 20%	£1600
From 6% to 10%	For the next 5–10% (£10 000), the contractor is paid 40%	£4000
Greater than 10%	The saving does not exceed 10%	£0
	Total 'contractor's share' paid by the client	**£5800**

Excess example

o The 'difference' between the 'total of the prices' (which is £4.5 M) and the 'Price for Work Done to Date' (which is £4.8 M) is (£300 000).

o Applying the same share ranges as the 'Saving example' shows an excess of 6.67% the calculation of the 'contractor's share', as follows:

Share range:		£
Less than 1%	Up to 1% (£3000), the client is paid 10%	(£300)
From 1% to 5%	For the next 1–5% (£12 000), the client is paid 20%	(£2400)
From 6% to 10%	For the next 5–10% (£5100), the client is paid 40%	(£2040)
Greater than 10%	The excess does not exceed 10%	£0
	Total 'contractor's share' paid to the client	**(£4740)**

If Option X22 – Early Contractor Involvement (see Section 5.4.3)[382] applies, then the project manager will also include his assessment of the contractor's 'budget incentive' in the 'final amount due' under clause 53.1. The 'budget incentive' will only be payable if the 'final Project Cost[383]' is less than the Budget.[384] The project manager calculates the 'budget incentive' as follows: ([Budget] – [final Project Cost]) × % in Contract Data Part One[385] = 'budget incentive'

---

[382] NEC4, Option X22 Early Contractor Involvement is only used with Option C (Target Contract with Activity Schedule) or Option E (Cost Reimbursable Contract).

[383] NEC4, Option X22 (Early Contractor Involvement), clause X22.1(2) defines 'Project Cost' as the total paid by the client to the contractor (and Others – as defined in clause 11.2(12)) for the items included in the 'Budget'.

[384] NEC4, Option X22 (Early Contractor Involvement), 'Budget' is defined in clause X22.1(1).

[385] NEC4, Contract Data Part One – entry under Option X22 for the applicable percentage of the saving as calculating the 'budget incentive'.

Once the final assessment is provided (either by the project manager under clause 53.1 or by the contractor under 53.2 – hereafter referred to as the 'relevant assessment'), it becomes 'conclusive evidence' of the final amount due, unless the party disagreeing with the relevant assessment of the final amount takes one of the dispute resolution options under clause 53.3. Clause 53.3 sets out the appropriate procedure under whichever of the dispute resolution Options W1, W2, and W3 has been selected to apply in Contract Data Part One.[386] The selection of Option W1, W2, or W3 will depend upon whether the HGCRA 1996 applies, and what the parties select as the 'tribunal' (in Contract Data Part One).

Option W2 will be selected if the HGCRA 1996[387] applies – which will be in the majority of cases if the project is in the UK and subject to UK law, and the construction contract is a contract for 'construction operations' within the meaning of s105 of the HGCRA 1996.

The definitions of 'construction operations' in ss105(a) to (f) are extensive, so s105(2) helpfully sets out the excluded contracts – that is, what are not 'construction operations'. Many of these excluded contracts include works of a civil or process engineering nature and are summarised below:

o Extraction of oil or natural gas, including drilling.[388]
o Extraction of minerals (from the surface or underground), including boring, tunnelling, and underground work to facilitate their extraction.[389]
o For sites where the primary activity is:
 – Power generation, nuclear processing, or water or effluent treatment[390]; or
 – The production, processing, transmission, or bulk storage (except warehousing) of chemicals, pharmaceuticals, oil, gas, steel, or food and drink.[391]

The excluded work includes the assembly, installation, or demolition of plant or machinery, or erection or demolition of steelwork for supporting or providing access to plant or machinery.[392]

o Manufacture or delivery to the site of equipment or components for building or engineering,[393] materials, plant or machinery,[394] or components for heating, venting, air-conditioning, drainage, water supply, sanitation, power supplies, lighting, communications, security, and fire protection[395]; and
o Making, installing, and repairing artistic works.[396]

---

386 NEC4, Contract Data Part One, Section '1 General' entry against 'Option for resolving and avoiding disputes'.
387 Housing Grants Construction and Regeneration Act 1996 (as amended) – refer to Section 8.7.
388 HGCRA 1996, s105(2)(a).
389 HGRA 1996, s105(2)(b).
390 HGCRA 1996, s105(2)(c)(i).
391 HGCRA 1996, s105(2)(c)(ii).
392 HGCRA 1996, s105(2)(c).
393 HGCRA 1996, s105(2)(d)(i).
394 HGCRA 1996, s105(2)(d)(ii).
395 HGCRA 1996, s105(2)(d)(iii), unless the contract also provides for their installation.
396 HGCRA 1996, s105(2)(e) – including murals, sculptures, and other works of a wholly artistic nature.

Option W2 permits the parties to refer a dispute to 'senior representatives'[397] under clause W1.1, but as it must comply with s108(2)(a) of the HGCRA 1996, the parties have the right to refer a dispute to adjudication at any time. Therefore, any attempt by option W2 to impose negotiations between senior representatives as a precursor to adjudication will be ineffective. It is therefore perfectly feasible that work being undertaken under NEC4 in the UK may include excluded contracts that are not 'construction operations' within the meaning of s104(1)(a) of the HGCRA 1996. Consequently, dispute resolution options W1 and W3 may also apply to works in the UK, as well as in other jurisdictions.

Option W1 allows for adjudication, but only to resolve the matters that were not agreed upon by the 'senior representatives'.[398] Option W3 neither involves the 'senior representatives' nor adjudication, as it only involves the Dispute Avoidance Board (DAB).[399] The DAB takes an active role in visiting the site and reviewing all potential disputes, providing the parties with their recommendations for how to resolve the potential dispute.[400]

If either of the parties are dissatisfied with the adjudicator's decision (Options W1 and W2) or with the DAB (Option W3), they may refer the remaining dispute to the 'tribunal' within the prescribed time. The parties agree on the type of tribunal, and this can be the Courts (litigation), an arbitrator, or expert determination.[401] If the tribunal is the Court, then the prevailing court rules for the particular jurisdiction will apply (e.g. the Civil Procedure Rules in the UK). Whilst the NEC4 *User's Guide*[402] refers to expert determination under Option W3, the Contract Data does not contain any guidance for either appointing an expert or the procedure under which his determination will be carried out.

The procedures applicable to Options W1, W2, and W3 are shown in Table 9.10.

Clause 53.4 provides that changes to the relevant assessment of the final amount due only arise after:

o The parties have agreed; or
o As provided in clause 53.3 if the following are not referred to the tribunal within four weeks:
   – An adjudicator's decision under W1 (stage 2) or W2 (stages 1 and 2), or
   – A recommendation from the DAB under W1 (stage 1).

Further details of dispute resolution methods are set out in Section 9.7.

---

397 NEC4, Contract Data Part One, entry against Options W1 or W2 under 'Resolving and avoiding disputes'.

398 NEC4, Option W1, clause W1.3(1).

399 NEC4, Contract Data Part One, entries against Option W3 under 'Resolving and avoiding disputes' for (i) the DAB being comprised of either one or three members, (ii) the client's DAB nomination, (iii) the frequency of DAB site visits, and (iv) the DAB nominating body.

400 NEC4, Option W3, clause W3.2(5).

401 NEC4, Contract Data Part One, entries against 'Resolving and avoiding disputes' for (i) the tribunal, and (ii) if the tribunal is an arbitrator (or panel of arbitrators), the arbitration procedure, and the place (the seat) where the arbitration is to be held.

402 *NEC4: A User's Guide – Managing an Engineering and Construction Contract*, vol. 4, (2020), p. 84.

Table 9.10 NEC4 clause 53.3 Options W1, W2, and W3.

	Option W1	Option W2 (if HGCRA 1996 applies)	Option W3
Stage 1	Refers a dispute to the senior representatives[a)] within 4 weeks of the relevant assessment;	Refers a dispute to the senior representatives or to the adjudicator within 4 weeks of the relevant assessment;	Refer to dispute to the Dispute Avoidance Board; and
Stage 2	Refers any remaining issues of disagreement to the adjudicator within 3 weeks of such disagreement[b)]; and	Refers any issues of not agreed by the senior representatives to the adjudicator within 3 weeks of such disagreement	Within 4 weeks of the Dispute Avoidance Board's decision, referring its dissatisfaction to the 'tribunal'.
Stage 3	Within 4 weeks of the adjudicator's decision, referring its dissatisfaction to the 'tribunal'	Within 4 weeks of the adjudicator's decision, referring its dissatisfaction to the 'tribunal'	None
	Likely overall duration = 15 weeks Stage 1 = 4 weeks Stage 2 = 3 weeks Stage 3 = 8 weeks (4 + 4)[c)]	Likely overall duration = 15 weeks Stage 1 = 4 weeks Stage 2 = 3 weeks Stage 3 = 8 weeks (4 + 4)[c)]	Likely overall duration = 8+ weeks Stage 1 = 4 weeks[d)] Stage 2 = 4+ weeks[e)]

a) NEC4, Senior Representatives are set out in Contract Data Part One, against Options W1 or W2 under 'Resolving and avoiding disputes'.
b) NEC4, clause 53.3 (Option W1), within three weeks of a list of the disagreed issues either being produced or when the list should have been produced.
c) NEC4, clause 53.3, Likely overall duration includes four weeks (28 days) for the adjudicator's decision.
d) NEC4, Option W3, clause W3.2(3), a dispute is referred to the Dispute Avoidance Board within two to four weeks after one party notifies the other and the project manager.
e) NEC4, Option W3. No duration for the Dispute Avoidance Board providing its recommendations is given.

9.7 Resolving Disputes

Construction contracts invariably involve the parties being in dispute to some degree, although the magnitude, frequency, and effects of most disputed matters are often relatively minor. Most disputes ultimately result in one party incurring a cost, even though the source of the dispute may not immediately be about money. In the UK, the wealth of reported cases since adjudication began in 1999 has explored the meaning of a 'dispute' in detail. However, for the purposes of this book (and particularly in relation to adjudication in the UK), the definition provided by the HGCRA 1996[403] that a dispute is 'any difference' is appropriate. In this context, a dispute could be 'any difference' over:

o Financial matters:
  – Valuation of work (including variations, particularly those involving design changes)

---

403 HGCRA 1996, s108(1).

- Late payments
- Loss and expense (additional costs)
o Time (or programme) matters:
  - Extensions of time
  - Disrupted sequence of working
  - Liquidated (delay) damages
  - Costs of late completion
o Quality matters:
  - Does the contractor's work comply with the contract?
  - Costs of rectifying defective work
o Administrative matters:
  - A party's failure to provide a document, reply, notice, or carry out a function within a prescribed period of time.

The definition of a dispute being 'any difference' is very broad. On one hand it can include the parties' respective valuations of work being millions of pounds apart, or on the other hand, it could mean their valuations of a significant, multi-million-pound item of work are only a few pounds apart. In most instances, the parties resolve the dispute themselves, without resorting to one of the formal means of resolving disputes.

The three design and build contracts all contain a range of dispute resolution methods, such as:

o Informal: The parties (or their designated representatives) meet with the aim of resolving a dispute.
o Contractual: The building contract provides a procedure by which the employer (or his representative) makes a 'decision' that is binding upon the parties until it is overturned by one of the formal methods.
o Formal: Formal methods of dispute resolution involve one or both parties appointing an individual or a tribunal to either help them resolve the dispute or to resolve it on the parties' behalf by making a decision (or ruling) that is binding upon them. Formal methods of dispute resolution are either available to the parties as a statutory right, or they can agree to use them by choosing a provision of the contract – summarised in Table 9.11.

This Section 9.7 departs from the regular format of dealing with the three design and build contracts in turn, as it considers each of the three dispute resolution methods described above and how they apply to each design and build contract.

9.7.1 Informal Dispute Resolution Methods

Informal methods of resolving disputes usually involve discussions between the parties in an effort to reach agreement. This can involve personnel from each party who have a significant and direct involvement in the project, but as the dispute may have arisen between these individuals, it may be better for other (often, more senior) staff from each party to meet to discuss the dispute. The usual advantages of involving more senior staff are that they may not be encumbered by any 'personal' issues that may have arisen, and they should have greater authority than the project staff to agree on terms of settlement.

Table 9.11 Dispute resolution methods.

Method	Person in charge	Person in charge gives binding decision?	Statutory right?	Right of appeal?
Mediation	Mediator	N/A	No	N/A
Adjudication	Adjudicator	Yes – temporarily	Yes (UK)	Yes
Dispute Avoidance Board	DAB	No	No	N/A
Dispute Avoidance/ Adjudication Board	DAAB	Yes	No	Yes
Arbitration	Arbitrator or tribunal	Yes	No	Yes
Litigation	Court	Yes	Yes (UK)	Yes
Expert determination	Expert	Yes	No	No

For example, a senior member of the employer's organisation may have the authority to agree a settlement that may alter or waive entitlements, rights, or procedures in the building contract. The employer's representative is unlikely to have such authority unless the employer grants it to him. Similarly, the contractor's project staff may not have the authority, as only a director may be able to agree on a settlement that sits outside the strict terms of the contract.

Such informal procedures are found in the design and build contracts as follows:

o JCT DB 2016's procedure is very short and simple. If the Contract Particulars state that Supplemental Provision 10[404] (Notification and Negotiation of Disputes) applies, then the following procedure applies:
  – Firstly, each party notifies the other of the dispute, which seemingly can be done at any time.
  – Secondly, the parties' nominees[405] ('senior executives' or their equivalent) are to meet as soon as is practicable, and enter into 'direct, good faith negotiations' with the aim of resolving the dispute.
o FIDIC Yellow Book 2017 contains two procedures, both of which may be used in respect of clause 21.5 (Amicable Settlement). The first is set out in clause 21.5 of the 'General Conditions', and the second version is an alternative, expanded version of clause 21.5 set out in the 'Special Provisions':
  – The first (General Conditions) version of clause 21.5 is very similar to JCT DB 2016's 'short and simple' procedure, whereby both parties try to settle the dispute 'amicably' within a 28-day, 'cooling off' period.[406]

---

404 JCT DB 2016, Schedule 2 contains the Supplemental Provisions – see paragraph 10 (Notification and negotiation of disputes).
405 JCT DB 2016, the entries in the Contract Particulars against para. 10 of the Supplemental provisions should be completed by both the employer and contractor to set out.
406 FIDIC Yellow Book 2017, refer to both the General Conditions' and the Special Provisions' versions of clause 21.5. The 28-day period commences when one party issues a Notice of Dissatisfaction under clause 21.4.4 to the other.

- However, unlike their JCT equivalent, both the first and second (Special Provisions) versions of clause 21.5 cannot be commenced at any time, as a NOD[407] must have been given beforehand, pursuant to clause 21.4 (Obtaining DAAB's Decision)[408] – see Section 9.7.2.1.
- The second (Special Provisions) version of clause 21.5 encourages the parties to 'actively engage' with each other to resolve the dispute by one (or perhaps a combination) of the following:
  (1) Direct negotiations between the parties' senior staff,
  (2) Mediation[409] (see Section 9.7.1),
  (3) Expert determination[410], or
  (4) Some other form of alternative dispute resolution ('ADR') that is swifter and cheaper than arbitration.
  The parties are also encouraged to extend the 28-day period[409] to facilitate fuller negotiations under the recommended procedures.
o NEC4 provides options W1 and W2 for 'senior representatives' to meet in an attempt to resolve the disputes. As Option W2 is selected for projects where the HGCRA 1996 applies (see Section 9.6.3), the meeting of senior representatives cannot be a mandatory precursor to adjudication, as the parties have the right to adjudication at any time, and is purely optional. Under Option W1, however, any adjudication is intended to take place after the senior representatives have met, in order to resolve any remaining disputes.

9.7.2 Contractual Dispute Resolution Methods

JCT DB 2016 does not contain any contractual methods of resolving disputes; however, FIDIC and NEC4 provide for the following:

o FIDIC Yellow Book 2017 contains the 'Agreement or Determination' provisions in clause 3.7, whereby the engineer provides his 'determination' of a disputed matter; and
o NEC4 Option W3 provides for the DAB to attend the site and take an active role in recommending how the parties resolve potential disputes.

---

407 FIDIC Yellow Book 2017, clause 1.1.57 defines 'Notice of Dissatisfaction' or 'NOD' as the Notice (i.e. complying with clause 1.3) that the party who is dissatisfied with either an engineer's determination under clause 3.7 (Agreement or Determination) or the decision made by the Dispute Avoidance/Adjudication Board (or DAAB) under clause 21.4 (Obtaining DAAB's Decision).
408 FIDIC Yellow Book 2017, clause 21.4.4 (Dissatisfaction with DAAB's decision).
409 FIDIC Yellow Book 2017, Special Conditions clause 21.5 suggests 'Example Mediation' rules covering (i) the mediator's appointment, (ii) the parties' obligations during the mediation, (iii) the Mediation Timetable, (iv) confidentiality, (v) the written settlement agreement to be binding, (vi) the Mediator's Opinion in the event that no agreement is reached, and (vii) the Mediation Costs.
410 FIDIC Yellow Book 2017, Special Conditions clause 21.5 suggests the Expert Determination is carried out in accordance with the International Chamber of Commerce ('ICC') Expert Rules (current version published in August 2019).

9.7.2.1 FIDIC Yellow Book 2017 – 'Agreement or Determination'

Turning first to FIDIC Yellow Book 2017's 'Agreement or Determination' procedure in clause 3.7, this provides an initially temporary resolution in the following circumstances:

o The following categories of 'Claim' (see Sections 8.10.4.1 and 8.11.3):
   - A 'Claim' under clause 20.1(c)[411];
   - A 'Fully Detailed Claim' under clause 20.2.4[412]; or
   - A 'Claim of continuing effect' under clause 20.2.6[413]; or
o Under the relevant subclause set out in the conditions[414] (as summarised in Table 9.12).

Table 9.12 FIDIC Yellow Book 2017, clause 3.7.3(a) determination or agreement subclauses. Adapted from FIDIC Yellow Book 2017, clause 3.7.3(a).

Subclause in the conditions	Description of subclause leading to the engineer's determination
1.9	Errors in the employer's requirements
4.2.2(b)	Claims under the performance security
4.7.3	Rectification measures, delay, and/or cost
4.12.5	Unforeseeable physical conditions, delay, and/or cost
8.5	Extension of time for completion
10.2	Taking over parts, reduction in delay damages
11.2	Cost of remedying defects
13.3.1	Variation by instruction
13.5	Daywork
14.3(vi)	Inclusion of sums arising from agreement or determination in IPC
14.4	Schedule of payments, revised instalments
14.5	Amount to be added for plant and materials
14.6.1(b)	The IPC
14.6.3(b)	Correction or modification
14.13	Issue of FPC
15.3	Valuation after termination for contractor's default
15.6	Valuation after termination for the employer's convenience
18.5	Optional termination
20.1(c)	Claims
20.2.2	Engineer's initial response, review of disagreement
20.2.4	Fully detailed claim
20.2.5	Agreement or Determination of Claim
20.2.6	Claims of continuing effect

---

411 FIDIC Yellow Book 2017, clause 3.7.3(b).
412 FIDIC Yellow Book 2017, clause 3.7.3(c)(i).
413 FIDIC Yellow Book 2017, clause 3.7.3(c)(ii).
414 FIDIC Yellow Book 2017, clause 3.7.3(a).

The 'Agreement or Determination' procedure is intended to last for 42 days (or some other period that the engineer proposes and the parties subsequently agree upon)[415] and commences when one of the three circumstances in clause 3.7.3 (as described above) occurs. The 42-day (or extended) period is referred to as the 'time limit for agreement', during which the engineer is to consult with the parties for the purposes of them reaching agreement. If no agreement is either likely[416] or achieved[417] within the time limit for agreement, then the engineer must proceed to make his determination – seemingly, within whatever 'time limit for agreement' remains,[418] after giving notice to the parties. This could put the engineer in a difficult position if no agreement is either likely or achieved quite late into the 'time limit for agreement', as this does not leave much time for his determination. Clause 3.7.2 appears to offer some assistance as it states that when making his determination, the engineer shall take 'due regard of all relevant circumstances'. In such a case, the relevant circumstances would be having to give a determination in a considerably shorter period than the 'time limit for agreement'. The 'Agreement or Determination' procedure in clause 3.7 is shown in Figure 9.7.

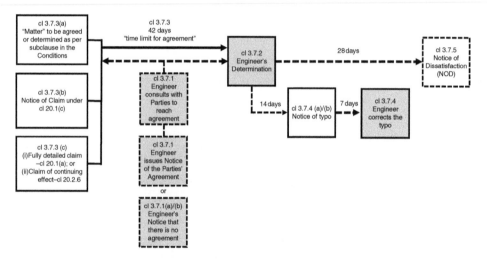

Figure 9.7 FIDIC Yellow Book 2017, clause 3.7, procedure for agreement or determination. Adapted from FIDIC Yellow Book 2017, clause 3.7

When consulting with the parties or making his determination, clause 3.7 provides that the engineer is to act 'neutrally' between the parties, and shall not be considered to be working for the employer. Clause 3.7.4 provides that his determination shall be final and binding[419] upon each party (and the engineer must comply with it) unless:

o There is a typographical, clerical, or arithmetical error that the engineer corrects under clause 3.7.4; or

---

415 FIDIC Yellow Book 2017, clause 3.7.3 (Time limits).
416 FIDIC Yellow Book 2017, clause 3.7.1(ii).
417 FIDIC Yellow Book 2017, clause 3.7(ii).
418 FIDIC Yellow Book 2017, clause 3.7.1.
419 FIDIC Yellow Book 2017, clause 3.7.5.

o The determination is revised by arbitration under clause 21.6. However, in order to do so, a party must firstly express its dissatisfaction with the determination under clause 3.7.5, by issuing a NOD[420] not later than 28 days after the engineer either has or should have issued his determination[421] before commencing arbitration[422].

In the event that the engineer does not provide his determination within the 'time limit for agreement', then he will be deemed to have given[423]:

o A determination rejecting a claim under clause 20; or
o For any other matter (see Table 9.12), it will be considered a dispute referred to the DAAB.

9.7.2.2 NEC4 Option W3 (Dispute Avoidance Board)

NEC4's Option W3 is only intended to be used where the HGCRA 1996 does not apply (see Section 9.6.3). It requires the involvement of the DAB, whose role is to assist the parties in resolving potential disputes.[424] avoiding the need to resolve them formally.[425] The DAB does this by visiting the site to inspect the works.[426] Then, acting impartially,[427] it gives a non-binding[428] recommendation[429] to the parties for resolving the potential dispute that has been referred to them.[430]

The DAB can be appointed in the following ways:

o If the DAB comprises one member,[431] it seems that the parties must either:
  – Agree on the DAB member's identity. This could simply be by either the client or the contractor inserting a name against Option W3 in the Contract Data Part One or Two (respectively); or
  – If no member of the DAB is identified in the Contract Data (or a DAB member is unable to act), either party can jointly choose the member or seek the DAB member's appointment by the DAB nominating body,[432] which must choose the DAB member within seven days.[433]

---

420 FIDIC Yellow Book 2017, clause 3.7.5(a) requires the NOD to be given to the other party with a copy to the engineer, and clause 3.7.5(b) stipulates it must state 'Notice of Dissatisfaction with the Engineer's Determination'.
421 FIDIC Yellow Book 2017, clause 3.7.5(c).
422 FIDIC Yellow Book 2017, clause 21.6 (Arbitration), conducted under the ICC Rules of Arbitration – clause 21.6(c).
423 FIDIC Yellow Book 2017, clause 3.7.3.
424 NEC4, Option W3, clause W3.2(1).
425 NEC4, Option W3, clause W3.2(5).
426 NEC4, Option W3, clause W3.1(5).
427 NEC4, Option W3, clause W3.1(3).
428 *NEC4: A User's Guide – Managing an Engineering and Construction Contract,* (2020), p. 84.
429 NEC4, Option W3, clause W3.2(5), fourth bullet point.
430 NEC4, Option W3, clause W3.2(2).
431 NEC4, Contract Data Part One, entry for Option W3 against 'Resolving and avoiding disputes' to select one or three members of the DAB.
432 NEC4, Contract Data Part One, entry for Option W3 against 'Resolving and avoiding disputes' to select the 'Dispute Avoidance Board nominating body'. It is not clear how the DAB member(s) could be appointed if (i) no DAB members are identified in Contract Data (either Part One or Part Two), and (ii) if no Dispute Avoidance Board nominating body is identified either.
433 NEC4, Option W3, clause W3.1(4).

o If the DAB consists of three members[431], then each party:
- Identifies one proposed DAB member in the Contract Data[434] and
- Jointly selects the third DAB member.[435]

o The DAB is appointed under the NEC Dispute Resolution Service Contract.[436]

The DAB should visit the site during the works,[437] and as often as is set out in the Contract Data[438] unless the parties agree that the DAB's visit is not required. To assist the DAB, the parties provide the following:

o Their suggested agenda for the DAB's site visit, although the DAB has the power to decide what (if any) part of the suggested agenda they may comply with[439];

o Further documentation including[440]:
- A copy of the contract,
- Progress reports, and
- Any other material consider the DAB will find relevant.

The DAB prepares a note of their visit,[441] but can only act (providing their recommendation) if the potential dispute is referred to them by one of the parties; however, the other party and the project manager must be notified of the potential dispute not later than four weeks beforehand.[442]

Clause W3.3(1) encourages the parties to utilise the DAB's assistance, as it prohibits any dispute being referred to the 'tribunal' (see Sections 9.7.3.3, 9.7.3.4, and 9.7.4.2) until the following has occurred:

o The DAB has provided its recommendation under clause W3.2(5); and

o A party has notified the other that it intends to refer the matter to the tribunal, and it must issue such notification not later than four weeks after the recommendation.[443]

The DAB (including their employees or agents) are not liable to the parties for any action (or failure to act) unless that action or failure to act was in bad faith.[444]

The terms of the DAB's appointment[445] are set out in NEC4's Dispute Resolution Service Contract (along with those covering adjudication[446] – see Section 9.7.4.1). It includes the

---

434 NEC4, the client's and contractor's nominees are set out in the entries against 'Resolving and avoiding disputes' for Option W3 in Contract Data Part One (for the client's nominee) and Contract Data Part Two (for the contractor's DAB nominee).

435 NEC4, Option W3, clause W3.1(1).

436 NEC4, Option W3, clause W3.1(2).

437 NEC4, Option W3, clause W3.1(5) – during the course of the works means from the 'starting date' (see 8.2.3) until the 'defects date' (see 9.5.3).

438 NEC4, Contract Data Part One, entry for Option W3 against 'Resolving and avoiding disputes' to set out the frequency (expressed in months) for the DAB's site visits.

439 NEC4, Option W3, clause W3.1(6). Also see clause W3.2(6), which enables the DAB to take the initiative in reviewing potential disputes.

440 NEC4, Option W3, clause W3.2(4). Surprisingly, clause W3.2(4) does not expressly refer to the programme (or any revised programme), although that could be provided as 'any other material'.

441 NEC4, Option W3, clause W3.2(5), third bullet point. Clause W3.2(5) does not specify that the DAB must provide this note to either of the parties or the project manager.

442 NEC4, Option W3, clause W3.2(3).

443 NEC4, Option W3, clause W3.3(2).

444 NEC4, Option W3, clause W3.1(7).

445 NEC4, Dispute Resolution Service Contract (2017), clause 3.

446 NEC4, Dispute Resolution Service Contract, clause 2.

DAB's terms of payment,[447] along with generic clauses dealing with payment (assessment and payment of the amount due to the DAB) and terminating the DAB's appointment.[448]

9.7.3 Formal Dispute Resolution Methods

The formal methods of resolving disputes are summarised in Table 9.11. These involve the parties referring a dispute to an individual or a tribunal (such as a Court) who either assist them in resolving the dispute or decide how the parties will resolve the dispute. The parties can proceed along the various formal dispute resolution routes as follows:

o They may agree to adopt a particular method, either by selecting that method in the contract or by subsequently agreeing to use it. Some examples include:
  – Mediation
  – Dispute advisory board
  – Arbitration
  – Expert determination
o The parties may have a statutory entitlement to a particular method, such as:
  – Adjudication
  – Litigation

The range of some methods of formal dispute such as adjudication, arbitration, and litigation is vast, particularly when considering that the FIDIC Yellow Book 2017 and NEC4 contracts may be used in legal jurisdictions outside the UK, so anything other than a very brief consideration is beyond the scope of this book.

9.7.3.1 Mediation

Mediation is perhaps the 'odd one out' when it comes to methods of finally resolving disputes. Whilst it shares many similarities with the other methods, such as the parties appointing a person (the mediator), submitting summaries of their cases to him (albeit brief ones), and meeting with the aim of resolving the dispute, the fundamental difference with mediation is that the mediator does not have any power to decide anything. Instead, the mediator's role is to assist the parties reaching agreement over the dispute. Most mediations take place over the course of a single day, and require a venue with at least three or four rooms: one for each party, one for meetings, and possibly one for the mediator. The parties may agree upon the mediator between them or may apply to a nominating body[449] to appoint the person. It is usual for each party to pay an equal share of the mediator's fees.

---

447 NEC4, Dispute Resolution Service Contract, clauses 3.7 to 3.9, and generic clauses (that apply to both the DAB or an adjudicator – both referred to as the 'Dispute Resolver') dealing with the assessment and payment of the amount due to the Dispute Resolver (i.e. DAB or adjudicator) in clause 4.1 to 4.7.
448 NEC4, Dispute Resolution Service Contract, clause 5. As with clauses 4.1 to 4.7, clause 5 is a generic clause that refers to the 'Dispute Resolver' (i.e. the DAB or an adjudicator).
449 Suitable professional bodies include the Royal Institution of Chartered Surveyors (RICS), the Chartered Institute of Arbitrators (CIArb) or the Technology and Construction Solicitors Association (TeCSA), who can appoint a mediator, along with publishing their own guidelines and/or rules for mediation.

Firstly, the parties (and their representatives) usually convene with the mediator in the meeting room, so the mediator can run through he mediation's procedure (such as side meetings between the parties that may narrow the issues in dispute, or with the mediator present) or request further information from the parties. The next stage is often called 'shuttle diplomacy', where the mediator goes back and forth between the parties, meeting them privately and attempting to get to the root of the dispute. The root of the dispute may not be purely commercial (about money). There may be other solutions that a skilled mediator discovers, which he can present to the parties to reach a settlement. If the dispute is resolved (entirely or in part), the mediator may help the parties draft a binding agreement that captures the settled issues.

Mediation is included as an option under clause 9.1 of JCT DB 2016, and under FIDIC Yellow Book 2017 as a 'Special Provision' whereby the parties can adopt mediation as part of the 'Amicable Settlement' procedure in clause 21.5.[450]

9.7.3.2 Dispute Avoidance/Adjudication Board

The 'dispute advisory board' is unique to FIDIC Yellow Book 2017, as clause 21 (Disputes and Arbitration) requires the parties to appoint the 'DAAB'. FIDIC's DAAB may initially seem very similar to NEC4's DAB (see Section 9.7.2.2), but there are two fundamental differences between them:

o FIDIC's DAAB:
  - Decides disputes[451]; and
  - Their decision is binding upon the parties.[452]
o NEC4's DAB:
  - Only provides a recommendation[453] (not a decision) of how to resolve the dispute; and
  - The recommendation is not binding[454] upon the parties.

Clause 21.1 provides that the DAAB comprises three members by default, unless the parties enter a single member in the Contract Data.[455] The parties must jointly appoint the DAAB members (including the 'chairperson' if there are three DAAB members) within either the time stated in the Contract Data or within the default period of 28 days after the contractor receives the Letter of Acceptance.[456] The parties must also agree if they are to terminate any DAAB member's appointment, and on the identity of his replacement. If the parties either fail to appoint a single member, or three members, or any replacement

---

450 FIDIC Yellow Book 2017, see 'Particular Conditions Part B – Special Provisions' (see also footnote 409), encouraging the parties to adopt the ICC Mediation Rules (1 January 2014).
451 FIDIC Yellow Book 2017, clause 21.1 (Constitution of the DAAB).
452 FIDIC Yellow Book 2017, clause 21.4.3 (The DAAB's Decision).
453 NEC4, Option W3, clause W3.2(5).
454 *NEC4: A User's Guide – Managing an Engineering and Construction Contract*, (2020), p. 84.
455 FIDIC Yellow Book 2017, Particular Conditions Part A – Contract Data, entries for clause 21.1, comprising three proposed DAAB members each given by the employer and the contractor.
456 FIDIC Yellow Book 2017, clause 21.1 (and see clause 1.1.50 – defines the 'Letter of Acceptance').

member, clause 21.2 (Failure to Appoint DAAB Member(s)) applies, so the appointing body named in the Contract Data[457] makes the relevant appointment, which is final and binding.

Clause 21.4 provides that the parties commence the DAAB proceedings by issuing their 'reference' to the DAAB as follows:

o The 'reference' must be in writing,[458] and confirm that it is a 'reference' under clause 21.4.1.[459]

o For the matter in dispute,[460] it must set out the case[461] made out by the party making the reference.

o Within 42 days of issuing a Notice of Dissatisfaction ('NOD' – under clause 3.7.5) with the engineer's determination under clause 3.7 (Agreement or Determination – see Section 9.7.2.1 and Figure 9.7).[462]

o If the DAAB comprises three members, they are deemed to have received the reference when it is received by the 'chairperson'.

The parties must assist the DAAB in reaching its decision, giving access to the site (and any facilities such as for testing) and all relevant information.[463] The DAAB must then provide its decision within 84 days of the clause 21.4.1 'reference' unless the parties agree otherwise,[464] and such decision must[465]:

o Be in writing, and contain the DAAB's reasons; and

o Be issued to both parties and to the engineer.

Irrespective of whether a party gives a NOD under clause 21.4.4 (Dissatisfaction with the DAAB's Decision), the decision is binding upon the parties, who shall 'promptly' comply with it, which may include:

o Immediate payment of a sum of money from one party to another[466];

o Paying security for costs if the DAAB's decision is reversed in arbitration[467]; and

o The employer being responsible for and ensuring that the engineer complies with the DAAB's decision.[468]

---

457 FIDIC Yellow Book 2017, Particular Conditions Part A – Contract Data, entries for clause 21.1, comprising three proposed DAAB members each given by the employer and the contractor.
458 FIDIC Yellow Book 2017, clause 21.4.1(d).
459 FIDIC Yellow Book 2017, clause 21.4.1(b).
460 FIDIC Yellow Book 2017, clause 1.1.29 defines 'Dispute'.
461 FIDIC Yellow Book 2017, clause 21.4.1(c).
462 FIDIC Yellow Book 2017, clause 21.4.1(a) (Reference of a Dispute to the DAAB).
463 FIDIC Yellow Book 2017, clause 21.4.2 (The Parties' obligations after the reference).
464 FIDIC Yellow Book 2017, clause 21.4.3(a). Clause 21.4.3(b) provides that a period other than 84 days is firstly proposed by the DAAB and then agreed on by the parties. The DAAB may postpone its decision if any DAAB member's invoice remains outstanding, but upon payment in full, shall issue the decision as soon as possible (clause 21.4.3).
465 FIDIC Yellow Book 2017, clause 21.4.3 (The DAAB's decision).
466 FIDIC Yellow Book 2017, clause 21.4.3(i).
467 FIDIC Yellow Book 2017, clause 21.4.3(ii), a party may request the DAAB orders payment of security of the arbitration costs as part of its decision, but only if the DAAB has reasonable grounds to believe that the party receiving the payment decided by the DAAB will be unable to repay such an amount if the DAAB's decision is reversed in arbitration (under clause 21.6).
468 FIDIC Yellow Book 2017, clause 21.4.3.

If neither party issues a clause 21.4.4 NOD ('clause 21.4.4 NOD') within 28 days[469] of the date of the DAAB's decision, then it is final and binding upon the parties.[470] A clause 21.4.4 NOD may be issued in the following circumstances:

o If the DAAB has failed to provide its decision within 84 days, then either party may issue a clause 21.4.4 NOD to the other[471];
o If a party disputes the entirety of the DAAB's decision, then the clause 21.4.4 NOD:
  – Is issued to the other party, with copies to the engineer and the DAAB,[472] clearly stating that it is a 'Notice of Dissatisfaction with the DAAB's decision'[473]; and
  – Shall set out the matter in dispute, and the disputing party's reasons.
o If a party only disputes part of the DAAB's decision, then the clause 21.4.4 NOD must clearly identify which parts are disputed.[474]

A clause 21.4.4 NOD disputing only part of the DAAB's decision shall have the following effects:

o The parts of the DAAB's decision that are not disputes by the clause 21.4.4 NOD are deemed to become final and binding on the parties[475]; and
o The parts of the DAABs decision that are affected by the clause 21.4.4 NOD shall be severable from the rest of the decision.[476]

If a party fails to comply with the DAAB's decision, then under clause 21.7 (Failure to Comply with DAAB's Decision), the other party may proceed immediately[477] to refer a dispute to arbitration under clause 21.6 (Arbitration). The arbitral tribunal is encouraged to give effect to the DAAB's decision by adopting an 'expedited procedure', or ordering some 'summary', 'interim', or 'provisional measure' as an arbitral award.[478]

9.7.3.3 Arbitration

Arbitration is a method of finally resolving disputes where a professional person (often from a technical or legal background) or a panel of professional people are appointed to decide the dispute. If the parties agree, arbitration can offer them a very flexible procedure, but usually its procedures often reflect those used in litigation. Arbitration remains a popular means of resolving international disputes; however, its use in the UK has diminished, being replaced by adjudication and the streamlining of construction litigation (including the Technology

---

469 FIDIC Yellow Book 2017, clause 21.4.4(c).
470 FIDIC Yellow Book 2017, clause 21.4.4 (Dissatisfaction with the DAAB's decision).
471 FIDIC Yellow Book 2017, clause 21.4.4.
472 FIDIC Yellow Book 2017, clause 21.4.4(a).
473 FIDIC Yellow Book 2017, clause 21.4.4(b).
474 FIDIC Yellow Book 2017, clause 21.4.4(i).
475 FIDIC Yellow Book 2017, clause 21.4.4(iii).
476 FIDIC Yellow Book 2017, clause 21.4.4(ii).
477 FIDIC Yellow Book 2017, clause 21.7 – seemingly, a party may proceed 'immediately' to arbitration without issuing (a) a clause 21.4.4 NOD, or (b) attempting to reach an 'Amicable Settlement' under clause 21.5 beforehand.
478 FIDIC Yellow Book 2017, clause 21.7. Where the DAAB's decision is binding (but not final) and the arbitral tribunal gives an 'interim' or 'provisional award' in the face of a party's failure to comply with the DAAB's decision, the parties' rights as to the merits of the Dispute are reserved until they are finally resolved by an arbitral award.

and Construction Court (TCC), along with pre-action protocols for construction and engineering disputes – see Section 9.7.4.2). One aspect of arbitration that parties often find attractive is that, unlike litigation, it is a private procedure. Therefore, if a dispute concerns issues that the parties may want to remain private (such as involving a party's conduct, its finances, or its intellectual property), they may select arbitration instead of litigation.

Arbitration in the UK is governed by the Arbitration Act 1996,[479] which empowers the parties to adopt flexible procedures. Unfortunately, this very progressive piece of legislation has been somewhat overshadowed by adjudication under the HGCRA 1996 (which came into force in 1999) and the litigation reforms resulting in the TCC. The decline in UK arbitration is also reflected in the JCT standard forms of building contract that, since their 2005 revisions, provide that the default method of finally resolving disputes has been litigation.[480]

Under the three design and build contracts, the parties may select Arbitration in the following ways:

o JCT DB 2016. The entry in the Contract Particulars against Article 8 (Arbitration) must state that 'Article 8 and clauses 9.3 to 9.8 (Arbitration) apply' [underlining added], along with identifying the body[481] that will appoint the arbitrator. The arbitration will be conducted in accordance with the JCT 2016 edition of the Construction Industry Model Arbitration Rules (CIMAR).[482]
o FIDIC Yellow Book 2016. Clause 21.6 provides for arbitration and states that the Rules of Arbitration of the International Chamber of Commerce (ICC) shall apply to:
  – The appointment of the arbitrator(s); and
  – The conduct of the arbitration.
o NEC4. The following entries in Contract Data Part One under 'Resolving and avoiding disputes' are completed:
  – Arbitration is selected as the 'tribunal',
  – The arbitration procedure,[483]
  – The place where the arbitration is to take place, and
  – The arbitrator nominating body (described as the 'person or organisation').

The arbitrator's decision (also often referred to as the 'award') is binding upon the parties, and in the UK it can be enforced by the Court. In the UK, the parties do have limited rights of appeal under the Arbitration Act 1996.

9.7.3.4 Expert Determination

Expert determination is another procedure in which a third party (the expert) determines a dispute; however, because it is viewed as a process whereby the parties effectively put

---

479 1996 c23.

480 Previously, JCT design and build contracts (CD81 and JCT WCD 1998 editions) included arbitration as the default method of finally resolving disputes.

481 JCT DB 2016, Contract Particulars' entry for clause 9.4.1. The parties shall select a President or a Vice President of the Royal Institute of British Architects (RIBA), the Royal Institution of Chartered Surveyors (RICS), or the Chartered Institute of Arbitrators (CIArb) to appoint the arbitrator. If nobody is selected, the default appointor is the President or a Vice President of the RIBA.

482 JCT DB 2016, clause 9.3 (Conduct of Arbitration).

483 NEC4. For example, the parties could adopt the CIMAR or the ICC rules of Arbitration.

the entire dispute in the expert's hands to resolve, the parties' rights of appealing an expert determination are practically non-existent, save for exceptional circumstances. This 'additional finality' often discourages parties from adopting expert determination. Like arbitration, expert determination is a private procedure, and one where the parties may choose flexible procedures. However, in practice, and perhaps given parties' relative unfamiliarity with expert determination, they are likely to adopt established rules[484] covering procedure and timetable.

There is no provision in JCT DB 2016 for expert determination. FIDIC Yellow Book 2017 suggests expert determination in the Special Provisions version of clause 21.5, using the ICC's Expert Rules (see Section 9.7.1). Expert determination can be selected as the 'tribunal' for NEC4 (see Section 9.7.1).

9.7.4 Statutory Methods of Dispute Resolution

This section only considers a party's statutory right to a method of dispute resolution if the law applicable to the contract is UK law. It briefly describes statutory adjudication and litigation, both of which, under most circumstances, are available to the parties.

9.7.4.1 Adjudication

A party's statutory right to adjudication was provided by the HGCRA 1996 (as amended) and came into force in the UK in 1999. Adjudication's principal aim was to provide a swift, relatively cheap and temporarily binding resolution of a dispute, in order to allow the project to continue relatively uninterrupted. The focus of this principal aim was to provide a means to preserve cash flow, so a party didn't have to wait until much later (or in some instances, after the project was completed) to commence dispute resolution proceedings.

A party's right to statutory adjudication is conditional upon the particular contract being a 'construction contract' for 'construction operations' within the meanings of sections 104 and 105, respectively, of the HGCRA 1996 (see Section 9.6.3). If the contract in question is a 'construction contract' within the meaning of s104 of the HGCRA 1996,[485] then s108 provides that a party has the right to refer a dispute for adjudication, under either:

o The adjudication terms of that contract, provided that in all cases, such terms provide:
  – That a party may give notice of his intention to refer a dispute to adjudication at any time[486];
  – A timetable with the object of securing the adjudicator's appointment and referral of the dispute to him within seven days of the such notice[487];
  – That the adjudicator reaches his decision within 28 days[488] (or extended by 14 days with the agreement of the party referring the dispute, or such longer period requiring the agreement of both parties[489]) after the date when the dispute was referred;

---

484 For example, see the Academy of Experts Rules for Expert Determination (2008) or the International Chamber of Commerce ('ICC') Expert Rules (current version published in August 2019).
485 But not a contract involving a residential occupier – s106 HGCRA 1996.
486 s108(2)(a) HGCRA 1996.
487 s108(2)(b) HGCRA 1996.
488 s108(2)(c) HGCRA 1996.
489 s108(2)(d) HGCRA 1996.

- That the adjudicator acts impartially,[490] can take the initiative when ascertaining the facts and the law,[491] and written terms that he is not liable for anything done (or omitted) when discharging his functions as adjudicator (unless that act or omission is in bad faith[492]); and
- Written terms that the adjudicator's decision is temporarily binding upon the parties until finally determined by litigation, by the parties' agreement, or by arbitration if the contract provides for arbitration[493]; or
o If the contract does not provide for all of the points above,[494] then the default adjudication procedure set out in the Scheme for Construction Contracts (England and Wales) Regulations 1998[495] as amended ('the Scheme') will apply.[496]

JCT DB 2016 adopts the Scheme as its adjudication procedure.[497] Presumably owing to its 'international' nature, there is no provision for adjudication in FIDIC Yellow Book 2017, so if the subject of that contract is 'construction operations' and the applicable law is UK law, then without any express adjudication provisions, the Scheme will apply. NEC4's Option W2 complies with s108 of the HGCRA 1996 and essentially mirrors the Scheme's procedure, but with the following adjustments:

o The adjudicator is appointed under the NEC Dispute Resolution Service Contract.[498]
o Additional timetable matters:
  - Within three days of receiving the notice of adjudication,[499] the adjudicator must inform the parties if he can or cannot act[500];
  - Not later than 14 days after it has referred the dispute to the adjudicator[501] (and the other party), the referring party must provide any further information to be considered by the adjudicator[502];
  - Instruct a party to provide further information within a specified period of time.[503]
o The ability to join a related dispute with a subcontractor to the adjudicator.[504]
o Regarding the adjudicator's decision:
  - He must also provide a copy of his decision to the project manager.[505]
  - It is understood that the adjudicator may open up and revise matters that become final and conclusive (unlike the restriction imposed by s20(a) of the Scheme); otherwise he

---

490 s108(2)(e) HGCRA 1996.
491 s108(2)(f) HGCRA 1996.
492 s108(4) HGCRA 1996. This protection extends to the adjudicator's employees or agents.
493 s108(4) HGCRA 1996.
494 As set out in ss108(1) to 108(4) of the HGCRA 1996.
495 The Scheme for Construction Contracts (England and Wales) Regulations 1998 SI 1998 No. 649, or The Scheme for Construction Contracts (Scotland) Regulations 1998 (as amended) 1998 SI No. 687.
496 s108(5) HGCRA 1996.
497 JCT DB 2016, clause 9.2 (Adjudication).
498 NEC4, Option W2, clause W.2.2(2).
499 The notice of intention to refer a dispute to adjudication, complying with s108(2)(a) of the HGCRA 1996.
500 NEC4, Option W2, clause W2.3.
501 NEC4, Option W2, clause W2.3(2), complying with s108(2)(b) of the HGCRA 1996.
502 NEC4, Option W2, clause W2.3(2).
503 NEC4, Option W2, clause W2.3(4).
504 NEC4, Option W2, clause W2.3(3).
505 NEC4, Option W2, clause W2.3(8).

would seemingly be unable to give a decision regarding implemented compensation events (see Section 9.8.3).
- The decision becomes final and binding upon the parties, unless a party notifies the other within four weeks of the adjudicator's decision[506] that he intends to refer the disputed matter to the 'tribunal'.

9.7.4.2 Litigation

Under UK law, a party has a right to have a dispute finally resolved by litigation (Court proceedings) unless it has already agreed on another method of finally resolving a dispute (e.g. by arbitration). Civil litigation in the UK is governed by the Civil Procedure Rules (CPR), and in most instances, cases are allocated to a particular procedural 'track' depending upon their complexity and the value in dispute:

o The small claims track if the amount in dispute does not exceed £10 000[507];
o The 'fast track' if
  - The amount in dispute does not exceed £25 000[508];
  - The trial length is likely to be no more than one day[509]; and
  - Expert evidence is limited to one expert per party to any expert field[510] and expert evidence in two fields[511]; and
o To the 'multi-track' if neither allocated to the small claims nor to the fast track.[512]

Most construction disputes allocated to the multi-track are litigated in the TCC. Since its current incarnation as the TCC in 1998, it has been led by several judges who have practised extensively in the fields of construction and engineering disputes.

Prior to embarking upon litigation, the parties are encouraged to resolve or clarify as much of the dispute[513] as they can by complying with the Pre-Action Protocol for Construction and Engineering Disputes[514] ('the Protocol'). The Protocol's objectives are:

o To exchange sufficient information about the proposed proceedings broadly to allow the parties to understand each other's position and make informed decisions about settlement and how to proceed[515]; and
o To make appropriate attempts to resolve the matter without starting proceedings and, in particular, consider the use of an appropriate form of ADR[516] in order to do so.[517]

---

506 NEC4, Option W2, clauses W2.3(11) and W2.4(2).
507 CPR r26.6(3).
508 CPR r26.6(4).
509 CPR r26.6(5)(a).
510 CPR r26.6(5)(b)(i).
511 CPR r26.6(5)(b)(ii).
512 CPR r26.6(6).
513 The dispute may also include professional negligence claims against architects, engineers, and quantity surveyors – *Protocol*, paragraph 1.1.
514 *Pre-Action Protocol for Construction and Engineering Disputes*, 2nd ed., (17 February 2017), www .justice.gov.uk/courts/procedure-rules/civil/protocol/prot_ced
515 *Protocol*, objective 3.1.1.
516 Alternative dispute resolution – such as mediation (see Section 9.7.3.1).
517 *Protocol*, objective 3.1.2.

Paragraph 4 of the Protocol dissuades the parties from 'flagrant or very serious disregard' for the Protocol's terms, and the Court may impose cost sanctions upon a party in the case of such failure.

The Protocol does not apply to the enforcement of an adjudicator's decisions.[518]

JCT DB 2016's default method of finally resolving disputes is litigation (legal proceedings). Therefore, if the parties select neither option in the Contract Particulars' entry for 'Article 8 – Arbitration' – leaving it as 'apply/do not apply', then litigation will apply.

FIDIC Yellow Book 2017 does not contain any provision for litigation (or legal proceedings), as clause 21.6 (Arbitration) will apply. NEC4 can adopt litigation if the entry in Contract Data Part One, under 'Resolving and avoiding disputes', states that the 'tribunal' is the Court.

9.7.5 Summary of Dispute Resolution Methods

Table 9.13 summarises the methods of dispute resolution available across the three design and build contracts.

Table 9.13 Summary of dispute resolution methods.

Dispute resolution method		JCT DB 2016	FIDIC Yellow Book 2017	NEC4
Informal	Negotiation	Supplemental Provision 10	Clause 21.5 (Amicable Settlement)	—
Contractual	Engineer's determination	—	Clause 3.7 (Agreement/ Determination)	—
	Recommendation (non-binding)	—	—	Option W3 (Dispute Avoidance Board)
Formal	Mediation	Clause 9.1	Clause 21.5 (Amicable Settlement)	—
	Dispute Avoidance/ Adjudication Board	—	Clause 21	—
	Arbitration	Article 8, Clauses 9.3–9.8	Clause 21.6	Tribunal is arbitration
	Expert determination	—	Clause 21.5 (Amicable Settlement)	Tribunal is expert determination
	Adjudication[a]	Clause 9.2	— (HGCRA 1996)	Option W2
	Litigation[a]	Applies by default unless arbitration selected	—	Tribunal is legal proceedings

a) A statutory method of resolving disputes under UK law.

---

518 Under r24 of the CPR. *Protocol,* paragraph 2.1 (Exceptions).

9.8 Concluding the Contract

This section considers the ways that the three design and build contracts effectively bring the contract to an end – when each party has exhausted its obligations to the other under that contract. By this stage, the parties should be in the following positions:

○ The contractor will have:
  – Completed the project's design and constructed the works, satisfactorily carried out all testing and commissioning, and rectified any defects;
  – Submitted all documents (either progressively during the course of the contract or by a specific date) for the employer (or his representative) to calculate the final sum due[519]; and
  – Only his statutory obligations remain (such as for latent defects).
○ The employer (or his representative) will have:
  – Issued formal documentation attesting to the works being completed, passing all tests and performance requirements, and that all defects have been either accepted (in return for a reduction in the contract price) or rectified; and
  – Reviewed and responded to the information provided by the contractor to calculate the final sum due.

Therefore, at this final stage, the works should be indisputably complete, defect-free, and fully functional, and each party should know the other's position in terms of the final sum due for the works. All that remains is to bring the contractual process to a conclusion.

9.8.1 JCT DB 2016

Clause 1.8.1 of JCT DB 2016 concludes the contract 'from the due date for the final payment' – which occurs one month after the relevant Final Statement[520] is issued (refer to Section 9.6.1 and Figure 9.5). From that date, and subject to the exceptions set out below, clause 1.8 (Effect of the Final Statement) provides that the relevant Final Statement has the effect of becoming 'conclusive evidence' in respect of the following:

○ The 'balance due' from either the employer to the contractor or from the contractor to the employer (as the case may be)[521];
○ That where the following are for the employer's approval; either expressly described in either the Employer's Requirements or in an instruction from the employer,
  – The quality of materials or goods and/or
  – The standard of any item of workmanship,

---

519 The final sum due (if any) will be either (a) from the employer to the contractor, or (b) from the contractor to the employer.

520 JCT DB 2016. The 'relevant Final Statement' means either (a) the Final Statement issued by the contractor to the employer under clause 4.24.1, or (b) the employer's Final Statement under clause 4.24.4, as the case may be.

521 JCT DB 2016, clause 4.24.6: referring to the 'sum due' under clause 4.24.2, which is the 'balance due' as calculated in clause 4.24.2 calculated as the difference between (a) the Contract Sum as adjusted under clause 4.2 (clause 4.24.2.1), and (b) the amount already paid by the employer to the contractor (clause 4.24.2.2).

are to the employer's satisfaction, and comply with the contract[522]:

o All extensions of time (under clause 2.25) have been given[523]; and
o Any loss and expense due to the contractor (irrespective of whether it arises from a Relevant Matter,[524] or breach of statutory duty, breach of contract, breach of a duty of care, or otherwise) is in final settlement of any claims.[525]

The 'exceptions' to the clause 1.8.1 matters becoming 'conclusive evidence' (as provided in that clause) are:

o In the case of fraud[526];
o If the employer notifies[527] the contractor that he disputes anything in the contractor's Final Statement[528] before the due date for the final payment[529]; and/or
o If proceedings (meaning adjudication, arbitration, or other proceedings[530]) are commenced:
  – Either before, or within 28 days after the date when the relevant Final Statement is issued[531]; or
  – If an adjudication began not later than 28 days after the relevant Final Statement is issued,[532] and the adjudicator's decision is due after the relevant Final Statement is issued, *and* legal proceedings (or arbitration) are commenced within 28 days after the date of the adjudicator's decision;

then any 'conclusive evidence' effect of the relevant Final Statement is[533]:

  – Firstly, suspended until such proceedings[530] are concluded[534]; and
  – Secondly, subject to the outcome of those proceedings.[530]

The 'conclusive evidence' effect of the relevant Final Statement essentially means that no contradictory evidence can be provided to cast doubt on or attempt to overturn the relevant Final Statement's effect under clauses 1.8.1 and 4.24.6. Therefore, once the 'conclusive evidence' provision takes effect, it is final and binding upon the parties, as there are relatively few grounds upon which a party may challenge the 'conclusive evidence'

---

522 JCT DB 2016, clause 1.8.1.1.
523 JCT DB 2016, clause 1.8.1.2.
524 JCT DB 2016, 'Relevant Matters' are set out in clause 4.21 (see Section 8.11.2).
525 JCT DB 2016, clause 1.8.1.3.
526 JCT DB 2016, clause 1.8.1.
527 JCT DB 2016, the employer's 'notice' under clause 4.24.6 must comply with clause 1.7.
528 JCT DB 2016, the 'contractor's Final Statement' is issued by the contractor pursuant to clause 4.24.1.
529 JCT DB 2016, clause 1.8.1 refers to the 'exception' as set out in clause 4.24.6 (the employer's notice disputing the contractor's Final Statement).
530 JCT DB 2016. It is unlikely that 'other proceedings' will include mediation (under clause 9.1) as clauses 1.8.2 and 1.8.2.2 refer to a 'decision', 'award' or 'judgement', whereas a mediator does not provide any decision (or the like), but merely assists the parties in coming to an agreement.
531 JCT DB 2016, clause 1.8.2.1.
532 JCT DB 2016, the 28-day period referred to in clause 1.8.2.1.
533 JCT DB 2016, clause 1.8.2.2.
534 JCT DB 2016, clause 1.8.3 provides that if neither party takes any steps pursuant to any proceedings for a period of 12 months commencing after the relevant Final Statement has been issued, then those proceedings are deemed to have been 'concluded'.

effect of the relevant Final Statement. It is suggested that aside from a party commencing proceedings not later than 28 days after the relevant Final Statement is issued,[535] the parties' only realistic grounds to mount a challenge are to:

o Provide evidence of fraud[536]; or
o Demonstrate that there was some irregularity in issuing the relevant Final Statement, such as:
  – The employer failed to notify the contractor under clause 4.24.3, either properly or at all, before issuing the Employer's Final Statement under clause 4.24.4; or
  – The relevant Final Statement is deficient in some way – e.g. it is not in the form and with the details required by clause 4.24.2, although if it is issued late, this may not prevent it from becoming 'conclusive evidence'.[537]

It is vitally important that if a party intends to challenge the relevant Final Statement that they commence proceedings not later than 28 days after the statement is issued. Whilst the statement's 'conclusive evidence' effect does not manifest itself until the 'due date for the final date for payment', and clause 4.9.5 enables the paying party to issue a 'pay less notice', the effect of a pay less notice is unlikely to be permanent, as a Court, arbitrator ,or adjudicator is unlikely to consider it has any merit, if (and by virtue of clause 1.8.1) they have already agreed that the relevant Final Statement is 'conclusive evidence', thereby prohibiting any dissenting evidence (that could support a pay less notice) being permitted.

9.8.2   FIDIC Yellow Book 2017

FIDIC Yellow Book 2017 does not provide for a single document that concludes the contract in the same way as JCT DB 2016 does (with the relevant Final Statement). Instead, FIDIC Yellow Book 2017 has the following clauses that confirm each party's liability to the other is at an end:

o The contractor's obligations to the employer end with the Performance Certificate[538] under clause 11.9, certifying that the contractor has completed its obligations under the contract; and
o The employer's obligations to the contractor end with:
  – The FPC[539] (under clause 14.13); and
  – Clause 14.14 (Cessation of Employer's Liability) provides that the amounts certified in the FPC become conclusive (or are 'deemed to be accepted' by the contractor) if the contractor does not make a claim under clause 20.2 (Claims for Payment and/or EoT) within 56 days of receiving the FPC.

---

535 JCT DB 2016, under clause 1.8.2.1.
536 JCT DB 2016, as per clause 1.8.1.
537 *Keating on Construction Contracts*, 10th ed., (2016), paragraph 20-067.
538 FIDIC Yellow Book 2017, clause 1.1.64 defines 'Performance Certificate'.
539 FIDIC Yellow Book 2017, clause 1.1.40 defines the 'Final Payment Certificate' or 'FPC'.

In all cases, the party's liabilities to the other may only continue in respect of the following:

o 'Unfulfilled Obligations' to each other[540] as set out in clause 11.10; and/or
o In the cases of fraud, gross negligence, deliberate default, or reckless misconduct.[541]

9.8.3 NEC4

NEC4 concludes the contract differently from JCT DB 2016 and FIDIC Yellow Book 2017, as the NEC ethos is to progressively resolve virtually all matters under the contract. It achieves this by providing for the following:

o For changes to the Prices[542] or to the Completion Date[543]:
   – These are dealt with via the Compensation Event procedure as per clause 63.6 (see Section 8.8.5.2); and
   – Once implemented, a Compensation Event cannot be changed as per clause 66.3 (see Section 8.9.3).
o The Defects Certificate[544] under clause 44.3[545] (see Section 9.5.3) sets out that either:
   – The contractor has corrected all the defects notified before the 'defects date'; or
   – If the contractor has failed to correct any remaining defects within the relevant 'defects correction period', clause 46.1 obliges the contractor to pay (to the client) the cost of others correcting the defect(s) as assessed by the project manager.

The only other matter is for the project manager to assess the 'final amount due' under clause 53 (see Section 9.6.3), which will include:

o The 'contractor's share' under Options C (Target Contract with Activity Schedule) and D (Target Contract with Bill of Quantities); and/or
o The 'budget incentive' if Option X22 (Early Contractor Involvement) applies.

9.8.4 Summary

Table 9.14 summarises the various ways in which the three design and build contracts conclude the contract.

---

540 FIDIC Yellow Book 2017, clause 11.10 (Unfulfilled Obligations). The contractor should not have any further obligations to perform after the engineer issues the Performance Certificate save for (i) clearing the site under clause 11.11 and (ii) issuing the draft Final Statement and agreed Final Statement under clause 14.11 and the 'discharge' under clause 14.12.

541 FIDIC Yellow Book 2017, the parties' liabilities to the other are not restricted in the cases of fraud, gross negligence, deliberate default, or reckless misconduct under clause 11.10 (Unfulfilled Obligations) in the case of the contractor, or under clause 14.14 (Cessation of the Employer's Liability) for the employer.

542 NEC4, clause 63.1 – akin to JCT DB 2016's clause 4.24.2 (final adjustment of the Contract Sum), and clause 1.8.1.3 (final calculation of loss and/or expense).

543 NEC4, clause 63.5 – akin to JCT DB 2016's clause 1.8.1.2 (finals assessment of extensions of time).

544 NEC4, clause 11.2(7) defines the 'Defects Certificate'.

545 NEC4, clause 44.3 confirms that the Defects Certificate does not affect the Client's rights for latent defects (i.e. those not discoverable until after the last 'defect correction period').

Table 9.14 Summary of methods of concluding the contract.

Description	JCT DB 2016	FIDIC Yellow Book 2017	NEC4
Final document(s) to conclude the contract	Yes, relevant Final Statement (cl 4.24)	Final Payment Certificate (cl 14.13), Performance Certificate (cl 11.9)	Only[a)] in respect of the 'final amount due' (cl 53), and the Defects Certificate (cl 44.3)
'Conclusive effect'	Yes (cl 1.8.1)	Yes (cl 14.14)	Only in respect of Compensation Events (cl 63.6, 61.3)
Time limit to commence dispute resolution proceedings	28 days (cl 1.8.2)	56 days (cl 14.14)	No

a)   NEC4, clause 66.3 provides that once a 'compensation event' is implemented, it is not revised.

9.8.5   Latent Defects

Under UK law, the period of limitation (for which a party may be liable for a breach of contract) is commonly either 6 years if the contract is signed 'under hand', or 12 years if the contract is signed as a deed. In either case, the period of limitation runs from completion of the works.[546] This is set out in the Limitation Act 1980.[547]

Broadly speaking, the three design and build contracts reflect or accommodate this approach:

o   JCT DB 2016 only limits the contractor's liability for the quality of materials or goods, or the standard of workmanship that was (i) set out in the Employer's Requirements, and (ii) was said to be to the employer's satisfaction[548] (see Section 9.8.1);

o   FIDIC Yellow Book 2017, unless prohibited by law, the contractor's liability for plant is limited to two years.[549] Its liability for latent defects for work other than plant is likely to be unaffected; and

o   NEC4 confirms that the client's right in respect of a latent defect (i.e. one that has not been found or notified) is unaffected by the supervisor issuing the Defects Certificate.[550]

---

546 'Completion' in this case means (i) 'practical completion' under JCT DB 2016, (ii) 'Taking Over' under FIDIC Yellow Book 2017, and (iii) 'Take over' under NEC4.
547 Limitation Act 1980. C58.
548 JCT DB 2016, clause 1.8.1.1.
549 FIDIC Yellow Book 2017, clause 11.10.
550 NEC4, clause 44.3.

Part II

10

Design and Build in Its Current Form

Section 1.4 in this book – on recent developments in design and build – briefly discussed the employer's increasing selection of design and build contracts along with a general summary of how employers use bespoke contract amendments to reduce their risk.

These recent developments make the study and understanding of design and build procurement problematic for both students and practitioners. It is almost as if one has to firstly learn the rules of design and build in order to break them! Part One of this book considered how design and build was firstly conceived and then intended to be operated – all in accordance with the procedures intended by the JCT DB 2016, FIDIC Yellow Book 2017, and NEC4 contracts. Many of these procedures are maintained in the more recent use of design and build, but there are significant differences. It is therefore suggested that in order to understand the changes in the way that design and build contracts are often now operated, it is important to firstly understand the original (or intended) procedures and compare them with the changes to design and build that are now being made in current practice.

Part Two of this book explores how these changes are now altering the way design and build contracts are being operated. The changes can be stark – often completely altering the principles upon which design and build was originally established in the early 1980s.

The main changes in how design and build contracts are operated, in particular in the construction sector, are essentially twofold:

o The employer wants to exert control over the project's design (which is covered in this chapter); and
o The employer wants to take on less risk (covered in Chapter 11).

10.1   Reduction of the Contractor's Design

In order to establish a greater control of the project's design, the employer must effectively reduce the amount of the design that the contractor carries out. This can partly be achieved without amending any terms of the three design and build contracts, by simply providing a very comprehensive Employer's Requirements document.[1] The following examples (broadly showing how the extent of the contractor's design can be reduced over three 'levels') illustrate how the detail contained in the Employer's Requirements has evolved:

o **Level 1.** The Employer's Requirements may contain specific descriptions of each element of work, broken down into the following:
  – Precise specifications for materials, products, and goods;
  – Specific standards for workmanship (such as tolerances); and
  – Precise performance specifications, and required outputs for plant and equipment.
  This would leave the contractor to carry out the design through RIBA Stages 3 (Concept Design), 4 (Spatial Coordination), and 5 (Technical Design). In the above example of *only* precisely specifying the construction components and performance criteria, the contractor would still be producing the project drawings and schedules. Therefore, despite the precise specifications, the employer may still be dissatisfied with elements of the contractor's design.
o **Level 2.** In addition to the Employer's Requirements containing the precise construction and performance specifications from the Level 1 example, they set out rudimentary design parameters such as:
  – RIBA Stage 2 (Concept Design): drawings showing the architectural concept, strategic engineering, project strategies, and design reviews; and
  – RIBA Stage 3 (Spatial Coordination): design studies, engineering analysis, and drawings showing a spatially coordinated design.
o **Level 3.** RIBA Stage 4 (Technical Design): produced in addition to the design described in the Level 1 and 2 examples (above). This includes the technical designs for architectural and engineering elements.

In providing a design up to the Level 3 example (i.e. including the detailed specifications and designs for RIBA Stages 2, 3, and 4), the employer would essentially have an almost fully developed design; effectively leaving the contractor with responsibility for what is often referred to as 'design development', which involves:

o Resolving any inadequacies, inaccuracies, divergences, or discrepancies in the Employer's Requirements;
o Coordination of the design in the Employer's Requirements:
  – Ensuring the architectural and engineering details (such as for structural, mechanical, and electrical elements) are coordinated; and
  – Integrating and coordinating any subcontractor's design.
o Additional design:
  – Completing any design 'gaps' that were absent from the Employer's Requirements' design yet are indispensably necessary to complete the project's design; and
  – Further design arising from variations.

---

1 'Employer's Requirements' set out the design that the contractor must complete in JCT DB 2016 and FIDIC Yellow Book 2017, whereas in NEC4, the extent and degree of the contractor's design is set out in the Scope.

Carrying out 'design development' clearly falls within the contractor's general obligations for design under the following (see Section 3.1.2):

o JCT DB 2016 – clause 2.1.1
o FIDIC Yellow Book 2017 – clause 4.1; and
o NEC4 – clauses 20.1 and 21.1.

The extent and detail of design included in the Employer's Requirements is often up to RIBA Stage 4 (Technical Design), which is more aligned to that provided by the employer under the traditional procurement methods (e.g. using JCT SBC 2016, FIDIC Red Book 2017,[2] or NEC4 without any contractor's design).

The employer can then obtain the benefit of an established design (from which the contractor cannot significantly deviate), and pass the entire responsibility for that design onto the contractor. At first glance, this may seem to place an additional and potentially unfair burden on the contractor; who must thoroughly check through the Employer's Requirements in their entirety, checking for any ambiguities, discrepancies, and inadequacies in order to satisfy himself that his price for the works and his contractor's proposals sufficiently deal with any issues that may arise. However, any 'unfairness' should only be perceived where the employer does not allow the contractor a sufficient period of time to make such checks. As is described in Section 5.4 (Early Contractor Involvement), employers are increasingly likely to engage the contractor from one of the earlier RIBA stages (between Stage 0 – Strategic Definition, and Stage 4 – Technical Design), so the contractor is often not only given a greater opportunity to scrutinise and check the design they are given (the Employer's Requirements), but also are often hand-in-hand, liaising with the employer's design team, and often advising them regarding the production and implications of that design.

10.2 Novation of Design Consultants

Novation means the substitution of a contract with another, or the transferral of a contract. It is a concept that has been applied to design and build contracts for some time – when substituting or transferring a contract with a design consultant that is initially between the designer and the employer, and is then between the design consultant and the contractor.

The advantage of novation, as perceived by the employer, is that by ensuring that the same design consultant is engaged (initially by himself, and subsequently by the contractor), continuity of the employer's design will be maintained, from the early RIBA Stages 2 (Concept Design) to 4 (Technical Design), into the construction phase (RIBA Stage 5 – Manufacturing and Construction).

Novation tends to be used more for construction projects, as this provides the employer with a greater degree of control over the project's aesthetic design. It is less common in engineering projects, as the contractor is often engaged during the early RIBA stages (Stage 0 – Strategic Definition to Stage 2 – Concept Design) to undertake design studies,

---

2 FIDIC Red Book 2017 – Conditions of Contract for Construction, 2nd ed., (2017), for Building and Engineering Works designed by the Employer.

because he is undertaking the majority of design work in any event, and achieving the project's function or performance level (rather than its aesthetic quality) is his principal aim.

10.2.1 How to Achieve Novation

Novation involves three parties – the employer, the consultant, and the contractor – so there must be an agreement between all three parties that transfers the design consultant's appointment from the employer to the contractor. Clearly, the design consultant and the contractor have to agree to this transferral, as it cannot simply be unilaterally imposed by the employer. Therefore, the employer must ensure that the terms under which the design consultant's appointment will be transferred (hereafter referred to as the 'Novation Agreement'), are expressly incorporated into the following documents:

o The design consultant's conditions of engagement (or appointment). This will contain an 'enabling clause' stating that the conditions of engagement will transfer from the employer to the contractor under the terms of the Novation Agreement, a copy of which will be appended to the conditions of engagement. A typical enabling clause in the conditions of engagement will state:

> *'Following the appointment by the Employer of the contractor under the building contract, the Employer shall, in writing, request the Consultant to execute and deliver to the Employer the Novation Agreement appended hereto. The Consultant shall comply with this clause within 14 days of the Employer's written request.'*

o The building contract. This will similarly include an enabling clause, obliging the contractor to appoint the design consultant under the terms of the Novation Agreement, a copy of which will be appended to the building contract. A typical enabling clause in the building contract will state:

> *'Not later than 14 days after execution of the building contract, the Contractor shall enter into the Novation Agreement in the form appended thereto, with the Employer and each of the design consultants listed below:'*

Given the popularity of novating the employer's consultants to the contractor, it is surprising that none of the publishers of the standard forms of contract (JCT, FIDIC, or NEC) offer a novation agreement. This is particularly surprising in the case of NEC, who publish an otherwise almost comprehensive suite of contracts – both for building contracts (NEC4 ECC Options A–F) and for professional services (NEC4 PSC[3]). The only and somewhat indirect exception would be the RIBA PSC,[4] which (at clause 4.6) enables

---

3 NEC4 Professional Services Contract (see Section 4.1.1).
4 Refer to RIBA Standard Professional Services Contract 2020 – Sections 4.1 and 4.2.1. The RIBA is a member of the Joint Contracts Tribunal (JCT) which publishes JCT DB 2016. The other JCT members are the British Property Federation, Build UK Group Ltd., Contractors Legal Grp Ltd., Local Government Association, the RICS, and the Scottish Building Contract Committee Ltd.

the client to novate the architect/consultant to the contractor under 'a separate Novation Agreement' that must include the wording set out in clauses 4.8.1–4.8.4 (inclusive)[5] of the RIBA PSC.

The Construction Industry Council ('CIC') publishes the following two novation agreements:

o Novation Agreement – Ab Initio[6] (dated 2018). This can be used where a design consultant is appointed very early in the project, and carries out minimal services (e.g. RIBA Stage 0 – Strategic Definition, or Stage 1 – Preparation and Briefing), two possible examples being:
  – The consultant is engaged by a developer, for whom he undertakes services for a particular project (e.g. a hotel) to be built on the developer's land. The developer may subsequently sell the site to another party – who will become the employer (engaging the contractor under the building contract to build the hotel), to whom the consultant's appointment can be novated; or
  – Where the employer engages the contractor early in the project, and novates the design consultant's services to the contractor (see Section 5.4 – Early Contractor Involvement).
o Novation Agreement – Switch (2nd ed., dated 2021). This is used where the design consultant has carried out detailed design work (up to RIBA Stage 4 – Technical Design) and is then novated from the employer to the building contractor, either:
  – Prior to the employer entering into the building contract with the contractor but after the design consultant has undertaken a significant amount of design (e.g. during RIBA Stage 4 – Technical Design – if the contractor is also engaged to carry out some pre-contract works under a PCSA[7]); or
  – After the building contract has been executed.

The terms of the CIC's two novation agreements are similar. The City of London Law Society also publishes a novation agreement,[8] which is similar to the CIC's 'Switch' novation agreement.

Key characteristics of novation agreements are as follows[9]:

o The novation agreement will be between three parties: the employer, the design consultant, and the contractor.
o It will refer to the design consultant's conditions of engagement (or appointment), which, pre-novation, is between the consultant and the employer, then post-novation is transferred from the employer to the contractor.

---

5 RIBA PSC, clauses 4.8.1–4.8.4 (inclusive) provide the 'key characteristics' of the CIC's novation agreements – as described in Section 10.2.1.
6 Broadly meaning 'from the start'. The CIC's 'Ab Initio' novation agreement is recommended by the RIBA's Standard Professional Services Contract 2020 (clause 4.8).
7 See Sections 5.4 (Early Contractor Involvement) and 5.4.6 (Pre-construction Services Agreements – 'PCSA').
8 www.citysolicitors.org.uk/clls/clls-precedent-documents/construction-law-novation-agreement which also includes the guidance notes.
9 Also refer to clause 4.8 of the RIBA's Standard Professional Services Contract (2020).

- o It will contain the following key rights and obligations that arise from the date of novation:
  - The design consultant agrees to release the employer from any further or continuing obligations it would otherwise have had under the conditions of engagement (or appointment); and
  - The contractor agrees to be responsible for any further or continuing obligations that the employer would have had to the design consultant, but for the novation agreement.
- o The design consultant's responsibility for pre-novation services is dealt with as follows:
  - It is transferred from the employer to the contractor; and
  - The design consultant remains liable to the contractor for breaches of the pre-novation services, notwithstanding the employer may have suffered no loss (see below).

10.2.2 Problems with Novation

Novating design consultants can go very smoothly. Employers have the comfort that their design intent is maintained and not compromised by the contract, and contractors often enjoy working with a designer who has been familiar with the project, often from its outset. However, novation is not without its problems.

Firstly, there is the issue of conflicts of interest arising. At its simplest, a conflict of interest arises when one party attempts to 'serve two masters'. When that party is a professional design consultant, and one who is bound by rules of professional conduct (such as those applying to members of the RIBA or RICS), then that person can only serve one 'master' and act in his best interests. The design consultant cannot say that he can serve two parties' best interests, as those interests are likely to conflict. Where the design consultant is novated from the employer to the contractor, the following conflicts of interest can arise:

- o The conditions of engagement (or appointment) or the novation agreement itself (both of which are likely to be drafted by the employer) may expressly provide that after novation to the contractor, the design consultant must still carry out 'residual' duties for the employer (such as reporting on the contractor's quality of work); and
- o The design consultant may have a long-standing business relationship with the employer, and whilst neither his conditions of engagement nor his novation agreement may contain any 'residual' duties to the employer, he may nevertheless (and presumably, out of loyalty to the employer) continue acting partly in the employer's interests after novation, when he should be working exclusively for the contractor.

The RIBA's guidance in the case of conflict of interest arising is clear. The consultant must either:

- o If a conflict of interest occurs, the RIBA member must declare it to all parties concerned. This would include the contractor.[10]
- o Remove the conflict of interest's cause[11]; or
- o Withdraw from the situation.[12]

---

10 RIBA Code of Professional Conduct (2018), Principle 1: Integrity, rule 3.1.
11 RIBA Code of Professional Conduct (2018), Principle 1: Integrity, rule 3.1(a).
12 RIBA Code of Professional Conduct (2018), Principle 1: Integrity, rule 3.1(b).

The most common way of avoiding a conflict of interest is for the designer's conditions of engagement or novation agreement to confirm that the following occurs upon novation:

o The employer releases and discharges the design consultant from any further performance of its obligations under its conditions of engagement;

o The contractor and design consultant agree that from the outset, the design consultant's conditions of engagement are deemed to always have been and will then be between the design consultant and the contractor (at the exclusion of the employer);

o The design consultant releases the employer from any further obligations under his conditions of engagement; and

o The contractor agrees to perform all the obligations of the employer under the terms of the design consultant's conditions of engagement.

The second problem with novation relates to the scope of the design consultant's services (and his fees for them). It is often the case that when the contractor receives the tender information from the employer, that a copy of the novation agreement is appended to the proposed terms of the building contract, but a copy of the relevant design consultant's conditions of engagement is often not provided. This means that whilst the prospective contractor is aware that a consultant or consultants may be novated, he has no idea *what* is intended to be novated to him! A prudent contractor should always request copies of any conditions of engagement for proposed novated consultants.

When engaging professional consultants (such as an architect) for a design and build project, the employer is normally only concerned with the design services that the consultant will carry out, pre-novation, on the employer's behalf. He is unlikely to give much thought to the post-novation services that the design consultant will carry out for the contractor during RIBA Stages 5 (Manufacturing and Construction) and 6 (Handover). The scope of the design consultant's services (as agreed with the employer) may not include many of the services that the contractor requires the consultant to undertake on his behalf, which may include:

o Acting as 'lead designer' (see Sections 4.2.3 and 4.2.6) and being responsible for coordinating and integrating:
  - All designs produced by other consultants (such as architectural, structural engineering, mechanical, electrical and building services engineering, etc.); and
  - All design provided by specialist subcontractors.
o Administrative tasks:
  - Attending project/site meetings;
  - Appraising/reporting on the performance of other consultants;
  - Advising on the scope of subcontractors' design and their applications for payment.
o Monitoring quality:
  - Appraising subcontractors' designs, materials, and workmanship;
  - Snagging the contractor's and subcontractors' work.

The contractor should therefore ensure that either (i) the scope of consultant's services agreed upon with the employer oblige the consultant to carry out all the services that the contractor requires of him, *and* includes the consultant's fees for doing so, or (ii) the design consultant agrees to carry out such services for the contractor and the contractor includes the additional consultant's fees for doing so within the contract price with the employer when agreeing on the building contract.

The third difficulty in respect of novated consultants is how the contractor may recover any additional costs that he incurs during the construction stage (RIBA Stage 5) that arise from pre-novation breaches committed by the consultant (i.e. when he was engaged by the employer and not by the contractor). The 2001 Scottish case of *Blyth & Blyth Ltd v Carillion Construction Ltd*[13] gave a sobering example of how this problem could affect contractors under design and build contracts. Blyth & Blyth were consulting engineers who were initially engaged to provide structural engineering advice and designs to the employer. The contractor (Carillion) was engaged by the employer under an amended form of the JCT CD81 design and build contract,[14] and Blyth & Blyth were novated to them from the employer. Unfortunately, Carillion faced the following problems:

o When Blyth & Blyth carried out the structural engineering design for the employer, their design showed far less reinforcement than was actually required to construct the building in accordance with the building contract; and

o The terms of the building contract had been amended, so that 'any mistake, in accuracy, discrepancy or omission in…the design contained in the Employer's Requirements… shall be corrected by the Contractor but there shall be no addition to the Contract Sum…'.

In order to correct the reinforcement 'mistake' in the Employer's Requirements, Carillion had to expend significant monies on purchasing and installing the necessary additional quantities of reinforcement in order to comply with the contract. As they were unable to recover the cost of the additional reinforcement from the employer (partly owing to the amended term, but it was highly doubtful whether it would constitute a 'Change' under the contract), Carillion then claimed the additional expenditure for reinforcement from Blyth & Blyth. The problem for Carillion was that Blyth & Blyth's failure to include sufficient reinforcement in the structural design was a breach they committed pre-novation, when they were working for the employer. Therefore, Blyth & Blyth had not committed a breach of novation agreement with Carillion. Furthermore, the employer could not claim any money from Blyth & Blyth because he had not suffered any loss; by virtue of (i) in any event, Carillion were obliged to provide him with a building that complied with the terms of the contract, and (ii) by agreeing to the contract amendments, Carillion had agreed to rectify any 'mistake, discrepancy, or omission' at their own cost.

The problem of to whom the consultant is liable, particularly for any breach committed during the pre-novation stage (i.e. when working for the employer), can be resolved by ensuring that the consultant's appointment and/or novation agreement contains one of the two options illustrated in Figure 10.1.

---

13 *Blyth & Blyth Ltd. v Carillion Construction Ltd,* ScotCS 90, (18 April 2001).
14 See Section 1.3. JCT CD81 is a predecessor of the JCT DB 2016 contract and contains many provisions of its 2016 successor.

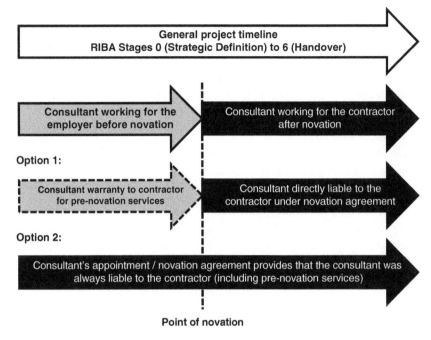

Figure 10.1 Novated consultant's liability transfer.

The two options are as follows:

o Option 1 provides the contractor with a warranty from the consultant to the contractor, promising that it is liable for any loss or damage suffered by the contractor that was caused by any default or negligent act by the consultant prior to the date of the novation agreement (i.e. pre-novation, when the consultant was engaged by the employer).[15]

o Option 2 provides the contractor with two limbs of protection:
  – Firstly, the consultant warrants to the contractor that all pre-novation services have been carried out correctly[16]; and
  – Secondly, the consultant is not absolved of liability to the contractor on the ground that the employer has suffered no loss.[17]

Increasingly for construction projects where the employer's representative[18] does not have a design background (for example, he is a quantity surveyor and not an architect or engineer), employers are being advised to retain their building services consultants and not novate them to the contractor. Two of the reasons for doing so are as follows:

o The employer's representative is not qualified (or possibly insured) to approve any building services design. This seems particularly short-sighted as it is unlikely that a professional person (such as a quantity surveyor) is suitably qualified to approve

---

15 As per clause 1.4 of the City of London Law Society Novation Standard Form of Novation Agreement (October 2007).
16 CIC Novation Agreement – Switch, 2nd ed., (2021), clause 4(a).
17 CIC Novation Agreement – Switch, 2nd ed., (2021), clause 4(b).
18 The Employer's Agent (JCT DB 2016) or Project Manager (NEC4).

architectural or structural engineering matters, but nevertheless, the architectural and structural engineering consultants are readily novated to the contractor; and

o The design produced by the building services consultant may be fairly simple. It may only consist of a performance specification so the detailed building services design will subsequently be produced by the contractor (often, by the building services subcontractors). Therefore, aside from a basic set of performance requirements, there is little in the way of detailed design for which the building services consultant may become liable.

However, the contractor should exercise caution in circumstances where the contract makes the contractor liable for all of the project's design (refer to Section 11.3) so any consultant contributing to the Employer's Requirements should either be:

o Novated to the contractor; or

o If the consultant is retained by the employer, provide the contractor with a collateral warranty that, before entering into the building contract (i.e. between the employer and the contractor) he performed his design duties in accordance with his conditions of engagement (between the employer and the designing consultant) correctly, or by exercising reasonable skill and care.

11

Common Amendments to Design and Build Contracts

Amendments to the standard terms of construction and engineering contracts may be introduced for a number of reasons. The three design and build contracts covered in this book, whilst carefully drafted, essentially adopt a 'one size fits all' approach, leaving the project-specific information to the following parts of the contract or the contract documents:

o Contract Particulars[1] for JCT DB 2016, or Contract Data[2] for FIDIC Yellow Book 2017 and NEC4 (refer to Section 2.2.4).
o The relevant contract documents providing the Scope of Works (refer to Sections 3.4 and 3.5):
  – JCT DB 2016 – the Employer's Requirements, and the Contractor's Proposals;
  – FIDIC Yellow Book 2017 – the Contract Agreement (including any addenda), the Letter of Acceptance, the Letter of Tender, Employer's Requirements, and Contractor's Proposals;
  – NEC4 – The Scope and Site Information.
o The relevant pricing documents (refer to Section 3.6):
  – JCT DB 2016 – the Contract Sum Analysis;
  – FIDIC Yellow Book 2017 – Schedule of Rates and Prices[3];
  – NEC4 – Data for the Short Schedule of Cost Components (for Options A[4] and B[5]), or the Data for the Schedule of Cost Components (for Options C[4], D[5], and E[6]).

Despite the contract obviously including information specific to the project (such as drawings, specifications, etc.), the employer or the contractor may consider that they and the project may benefit from amendments to the contract's conditions.

---

1 Contract Particulars contains the 'specific information' for JCT DB 2016 (see Sections 2.2.2 and 2.2.3).
2 Contract Data contains the 'specific information' (see Sections 2.2.2 and 2.2.3) in the case of FIDIC Yellow Book 2017 and NEC4.
3 FIDIC Yellow Book 2017 – the 'Schedule of Rates and Prices' is defined in clause 1.1.75. It is one of the 'Schedules' (defined in clause 1.1.72) which will be attached to the Letter of Tender (forming part of the 'Contract' – as defined in clause 1.1.9).
4 NEC4 Options A and C include the 'Activity schedule'.
5 NEC4 Options B and D include the 'Bill of quantities'.
6 NEC Option E is a cost-reimbursable contract.

Design and Build Contracts, First Edition. Guy Higginbottom.
© 2024 John Wiley & Sons Ltd. Published 2024 by John Wiley & Sons Ltd.

In most cases, contract amendments are firstly proposed by the employer, and are usually set out in the tender documents or discussed with the contractor during any early involvement stage or during negotiations. The contractor may proffer his own contract amendments, but this is less common, and they are often restricted to generic changes to the contract, rather than amendments which are project-specific.

Most amendments to design and build contracts fall into the following categories:

o Project-specific amendments
o Amendments for a practical benefit
o Increased contractor's design liability
o Reduced employer risk/increased employer benefit
o Reduced contractor rights (not design-related)
o Increased benefits to the contractor
o Clarifications

11.1   Project-Specific Amendments

Project-specific amendments are usually made to assist the employer in administering the contract, initially covering his obligations to third parties (such as funders, landlords, tenants, or public authorities) and his dealings with the contractor. The employer may find some of the contract's standard conditions to be unsuitable or inadequate for his needs, and may seek to make some of the changes described below.

11.1.1   Revised (or Wider) Definitions

The employer may require more appropriate or wider definitions than are set out in the contract's standard terms. Some examples are:

o The definition of the 'Works' may be widened to include areas other than the main site (for JCT DB 2016[7]):
    – The contract now includes (subsumes) any previous work, such as that carried out under a PCSA or as part of any enabling works.
o New definitions of 'third parties', such as:
    – Funders, purchasers, tenants;
    – Other beneficiaries under collateral warranties or bonds.

11.1.2   Additional Measures to Assist the Employer

The employer may require the contractor to perform his existing obligations in a particular way that provides further assistance to the employer or ensures that the employer is not in breach of any terms of a wider agreement for the project. This may be to satisfy the

---

7 The 'Site' can include multiple areas under FIDIC Yellow Book 2017 (see clause 1.1.77) and NEC4 (see clause 11.2(20) for the 'Working Areas').

requirements of those permitting the employer to carry out the works (such as a local authority, regulatory body, or a funder) and can include:

o Greater or more-specific insurance requirements (including professional indemnity insurance for the contractor or subcontractors).
o Confidentiality (contractor not to publicly disclose information concerning the works).
o Subcontracting:
    – 'Specified' subcontractors (either identified by name or by the class of work they will be undertaking) who will be designing works;
    – The employer may insist upon and require approval of a particular form of subcontract that the contractor must use;
    – Collateral warranties and bonds to be provided by the subcontractor.
o Further obligations during the works (e.g. to comply with local laws or conditions), often requiring the contractor to indemnify the employer in respect of:
    – Site safety, security, and maintenance;
    – Maintaining the site and cleaning surrounding area and access routes (roads, pathways);
    – Prevention of nuisance, trespass noise, or environmental pollution;
    – Working with other contractors engaged by the employer.
o Payment provisions:
    – Final date for payment is often extended to allow the employer time to process payment for the contractor. If the employer is a large organisation, its accounts department may require longer than the period set out in the contract to process the contractor's payments. In many cases, payments to the contractor may be substantial, so there may be several signatories required to authorise payment.
    – The contractor is required to provide a tax invoice, either instead of or in addition to an application for payment.
o Additional requirements for 'completion'[8]:
    – Specific documents to be provided as a condition of achieving completion (certification, testing certificates, approvals, etc.), additional copies, specific formats, etc.
o Post-completion obligations:
    – Maintenance
    – Monitoring
    – Training the employer's staff
o Changing the dispute resolution method:
    – Usually deleting arbitration in place of litigation.

11.2 Amendments with a Practical Benefit

Some amendments are intended to have a practical benefit, such as providing more detailed procedures for:

o Programme:
    – The contractor provides a programme (JCT DB 2016 only);

---

8 See Section 9.3.

- Contractor revises the programme after extensions of time (JCT DB 2016 and FIDIC Yellow Book 2017).
o Statutory compliance:
  - Compliance with specified local laws;
  - Contractor to obtain permits or approvals;
  - Adjudication and payment provisions (compliance with HGCRA 1996 for FIDIC Yellow Book 2017[9]).
o Work carried out by statutory bodies:
  - Contractor to complete all administrative functions and notifications regarding statutory bodies;
  - Work by statutory bodies to be shown on the contractor's programme.
o Contractor to accommodate other contractors engaged by the employer:
  - Contractor does not have exclusive possession of the site;
  - Fit-out contractors or contractors installing particular plant and equipment will be working at the same time, and often in the same locations as the contractor;
  - Contractor's programme to show activities or periods for the employer's contractors.
o Payment provisions:
  - Compliant payment provisions with UK Law (HGCRA 1996[9]) for FIDIC Yellow Book 2017;
  - Periodic reviews of the 'contractor's share' during the course of the works, in addition to the 'preliminary assessment' but before the 'final assessment' (NEC4 Options C and D, see Section 9.6.3).
o Additional provisions regarding completion[8]:
  - Specific obligations to achieve completion (such as Building Control approval, provision of energy efficiency verification);
  - Employer is entitled to take over, or take possession of the works at an earlier stage, in order to use the works.
o Rectifying defects
  - Specifying categories of defects and periods of time for rectifying each category of defect.
o Dispute resolution
  - Adjudicator can decide more than one dispute.

11.3 Increased Liability for Contractor's Design

The employer may introduce amendments that increase the contractor's liability for the project's design. Chapter 10 discussed reducing the extent of the contractor's design. This reduction can be partial so the Contractor's Proposals include some design (as well as that contained in the Employer's Requirements), or it can be to the extent that the Employer's Requirements contain all of the design. In the latter example, no design remains for the contractor to undertake, so he is left to coordinate and integrate the employer's design.

---

9 See Section 8.7.2.

Contract amendments in this regard increase the contractor's responsibility for design, to the point of making him responsible for the design that the employer has produced:

o The contractor is responsible for all the design in the Employer's Requirements:
  – The contractor's proposals must satisfy the Employer's Requirements[10];
  – The contractor may have to warrant or verify[11] that the design contained in the Employer's Requirements is accurate;
  – There is no entitlement for a variation, extension of time, or loss and expense if there is a discrepancy, divergence, inadequacy, or inconsistency in the Employer's Requirements[12];
  – The employer decides how any discrepancy, divergence, inadequacy, or inconsistency in the Employer's Requirements is resolved, at the contractor's cost.
o The contractor's warranty for the standard and competency of design is greater than simply exercising reasonable skill and care:
  – Often the required level of skill and care is of a contractor experienced in designing and constructing works of the same type, nature, and complexity of the current project;
  – The contractor may be obliged to design and construct the works to ensure (i) that any approvals or permissions are satisfied (construction projects), or (ii) its full, efficient, economic, and safe commercial operation (for engineering projects);
  – The contractor shall not specify any materials stipulated to be 'deleterious' or 'prohibited'. This obligation is usually referenced to a particular publication known in the construction or engineering industries.[13]
o The contractor may have a higher obligation when selecting materials or goods:
  – To specify materials or goods that are of 'best quality', a higher obligation than ordinarily specifying 'satisfactory quality';
  – Approvals (by or on behalf of the employer or his representative) do not relieve the contractor of any of his responsibilities;
  – Concluding the contract does not provide conclusive evidence that materials and goods are in accordance with the contract.[14]

---

10 For example, JCT DB 2016 the Third Recital is either deleted or its effect is reversed.
11 Refer to (i) JCT DB 2016, clause 2.15 in relation to Statutory Requirements, including a divergence occurring after the contract has commenced (i.e. after the 'Base Date'), and (ii) FIDIC Yellow Book 2017, clause 5.1.
12 JCT DB 2016, deletion of clauses 2.10 (Divergence in Employer's Requirements and definition of the site boundary), 2.12 (Employer's Requirements – inadequacy), and 2.14.2 (Discrepancies in documents). FIDIC Yellow Book 2017, clause 5.1 (General Design Obligations) not only makes the contractor responsible for the Works' design, but also obliges him to scrutinise the Employer's Requirements to ascertain if they contain any 'error, fault or other defect'. If there are any errors in the Employer's Requirements, then clause 1.9 (Errors in the Employer's Requirements) may entitle the contractor to a Variation (and an extension of time or additional cost), provided that certain criteria are met. Therefore, to ensure that the contractor bears complete responsibility for the Works' design, the rights in clause 1.9 could be reduced, or clause 1.9 could be deleted. NEC4 – deletion of clause 63.10. Refer to Section 3.3.
13 Often referred to documents such as The British Council of Offices' 'Good Practice in the Selection of Construction Materials' (2011), or other documents published by the Building Research Establishment (BRE).
14 JCT DB 2016 – deletion of clause 1.8.1.1.

o Contractor to novate the consultants specified by the employer. Increasingly, employers are retaining the building services consultant, so the contractor should either[15]:
  - Insist that the building services consultant is novated; or
  - Require that the building services consultant provides the contractor with a collateral warranty for any design (no matter how basic) provided for inclusion in the Employer's Requirements.
o Contractor to obtain permits or approvals[16] or satisfy planning conditions (that the employer would otherwise obtain).
o The contractor is obliged to provide the works (or units within the works – such as dwellings) to particular dimensions (e.g. specific areas per section or unit):
  - The employer may deduct liquidated damages ($£/m^2$) if the works (or any of its constituent units) are constructed to smaller areas than specified.

11.4 Reduced Employer's Risk/Increased Employer's Benefit

These types of amendments (along with those described in Section 11.5) are often the most controversial. Whilst the effects of amendments in the previous categories (project-specific, practical benefit, and contractor's increased liability for design) can be onerous for the contractor, those amendments are often justified by the project's requirements, or the contractor has a reasonable opportunity to accommodate them[17] (in the Contractor's Proposals, his price, and the programme). Amendments that either reduce the employer's risk or increase his benefit are usually to the detriment of the contractor. Examples of these amendments are as follows:

o Any limitation of the contractor's liability is either significantly removed or reduced.
o The contractor is not entitled to rely upon statements or actions by the employer (or by his representative) and must satisfy himself that the contract and its documents are accurate:
  - There may be an entire agreement clause, so the contractor cannot rely upon any pre-contract statements made by the employer, as the written contract terms and conditions contain all matters relevant to the works, and any prior representations (made by the employer) have no effect;
  - The contractor cannot rely upon any documents (such as surveys, reports, or test results) that relate to the site that the employer has provided;
  - Any approvals or consents given by the employer do not relieve the contractor of any of his obligations or liabilities or can be relied upon as evidence that the contractor's works are in accordance with the contract.
o The contractor (and not the employer) has to ascertain the site boundary.

---

15 Refer to Section 10.2.2.
16 Notable examples are the contractor obtaining planning permission, and ensuring that the design and construction of the works satisfies those requirements. Other approvals to be obtained are likely to include Building Control approval or the Fire Officer's approval, along with any approvals for particular statutory or regulatory requirements.
17 Particularly if the contractor has been engaged (or involved) during the pre-contract stages – for example, under a PCSA (see Section 5.4.6).

○ No test results can be relied upon as evidence that the contractor's works are in accordance with the contract.

○ The employer can omit works from the contractor's scope and engage others to carry them out, and the contractor is not entitled to any claim for loss of profit or loss of opportunity.

○ Additional procedures or requirements for the contractor claiming extensions of time and/or loss and expense are as follows:

   – Notifications of delay and/or loss and expense have to be in a particular format, and submitted within a specific time period (imposing a time bar obligation[18]);

   – The contractor must submit a programme with his notification of delay (or as part of the supporting documents) that must show the critical path, delays (including those caused by the contractor);

   – There is no entitlement for an extension of time for concurrent delays[19] (i.e. reversing the ordinary position that a contractor is entitled to an extension for time).

○ Assignment:

   – The employer may assign the benefit or rights under the contract without the contractor's prior written permission; or

   – The contractor is prohibited from any assignment.

○ Employer's additional rights of set-off or deduction from sums due to the contractor:

   – The time before the final date for payment for giving a pay less notice is often reduced to one day; and

   – The employer is entitled to set off or deduct monies from sums due to the contractor under other contracts, or then recover as a debt due.

○ Additional rights for the employer to terminate the contractor's employment under the contract[20]:

   – The employer may terminate following any breach by the contractor, irrespective of how minor that breach (or its effects) may be;

   – Upon termination, the employer may commandeer the contractor's plant, equipment, facilities, materials, and goods to complete the works;

   – The employer may engage others to complete the works, irrespective of whether termination of the contractor's employment was lawful or properly carried out; and

   – There may be reduced periods of notice for the contractor to comply with (i.e. cease a default or not repeat it) the employer's termination notice.

11.5 Reduced Contractor Rights (Not Design Related)

This category of contract amendments is similar to that described in Section 11.4 (Reduced Employer's Risk/Increased Employer's Benefit), albeit the amendments are often drafted

---

18 See Section 8.8.5.2 for NEC4 (time bar for notifying compensation events) and Section 8.10.4.1 for FIDIC (relating to the 'Notice of Claim').

19 See Section 8.10.1. The ordinary position is that the contractor is entitled to an extension of time, but not additional costs for concurrent delays.

20 See Section 9.1.

to remove or reduce a contractor's right, instead of providing an additional right to the employer.

o The contractor is responsible for the site conditions, including:
  – Ground and subsurface conditions (JCT DB 2016 only);
  – Conditions of existing buildings or structures; and
  – Preceding works (carried out by others).
o Variations:
  – Instructions for provisional sums (see Section 3.6.4) do not constitute variations, and the contractor is not entitled to an extension of time or additional costs;
  – The contractor cannot object to instructions from the employer, regardless of whether they may adversely affect the integrity of the contractor's design or possibly affect health and safety;
  – The contractor cannot object or refuse to provide estimates (for time and money) requested by the employer.
o Reduced entitlement for extensions of time and/or additional costs, by deleting the following events:
  – Encountering antiquities;
  – Adverse weather conditions;
  – Unexpected or adverse ground or site conditions (FIDIC Yellow Book 2017 and NEC4);
  – Fire, lightning, explosion, storm, flood, escape of water from any water tank, apparatus or pipe, earthquake, aircraft,[21] and other aerial devices or articles dropped therefrom[22];
  – Terrorism, civil commotion, or industrial action (strike, or lock-out);
  – Exercise of statutory powers by the government or local authority;
  – Force majeure (particularly in terms of any restrictions imposed by the government following the Covid-19 pandemic). Alternatively, the contractor must comply with additional notification provisions if the progress of work is likely to be affected by a force majeure event.
o Payments:
  – Interim payments may not become due 'periodically' (e.g. becoming due each month), but may only become due upon completion of a stage (or milestones) – and the conditions for achieving that stage may be numerous and difficult to achieve.
  – A further alternative is that interim payments become due according to a payment schedule, usually in the form of a table showing the dates (often occurring on a monthly basis) for the contractor's application, employer's payment, and pay less notices and the final date for payment. However, the schedule may not show a sufficient number of interim payments – particularly if the works become delayed. In such cases, the contractor may not be entitled to any further payments on an interim basis after the dates in the payment schedule have expired. He may only be entitled to the final payment, which is likely to occur some amount of time (often up to two years) after the works are completed.[23]

---

21 Often excluding pressure waves from aircraft travelling at supersonic speeds.
22 Referred to as 'Specified Perils' (JCT DB 2016).
23 *Balfour Beatty Construction Limited v Grove Developments Ltd*, EWCA Civ 990, [2016].

- The contractor's application for interim payment may have to be in a specific manner and form – for example, containing specific information and substantiation, and sent to the employer (or his representative) in a particular way.[24]
- The period from the 'due date' for payment to the 'Final date for payment' may be significantly extended[25] (e.g. increased from 14 up to 60 or 90 days).
- The employer has no obligation to act as trustee for the retention money (no fiduciary duty) and may be able to deduct retention for longer periods (e.g. the second half of retention may only be included with the final account – see Section 9.8).
- Payment of additional costs (loss and expense) may be restricted to the matters set out in the contract, because the contractor's right to common-law damages has been deleted (additional costs via the contract are an 'exclusive remedy').
- The contractor's entitlement to interim payment is conditional upon him providing performance bonds and/or collateral warranties to the employer.

o Rectification of defects:
  - The contractor may have additional obligations, such as to 'maintain' the works during the defects period after completion.
  - Opening up for testing and inspections does not entitle the contractor to an extension of time or additional costs, even if the test or inspections shows that the contractor's work complied with the contract.
  - The defects rectification period may run for a longer period – until the employer (or his representative) issues the relevant certificate or notice that all defects have been made good.

11.6   Increased Benefits to the Contractor

Amendments that give increased benefits to the contractor are relatively rare, as it is usually the employer who enjoys the strongest bargaining position before agreeing on the contract and can impose his amendments on the contractor. However, there may be circumstances in which the contractor enjoys a stronger bargaining position, perhaps because the economic conditions lean more towards a 'seller's market' rather than a 'buyer's market'. There may be only one contractor available to carry out the employer's project, perhaps in commercial or logistic terms.

If the contractor can negotiate the contract to include amendments that benefit him, they often take the form of removing, reversing, or reducing the effect of amendments that the employer introduced for his benefit, effectively having the opposite effect to many of the amendments described in Sections 11.3, 11.4, and 11.5. Other amendments that benefit the contractor may include:

o For larger projects (particularly engineering projects), and in addition to the costs that the contractor may incur for suspending work, he will also be entitled to costs of restarting work (or re-mobilisation).

---

24 The contractor's interim payment application may only be valid if sent in hard copy (by hand or post) and not by email.
25 Beyond (i) 14 days for JCT DB 2016 (see Section 8.7.1), (ii) 56 days for FIDIC Yellow Book 2017 (see Section 8.7.2), and (iii) 14 days for NEC4 (see Section 8.7.3).

o The contractor may have additional rights of termination:
  - He may be able to terminate following *any* default by the employer;
  - Under a wider definition of the employer's insolvency;
  - The contractor may be able to use the employer's facilities;
  - If the employer assigns the benefit or rights under the contract without the contractor's prior permission.
o Imposing specific periods for the employer to carry out an action, failing which, the contractor is entitled to an extension of time and/or additional costs.

11.7 Clarifications

This final category includes amendments that, for the most part, are intended to provide the parties with greater certainty over their respective rights and obligations, such as the following:

o Additional information to expand or better describe the existing terms in a contract;
o Confirmation of the parties' common-law obligations if not expressly stated within the contract terms; or
o Introducing a more precise, measured, or objective test (or standard) instead of a subjective test – such as one of 'reasonableness' – or applying it in particular circumstances.

Examples of these 'clarification' amendments may include the following.

o Confirming 'common law' matters such as:
  - The limitation period is 12 years (UK).
  - No third-party rights are conferred (see Section 7.1.2).
  - Materials and goods are to be of 'satisfactory quality' (SGA82 – see Section 6.3.2) but must be new and appropriate for their intended use.
  - The contractor must exercise a standard of 'reasonable skill and care' when carrying out the design.
  - The contractor cannot subcontract the whole of the works.
  - The parties are obliged to cooperate with each other.
  - No payments to the contractor are conclusive evidence that the contractor's works and design are in accordance with the contract unless otherwise stated.
  - A party cannot benefit from its own breach of contract (e.g. if the contractor is responsible for a delay, he is not entitled to an extension of time or additional costs[26]).
  - The contractor ought to mitigate any delays to progress or the incurrence of additional costs.
  - 'Concurrent delays' entitle the contractor to an extension of time but not to additional costs.
  - 'Completion' has its ordinary meaning (see Section 9.3).
  - The employer can recover any overpayment made to the contractor.

---

26 Some examples: (i) if the contractor's workmanship, materials or goods are not in accordance with the contract, (ii) a delay caused by a statutory body could have been avoided if it was reasonable for the contractor to place the contract for those works earlier, or (iii) if any industrial action (strike, lock-out) was caused by the contractor.

o Additional clarification points:
  – Variations do not arise from the contractor's default.
  – The contractor must provide reasonable access for the employer to areas for which 'partial possession' (JCT DB 2016) has been taken or which have been 'taken over' (FIDIC Yellow Book 2017 and NEC4).
  – Subsequent award of extension of time does not invalidate a previous notice to pay less or deduct liquidated damages, although the employer must repay liquidated damages commensurate with any subsequent extension of time.
  – The contractor must also 'rectify' as well as remove any non-compliant work (JCT DB 2016).
  – The contractor must provide the required documents (including 'as built' drawings) before the period for making good defects commences.
  – 'Tests after Completion' (FIDIC Yellow Book 2017) must be carried out 182 days after the Taking Over Certificate.
  – The employer is not obliged to issue (i) any instruction or schedule of rectification work, or (ii) any certification (or notice) that the contractor has rectified any defects until the expiry of the 'Rectification Period' (JCT DB 2016), the 'DNP' (FIDIC Yellow Book 2017), or the 'defects correction period' (NEC4).[27]
  – The contractor's liability for personal injury or death or damage to property continues during the period for making good defects.
  – The contractor must leave the site upon termination.
o Wider or expanded definitions:
  – 'Communications' may include instructions from the employer (or his representative), quotations from the contractor, and particular notices.
  – 'Permitted uses' of information provided by the contractor (such as design information or quotations) may be restricted to its intended purpose and not used for any other reason (such as price comparisons on other projects or for changes in the price for other reasons).
  – 'Notices' issued under the contract must state the clause under which they're issued.
  – 'Completion' requires additional certification or approval.[28]
o Increased objectivity – amendments which attempt to establish a standard for the following:
  – 'Unforeseeable' (e.g. ground conditions) is redefined with an objective standard.
  – 'Exceptionally adverse weather conditions' (JCT DB 2016) is more prescriptive (akin to NEC4[29]).
  – 'Experienced contractor'.

---

27 Refer to Sections 9.5.1, 9.5.2, and 9.6.3.
28 Such as independent certification (e.g. by a statutory body such as a highways or rail authority, or in order to secure a grant or funding) or approval by the Building Control or Fire officer.
29 NEC4, clause 60.1(13) 'weather measurement' compensation event – when compared to the 'weather data'.

12

Possible Future Development in Design and Build

Part One of this book considered how design and build contracts were originally intended to operate, and Part Two examined ways in which the contracts' operation has changed over time. This chapter aims to explore possible future developments in design and build, from the perspective of how recent developments or current practice may be reflected in revised versions of the JCT, FIDIC, and NEC contracts.

12.1   Continuing Trend of Reducing the Extent of Contractor Design

Section 10.1 described the trend for employers to exert more control over the project's design by reducing the amount of design that the contractor carried out. This is often achieved by the employer providing a design to RIBA Stage 4 (Technical Design) in the employer's requirements, and in practical terms, leaving the contractor's design consisting of little more than coordinating the various elements of the project's design. The employer may also amend the contract's terms to make the contractor responsible for all of the design (including that prepared by or on behalf of the employer) – refer to Section 11.3. This trend clearly indicates that employers prefer to significantly reduce their contractual risks and responsibilities, over possibly benefitting from any 'buildability' expertise that the contractor may provide.

12.1.1   JCT Design and Build Contracts

The changes to JCT design and build contracts to reflect the reduction in contractor's design seem fairly obvious. If the majority of projects let under a JCT contract are based upon an advanced design (e.g. to RIBA Stage 4) for which the contractor must not only 'complete' but take responsibility for, then the following changes could be seen:

o A simple name change. Instead of being called a 'design and build' contract, which implies that the contractor will be carrying out some design, the contract's name could be altered to reflect that the contractor will only be coordinating and taking responsibility for the employer's design. The contract may be called something like 'Standard Building Contract with Contractor's Design Coordination'.

Design and Build Contracts, First Edition. Guy Higginbottom.
© 2024 John Wiley & Sons Ltd. Published 2024 by John Wiley & Sons Ltd.

o Contractual changes. To make the contractor responsible for the employer's design, along with its coordination, the following amendments could be made to the current JCT DB 2016 conditions:
- Removing the 'Contractor's Proposals', as there is simply no design remaining for the contractor to 'complete'[1] (see Section 3.1.2); and
- Removing the contractor's right to a Change (variation) for inaccuracies, inadequacies, inconsistences, or incompleteness in the employer's design. By doing so, the contractor would not be entitled to any additional payment, extension of time, or loss and/or expense for resolving any of the aforementioned deficiencies in the employer's design.[2]
o In the event that the employer does require the contractor to carry out some or all of the project's design, and/or is content not to make the contractor responsible for deficiencies in the employer's design, then the current clauses in JCT DB 2016 that entitle the contractor to a Change in such circumstances, could be taken out of the Conditions, and included as a 'Supplemental Provision', in similar form to the 'Contractor's Designed Portion' in JCT SBC 2016.[3]

12.1.2 FIDIC Yellow Book Contracts

The trend for reducing the extent of contractor's design is less likely to apply to engineering contracts procured under the FIDIC Yellow Book 2017 than to construction projects procured under JCT DB contracts. This is mainly because the employer is likely to rely upon the contractor's expertise when designing and constructing plant, and often, the employer's requirements only contain a specification for the plant's performance (a performance specification). The contractor may be actively working on design studies and developments under a PCSA-type arrangement for a significant period of time before entering into the contract with the employer.

12.1.3 NEC4 Contracts

NEC contracts are often used for both construction and engineering projects, so they may be considered as a 'mid-point' between JCT DB 2016 and FIDIC Yellow Book 2017. It is relatively simple to reduce the scope of the contractor's design under a NEC4 contract, as the Scope (set out in Contract Data Part One[4]) can contain the detailed design (e.g. to RIBA Stage 4 – Technical Design) prepared by or on behalf of the client, but state that the contractor is responsible for that design, and take the following steps:

o In Contract Data Part Two (against '2 The Contractor's Main Responsibilities'), either:
- Make no entry (refer to Section 3.3.3), so no design remains for the contractor to carry out, but he remains liable for that design; or
- Enter an obligation for the contractor to complete and coordinate the design set out in the Scope in Contract Data Part One[4]; and

---

1 JCT DB 2016, clause 2.1.1.
2 Refer to Section 11.3.
3 JCT SBC/Q 2016, refer to the Ninth, Tenth, Eleventh, and Twelfth Recitals, and clauses 2.13–2.20.
4 Refer to Section 3.3.3, which refers to this part of the Scope as the 'Part One Scope' (i.e. the entry in '1 General' against 'The Scope is in...').

o Deletion of clause 63.10, to have the following effects:
- – That no 'ambiguities or inconsistencies' would be corrected by an instruction (which could become a 'compensation event' under clause 60.1[1]); and
- – The project manager must accept the contractor's design (under clause 21.2) before the contractor can proceed with the work.

12.2 Novation of Design Consultants

Considering that novation has been popular for some time, it is particularly surprising that none of the three design and build contracts include any express provisions for it. As mentioned in Section 10.2.1 (Novation of Design Consultants), it seems peculiar that in cases of FIDIC and NEC4, which both provide suites of coordinated contracts (including the main contract, a subcontract, and contracts for design services) do not contain any provision for novation. Curiously, it is only the RIBA Professional Services Contract 2020[5] that contains (at clause 4.6) any provision for novation, but despite the RIBA being one of the members of the JCT (Joint Contracts Tribunal), novation is not found in JCT design and build contracts.

It should be relatively simple to incorporate novation into the three design and build contracts:

o Add an enabling clause in the relevant main contract whereby the contractor agrees to novation of the relevant consultant or consultant. This could be added as follows:
- – JCT – as another Supplemental Provision, possibly confirming that the novated consultant was engaged under RIBA PSC.
- – FIDIC – via a new 'Special Provision' for novation in Particular Conditions Part B (novating the relevant designer's contract that is based upon the FIDIC Model Services Agreement[5]).
- – NEC – as another Secondary Option Clause (e.g. 'X23 Novation'), which novates the relevant designer's appointment under the NEC Professional Services Contract[5].
o Adding a novation agreement:
- – JCT – the novation agreement could be included in one of the Schedules to the JCT contract.
- – FIDIC – the novation agreement could be included as (i) one of the 'Example Forms' (such as 'Annex H – Example form of novation agreement') appended to the Yellow Book Contract,[6] and (ii) appended to the FIDIC Model Services Agreement.[5]
- – NEC – the novation agreement could be appended to both (i) the NEC ECC contract, and (ii) NEC Professional Services Contract.[7]

---

5 Refer to Sections 4.1 and 4.2.1.
6 FIDIC Yellow Book 2017 currently includes Annex A (Example form of parent company guarantee), Annex B (Example form of tender security), Annex C (Example form of performance security – demand guarantee), Annex D (Example form of performance security – surety bond), Annex E (Example form of advance payment guarantee), Annex F (Example form of retention money guarantee), and Annex G (Example form of payment guarantee by Employer).
7 NEC4. Both the Engineering and Construction Contract (ECC) and the Professional Services Contract (PSC) append forms of a 'Trust Deed' and a 'Joining Deed', so a form of novation agreement could be appended there.

12.3 Detailed Pricing Documents

The various pricing documents that are used for the JCT DB 2016, FIDIC Yellow Book 2017, and NEC4 contracts were described in Section 3.6.1 (Pricing Document – Common Format and General Contents), which suggested that the pricing document's format could be based upon the following:

o For construction projects: NRM2's 'Template for pricing summary for elemental bill of quantities (condensed)'; or
o For engineering projects: CESMM4's 'Work classification' with a 'Grand Summary'.

The pricing document will depend upon the form of design and build contract to be adopted, and in the case of NEC4, which of its main contract options is selected (refer to Table 3.7).

The required amount of detail to be shown in the pricing document is generally set out in the Employer's Requirements (in the cases of JCT DB 2016 and FIDIC Yellow Book 2017) or in the Scope (for NEC4). The contract conditions for FIDIC and NEC4 lean towards more detail being provided, in the 'Schedule of Rates and Prices' (FIDIC Yellow Book 2017) or the 'Data for the Schedule of Cost Components'[8] (NEC4). However, there is no such stipulation as to detail in the 'Contract Sum Analysis' in JCT DB 2016.

Given that a significant number of the Employer's Agents (under JCT DB 2016) are quantity surveyors,[9] the apparent lack of prescribed detail for JCT's 'Contract Sum Analysis' is perhaps surprising. When publishing JCT DB 2016, it was further surprising that the provision for a bill of quantities (found in JCT DB 2011[10]) was deleted, given that the RICS had only replaced SMM7 by introducing NRM2 (Detailed measurement for building works) in April 2012. The current absence of a bill of quantities provision in JCT DB 2016 is probably explained by employers not wanting to go to the pre-contract expense of a quantity surveyor producing a bill of quantities, when the very nature of a JCT design and build contract reduced the scope for variations, compared to a contract let under its JCT SBC/Q counterpart.

However, this does not seem to quell the quantity surveyor's thirst for cost information. When the Employer's Agent (a quantity surveyor) contributes to drafting the Employer's Requirements under a JCT DB 2016 contract, he often asks the contractor to submit detailed breakdowns of his price; often into individual and quantified unit rates, akin to a traditional bill of quantities. The contractor's estimating team may have carried out a similar exercise, producing 'builder's quantities' from which to let the subcontracted elements of the project (unless a subcontractor's lump-sum price is sought). Detailed cost information such as this will assist the Employer's Agent (quantity surveyor) monitor costs against a cost plan and assist with any 'value engineering' scope changes and evaluation of the contractor's price if a competitive tender process is chosen.

---

8 NEC4. The 'Data for the Schedule of Cost Components' is used for Options C, D, and E, whereas the 'Data for the Short Schedule of Cost Components' is used for Options A and B (refer to Table 3.7).
9 Refer to the RICS Scope of Services for Employer's Agent Services (May 2022) intended for use with JCT design and build contracts.
10 JCT DB 2011 included the provision for the Employer to describe the Works in the Employer's Requirements by a bill of quantities, by applying paragraph 3 (Bills of Quantities) of the 'Supplemental Provisions' (contained in Schedule 2).

In the case of JCT design and build contracts, the specific detail to be provided by the contractor could be expressly set out in the following two ways:

o Reintroducing the 'Bills of Quantities' supplemental provision that was previously included in JCT DB 2011,[11] thereby describing the Employer's Requirements (or parts of it) by a bill of quantities, to which the contractor can enter his unit rates and prices; or
o Introducing an alternative provision, obliging the contractor to set out the Contract Sum Analysis to a particular format and level of detail (e.g. in accordance with NRM2 or CESMM4). This would probably be unpopular for competitively tendered projects, but where the employer and contractor are negotiating the contract terms (particularly, if doing so during the course of a PCSA), then the contractor may be willing to go to the expense of providing specific levels of pricing detail, in the knowledge that he will be awarded the main contract.

12.4 Detailed Procedures for Testing and Commissioning

In Section 9.3 (Meaning of 'Completion') and Section 9.5 (Rectification or Completion of Outstanding Works after Completion), the various procedures for rectifying non-compliant work, completing any work that was permitted to be incomplete, along with testing and trial operation were discussed. Table 9.7 summarised the various procedures for rectification, testing, and operations, and unsurprisingly, given that it is a contract often used for procuring engineering projects with complex functions, FIDIC Yellow Book 2017 contains the most detailed provisions.

The complexity of construction projects has increased in recent years. The introduction of computer-aided design has given rise to more complicated structural and aesthetic designs and computerised building management systems (such as those controlling the internal environment, energy performance, transportation systems, security systems, and the fire alarm and fire strategy). Obtaining certification or formal approval from a third party is often necessary to achieve a particular standard (such as a BREEAM rating) or a financial grant or incentive.

JCT DB 2016 generally operates on the premise that 'practical completion' (see Section 9.3.1) means that the contractor has completed all the physical work and it is free from patent defects. However, the physical installation of various building management systems and their operating components is only part of the contractor's work. In many cases, he must ensure that these systems are tested, commissioned, and operating correctly in order to achieve practical completion. Having to account for testing and commissioning under JCT DB 2016 may not be straightforward. Depending upon the complexities of the relevant building management systems and the conditions for their operation (e.g. they may need the building to be fully occupied), testing and commissioning may run for many weeks or months after the physical work is completed.

---

11 JCT DB 2011, Schedule 2 (Supplemental Provisions), paragraph 3 (Bills of Quantities) only runs to four subparagraphs (3.1–3.4).

In terms of accommodating a testing and commissioning period within the construction programme, the parties could consider adopting sectional completion as follows:

o The first (or preceding) section is for completing the physical works:
  - The employer can arrange to take possession of the completed section (or sections) and effectively operate the building; and
  - The contractor can organise his programme with the subcontractors to show the physical works being completed during this section (or sections).
o The second (or final section):
  - The employer will be in possession of all parts of the physical works, but will have to accommodate and cooperate with the contractor to facilitate testing and commissioning of the building management systems; and
  - The contractor can organise the relevant subcontractors to carry out to carry out the testing and commissioning work during this second (or final) section.
o The parties can also agree on:
  - Appropriate amounts of liquidated damages for the relevant sections. Given that proportionally, the greatest value of work[12] may be in the first (or preceding) section, the amount of liquidated damages is likely to be relatively higher than for the 'testing and commissioning' section; and
  - Different Rectification[13] periods for the building management system, and the remaining works.

For more complex construction projects that necessarily involve a lot of building services, the terms of the JCT design and build contract could adopt a similar procedure to 'Tests on Completion' found in clause 9.1 of FIDIC Yellow Book 2017 (refer to Section 9.2.2). For less complex construction projects, such a detailed testing and commissioning procedure is less likely to operate, so making the procedures optional, by including them as Supplemental Provisions, seems appropriate. Fewer stages of testing than are provided in clause 9.1 of FIDIC Yellow Book 2017 may be appropriate, so reducing the testing and commissioning to two stages, equivalent to FIDIC's 'Commissioning Tests',[14] and 'Trial Operation',[15] may be all that is required. The operation of such a testing and commissioning procedure could proceed in the same way as FIDIC's 'Tests on Completion' insofar as:

o The contractor provides a programme for the testing and commissioning works, an appropriate period (i.e. an adequate period to allow for the testing and commissioning works, along with any third-party verification, certification, or approval) before either:
  - The Completion Date; or
  - If different, the date on which the contractor estimates that the Completion Date will occur.
o Following each testing or commissioning stage, the contractor must submit sufficient details (such as test results or certification) to the employer, so the employer can:
  - Approve the contractor's testing and commissioning results; and
  - Notify the contractor to proceed to the subsequent stage.

---

12 JCT DB 2016. The 'Section Sum' – see clause 2.34 and Section 8.3.1.
13 JCT DB 2016, refer to clause 2.35 and Section 8.3.1.
14 FIDIC Yellow Book 2017, clause 9.1(b).
15 FIDIC Yellow Book 2017, clause 9.1(c).

This is similar to both the procedure in clause 9.1 of FIDIC Yellow Book 2017 and JCT DB 2016's Design Submission Procedure[16] (see Section 3.5.2), which prohibits the contractor from proceeding to the next stage without the employer's approval.

12.5 Collateral Warranties (FIDIC and NEC4)

In Section 7.2, the book suggested how collateral warranties could be introduced into the FIDIC Yellow Book 2017 and NEC4 contracts as follows:

o FIDIC Yellow Book 2017:
   - By including an enabling clause as a 'Special Provision' (in Part B of the Particular Conditions) for the contractor to provide the form of collateral warranty included in the contract[17];
   - Including similar enabling clauses in any relevant subcontract[18] and/or designer's conditions of engagement[19]; and
   - Similarly, incorporating the relevant forms of collateral warranty into the relevant subcontract or designer's conditions of engagement.
o NEC4:
   - Using Option X8 (undertaking to others) in the main contract (ECC), NEC4's subcontract and designer's 'Professional Services Contract', the 'undertaking' is to provide a collateral warranty to the designated beneficiary;
   - Including a copy of the relevant collateral warranty as part of the documentation specified in Option X8.

Both FIDIC and NEC could provide an example of a collateral warranty to be used under their main contract, subcontracts, and conditions for engaging a designer. The collateral warranty example may have to contain a number of selectable options to reflect the more 'international' usage intended for the FIDIC and NEC4 contracts (compared to JCT DB 2016).

---

16 JCT DB 2016, Schedule 1 contains the Design Submission Procedure, paragraph 6 prohibits the contractor from carrying out any work that has not been approved (i.e. marked as 'C'), whereupon the contractor shall re-submit the Design Document accounting for the Employer's comments.
17 FIDIC Yellow Book 2017, for example, the collateral warranty could be referred to under section 2 of the Contract Agreement, as a new point '2(j) – Collateral warranty'.
18 FIDIC Yellow Book 2017 – Conditions of Subcontract for Plant Design-Build, 1st ed., (2019).
19 FIDIC Client/Consultant Model Services Agreement, 5th ed., (2017).

Appendix A

JCT Design and Build Contracts Reconciliation

DB 2016 Description	DB 2016	DB 2011	DB 2005	JCT SBC/Q 2016 CDP clause	JCT ICD 2016	MWD 2016
RECITALS						
Description of the Works and Employer's Requirements	First	First	First	First, Tenth	First, Second, Fourth	First, Second, Third
Contractor's Proposals and Contract Sum Analysis	Second	Second	Second	Eleventh	Sixth	—
Employer is satisfied the CP's meet the ERs	Third	Third	Third	Twelfth	Seventh	—
CIS scheme	Fourth	Fourth	Fourth	Fourth	Eighth	Fifth
Sectional Completion	Fifth	Fifth	Fifth	Sixth	Eleventh	—
Framework Agreement	Sixth	Sixth	—	Seventh	Twelfth	Seventh
Supplemental Provisions	Seventh	Seventh	—	Eighth	Thirteenth	Eighth
ARTICLES						
Contractor's obligations	Article 1	Article 1	Article 1	Article 1	Article 1	Article 1
Contract Sum	Article 2	Article 2	Article 2	Article 2	Article 2	Article 2
Employer's Agent	Article 3	Article 3	Article 3	—	*Article 3*	Article 3
Employer's Requirements and Contractor's Proposals	Article 4	Article 4	Article 4	—	—	—
Principal Designer (CDM officer)	Article 5	Article 5	Article 5	Article 5	Article 5	Article 4
Principal Contractor	Article 6	Article 6	Article 6	Article 6	Article 6	Article 5
Adjudication	Article 7	Article 7	Article 7	Article 7	Article 7	Article 6
Arbitration	Article 8	Article 8	Article 8	Article 8	Article 8	Article 7
Legal Proceedings	Article 9	Article 9	Article 9	Article 9	Article 9	Article 8

Design and Build Contracts, First Edition. Guy Higginbottom.
© 2024 John Wiley & Sons Ltd. Published 2024 by John Wiley & Sons Ltd.

DB 2016 Description	DB 2016	DB 2011	DB 2005	JCT SBC/Q 2016 CDP clause	JCT ICD 2016	MWD 2016
CONDITIONS						
Section 1 – Definitions and interpretation						
Definitions / Interpretation	1.1	1.1	1.1	1.1	1.1	1.1
References to clauses etc	1.2	1.2	1.2	1.2	1.2	—
Agreement etc, to be read as a whole	1.3	1.3	1.3	1.3	1.3	1.2
Headings, references to persons, legislation etc	1.4	1.4	1.4	1.4	1.4	1.3
Reckoning period of days	1.5	1.5	1.5	1.5	1.5	1.4
Contract (Rights of Third Partys) Act 1999	1.6	1.6	1.6	1.6	1.6	1.5
Notices and other communications	1.7	1.7	1.7	1.7	1.7	1.6
(2005) Electronic communications	—	—	1.8	—	—	—
Effect of Final Statement	1.8	1.8	1.9	1.9	1.9	—
Effect of payment other than payment of Final Statement	1.9	1.9	1.10	1.10	1.10	—
Consents and approvals	1.10	—	—	1.11	1.11	1.7
Applicable law	1.11	1.10	1.11	1.12	1.12	1.8
Section 2 – Carrying out the Works						
Contractor's Obligations						
General obligations	2.1	2.1	2.1	2.1 2.2 (CDP)	2.1	2.1
Materials goods and workmanship	2.2	2.2	2.2	2.3.2	2.2	2.2
Possession						
Date of Possession – progress	2.3	2.3	2.3	2.4	2.4	2.3
Deferment of possession	2.4	2.4	2.4	2.5	2.5	—
Early use by Employer	2.5	2.5	2.5	2.6	2.6	—
Work not forming part of the Contract	2.6	2.6	2.6	2.7	2.7	—

DB 2016 Description	DB 2016	DB 2011	DB 2005	JCT SBC/Q 2016 CDP clause	JCT ICD 2016	MWD 2016
Supply of documents, Setting Out etc						
Contract Documents	2.7	2.7	2.7	2.8	2.8	—
Construction information	2.8	2.8	2.8	2.9.4, 2.9.5	2.10	—
Site boundaries	2.9	2.9	2.9	—	—	—
Discrepancies and Divergences						
Divergence in Employer's Requirements and definition of site boundary	2.10	2.10	2.10	—		—
Preparation of Employer's Requirements	2.11	2.11	2.11	2.13.2	2.13.1.4	—
Employer's Requirements – inadequacy	2.12	2.12	2.12	2.14.2	—	—
Notice of discrepancies etc	2.13	2.13	2.13	2.15.5	2.13.3.1	—
Discrepancies in documents	2.14	2.14	2.14	2.16	2.13.3.2	2.5
Divergences from Statutory Requirements	2.15	2.15	2.15	2.17	2.15	2.6
Emergency compliance with Statutory Requirements	2.16	2.16	2.16	2.18	2.16	—
Design Work – liabilities and limitation	2.17	2.17	2.17	2.19	2.34	—
(JCT SBC/Q) Errors and failures – other consequences	—			**2.20**	—	—
Fees, Royalties and Patent Rights						
Fees or charges legally demandable	2.18	2.18	2.18	2.21	2.3	2.7
Patent rights and royalties – Contractor's indemnity	2.19	2.19	2.19	2.22		—
Patent rights – instructions	2.20	2.20	2.20	2.23		—
Unfixed Materials and Goods – property, risk etc						
Materials and goods – on site	2.21	2.21	2.21	2.24	2.17	—
Materials and goods – off site	2.22	2.22	2.22	2.25	2.18	—

DB 2016 Description	DB 2016	DB 2011	DB 2005	JCT SBC/Q 2016 CDP clause	JCT ICD 2016	MWD 2016
Adjustment of the Completion Date						
Related definitions and interpretations	2.23	2.23	2.23	2.26	—	—
Notice by Contractor of delay to progress	2.24	2.24	2.24	2.27	2.19.1	2.8
Fixing Completion Date	2.25	2.25	2.25	2.28	2.19.2	2.8
Relevant Events	2.26	2.26	2.26	2.29	2.20	—
Practical Completion, Lateness and Liquidated Damages						
Practical Completion, Lateness and Liquidated Damages	2.27	2.27	2.27	2.30	2.21	2.10
Non-Completion Notice	2.28	2.28	2.28	2.31	2.22	—
Payment or allowance of liquidated damages	2.29	2.29	2.29	2.32	2.23, 2.24	2.9
Partial Possession by Employer						
Contractor's consent	2.30	2.30	2.30	2.33	2.25	—
Practical completion date	2.31	2.31	2.31	2.34	2.26	—
Defects etc – Relevant Part	2.32	2.32	2.32	2.35	2.27	—
Insurance – Relevant Part	2.33	2.33	2.33	2.36	2.28	—
Liquidated damages – Relevant Part	2.34	2.34	2.34	2.37	2.29	—
Defects						
Schedules of defects and instructions	2.35	2.35	2.35	2.38	2.30	2.11
Notice of Completion of Making Good	2.36	2.36	2.36	2.39	2.31	2.12
Contractor's Design Documents						
As-built Drawings	2.37	2.37	2.37	2.40	2.32	—
Copyright and use	2.38	2.38	2.38	2.41	2.33	—

DB 2016 Description	DB 2016	DB 2011	DB 2005	JCT SBC/Q 2016 CDP clause	JCT ICD 2016	MWD 2016
Section 3 – Control of the Works						
Access and Representatives						
Access for Employer's Agent	3.1	3.1	3.1	3.1	3.1	—
Site Manager	3.2	3.2	3.2	3.2	3.2	3.2
Subcontracting						
Consent to sub-contracting	3.3	3.3	3.3	3.7	3.5	3.3.1
Conditions of sub-contracting	3.4	3.4	3.4	3.9	3.6	3.3.2
Employer's Instructions						
Compliance with instructions	3.5	3.5	3.5	3.10	3.8	3.4
Non-compliance with instructions	3.6	3.6	3.6	3.11	3.9	3.5
Instructions other than in writing	3.7	3.7	3.7	3.12	—	—
Provisions empowering instructions	3.8	3.8	3.8	3.13	3.10	—
Instructions requiring Changes	3.9	3.9	3.9	3.14	3.11	3.6
Postponement of work	3.10	3.10	3.10	3.15	3.12	—
Instructions on Provisional Sums	3.11	3.11	3.11	3.16	3.13	—
Inspection – tests	3.12	3.12	3.12	3.17	3.14	—
Work not in accordance with the Contract	3.13	3.13	3.13	3.18	3.15	—
Workmanship not in accordance with the Contract	3.14	3.14	3.14	3.19	3.16	—
Antiquities	3.15	3.15	3.15 3.16 3.17	3.22		—
CDM Regulations	3.16			3.23	3.18	3.9
(2011) Undertakings to comply	—	3.16	3.18			
(2011) Appointment of successors	—	3.17	3.19			

DB 2016 Description	DB 2016	DB 2011	DB 2005	JCT SBC/Q 2016 CDP clause	JCT ICD 2016	MWD 2016
Section 4 – Payment						
Contract Sum and Adjustments						
Adjustment only under the Conditions	4.1	4.1	4.1	4.2	4.2	—
Items included in adjustments	4.2	4.2	4.2	4.3	4.3	4.3
Taking adjustments into account	4.3	4.3	4.3	4.4	—	—
Taxes						
VAT	4.4	4.4	4.4	4.5	4.5	4.1
Construction Industry Scheme (CIS)	4.5	4.5	4.5	4.6	4.6	4.2
Payments and Notices – general provisions						
Advance payment	4.6	4.6	4.6	4.7	4.7	—
Interim Payments – Contractor's Interim Payment Applications, due dates and Payment Notices	4.7	4.8	4.8	4.9, 4.10	4.8, 4.11	4.3, 4.4, 4.5
(2011) Issue and amount of Interim Payments	—	4.7	4.7			
Relevant statement and Final Payment Notice	4.8	4.12	4.12	—	—	
Interim and final payments – final date and amount	4.9	4.9	4.10	4.11	4.12	4.4
Pay-Less Notices and other general provisions	4.10	4.10	—	4.12	4.13	4.5.4
Contractor's right of suspension	4.11	4.11	4.10	4.13	4.14	4.7
Interim Payments – calculation of sums due						
Gross Valuation – Alternative A	4.12	4.13	4.13	4.14	4.9	4.3
Gross Valuation – Alternative B	4.13	4.14	4.14		—	4.3
Sums due as Interim Payments	4.14	—	—	4.15	4.9	4.3

DB 2016 Description	DB 2016	DB 2011	DB 2005	JCT SBC/Q 2016 CDP clause	JCT ICD 2016	MWD 2016
Listed Items	4.15	—	—	4.16	—	—
(2011) Off-site materials and goods	—	4.15	4.15		4.10	
Retention						
Rules on treatment of Retention	4.16	4.16	4.16	4.17	—	—
Retention bond	4.17	4.16	—	4.18	—	—
Retention – amounts and periods	4.18	4.18	4.17	4.19	—	—
Loss and Expense						
Matters materially affecting regular progress	4.19	4.20	4.19	4.20	4.15	—
Notification and ascertainment	4.20	—	—	4.21	4.16	—
Relevant Matters	4.21	4.21	4.20	4.22	4.17	—
Amounts ascertained – addition to Contract Sum	4.22	4.22	4.21	4.23	4.18	—
Reservation of Contractor's rights and remedies	4.23	4.23	4.22	4.24	4.19	—
Final Statement and Final Payment	4.24	4.12	4.12	4.25, 4.26	4.20, 4.21	4.8
Section 5 – Changes						
General						
Definition of Changes	5.1	5.1	5.1	5.1	5.1	—
Valuation of Changes and provisional sum work	5.2	5.2	5.2	5.2	5.2	3.6.2, 3.6.3, 3.7
Giving effect to Valuations, agreements etc	5.3	5.3	5.3	5.5	—	
The Valuation Rules						
Measurable Work	5.4	5.4	5.4	5.6	5.3	—
Daywork	5.5	5.5	5.5	5.7	5.4	—
(JCT SBC/Q) Contractor's Designed Portion – Valuation	—			5.8	5.7	—
Change of conditions for other work	5.6	5.6	5.6	5.9	5.5	—
Additional provisions	5.7	5.7	5.7	5.10	5.6	—

DB 2016 Description	DB 2016	DB 2011	DB 2005	JCT SBC/Q 2016 CDP clause	JCT ICD 2016	MWD 2016
Section 6 – Injury, Damage and Insurance						
Personal Injury and Property Damages						
Contractor's liability – personal injury or death	6.1	6.1	6.1	6.1	6.1	5.1
Contractor's liability – loss, injury or damage to property	6.2	6.2	6.2 6.3	6.2	6.2	5.2
Loss or damage to Existing Structures or their contents	6.3	6.3	—	6.3	6.3	—
Insurance against Personal Injury and Property Damage						
Contractor's insurance of his liability	6.4	6.4	6.4	6.4	6.4	5.3
Contractor's insurance of liability Employer	6.5	6.5	6.5	6.5	6.5	—
Excepted Risks	6.6	6.6	6.6	6.6	6.6	—
Insurance of the Works and Existing Structures						
Insurance Options and period	6.7	6.7	6.7	6.7	6.7	5.4A, 5.4B, 5.4C
Related definitions	6.8	6.8	6.8	6.8	6.8	—
Sub-contractors – Specified Perils cover under Works Insurance Policies	6.9	6.9	6.9	6.9	6.9	
Terrorism Cover – policy extensions and premiums	6.10	6.10	—	6.10	6.10	
Terrorism Cover – non-availability – Employer's options	6.11	6.11	6.10	6.11	6.11	
Evidence of insurance	6.12	B.1, C.3		6.12	6.12	5.5
Loss or damage – insurance claims and reinstatement	6.13	—	—	6.13	6.13	5.6
Loss or damage to Existing Structures – right of termination	6.14	—	—	6.14	6.14	5.7

DB 2016 Description	DB 2016	DB 2011	DB 2005	JCT SBC/Q 2016 CDP clause	JCT ICD 2016	MWD 2016
Professional Indemnity Insurance						
Obligation to insure	6.15	6.12	6.11	6.15	6.19	—
Increased cost and non-availability	6.16	6.13	6.12	6.16	6.20	—
Joint Fire Code – compliance						
Application of clauses	6.17	6.14	6.13	6.17	6.15	—
Compliance with Joint Fire Code	6.18	6.15	6.14	6.18	6.16	—
Breach of Joint Fire Code – Remedial Measures	6.19	6.16	6.15	6.19	6.17	—
Joint Fire Code – amendments/ revisions	6.20	6.17	6.16	6.20	6.18	—
Section 7 – Assignment, Performance Bonds and Guarantess, Third Party Rights and Collateral Warranties						
Assignment						
General	7.1	7.1	7.1	7.1	7.1	3.1
Rights of enforcement	7.2	7.2	7.2	7.2	—	—
Performance Bond and Guarantees	7.3	—		7.3	7.2	—
Clauses 7A to 7E – Preliminary						
Rights Particulars	7.4	—	—	7.4	7.3	—
(2005) References	—	—	7.3			
Notices	7.5	7.3	7.4	7.5	7.4	—
Execution of Collateral Warranties	7.6	7.4	7.5	7.6	7.5	—
Third-Party Rights from Contractor						
Rights for Purchasers and Tenants	7A	7A	7A	7A	7.6	—
Rights for a Funder	7B	7B	7B	7B	7.7	—

DB 2016 Description	DB 2016	DB 2011	DB 2005	JCT SBC/Q 2016 CDP clause	JCT ICD 2016	MWD 2016
Collateral Warranties from Contractor						
Contractor's Warranties – Purchasers and Tenants	7C	7C	7C	7C	7.4	—
Contractor's Warranty – Funder	7D	7D	7D	7D	7.5	—
Third-Party Rights and Collateral Warranties from Sub-contractors	7E	7E	7E 7F	7E	7.8	—
Section 8 – Termination						
General						
Meaning of insolvency	8.1	8.1	8.1	8.1	8.1	6.1
Notices under section 8	8.2	8.2	8.2	8.2	8.2	6.2
Other rights, reinstatement	8.3	8.3	8.3	8.3	8.3	6.3
Termination by Employer						
Default by Contractor	8.4	8.4	8.4	8.4	8.4	6.4
Insolvency of Contractor	8.5	8.5	8.5	8.5	8.5	6.5
Corruption and regulation 73(1) of the PC Regulations	8.6	8.6	8.6	8.6	8.6	6.6
Consequences of termination under clauses 8.4 to 8.6	8.7	8.7	8.7	8.7	8.7	6.7
Employer's decision not to complete the Works	8.8	8.8	8.8	8.8	8.8	—
Termination by Contractor						
Default by Employer	8.9	8.9	8.9	8.9	8.9	6.8
Insolvency of Employer	8.10	8.10	8.10	8.10	8.10	6.9
Termination by either Party and regulations 73(1)(a) and 73(1)(c) of the PC Regulations	8.11	8.11	8.11	8.11	8.11	6.10
Consequences of Termination under clauses 8.9 to 8.11 etc	8.12	8.12	8.12	8.12	8.12	6.11

DB 2016 Description	DB 2016	DB 2011	DB 2005	JCT SBC/Q 2016 CDP clause	JCT ICD 2016	MWD 2016
Section 9 – Settlement of Disputes						
Mediation	9.1	9.1	9.1	9.1	9.1	7.1
Adjudication	9.2	9.2	9.2	9.2	9.2	7.2
Arbitration						
Conduct of arbitration	9.3	9.3	9.3	9.3	9.3	7.3, Sch 1 para 1
Notice of reference to arbitration	9.4	9.4	9.4	9.4	9.4	Sch 1, para 2
Powers of Arbitrator	9.5	9.5	9.5	9.5	9.5	Sch 1, para 3
Effect of award	9.6	9.6	9.6	9.6	9.6	Sch 1, para 4
Appeal – questions of law	9.7	9.7	9.7	9.7	9.7	Sch 1, para 5
Arbitration Act 1996	9.8	9.8	9.8	9.8	9.8	Sch 1, para 6
SCHEDULES						
Schedule 1 – Design Submission Procedure	Sch 1	Sch 1	Sch 1	Sch 1	Sch 6	—
Schedule 2 – Supplemental Provisions				Sch 2	Sch 5	Sch 3
Part 1 – Supplemental Provisions						
(2011) Site Manager	—	Para 1	Para 1			
Named Sub-contractors	Para 1	Para 2	Para 2	Sch 8, para 9	3.7, Sch 2	—
(2011) Bills of Quantities	—	Para 3	Para 3			
Valuation of Changes – Contractor's estimates	Para 2	Para 4	Para 4	Para 1	—	—
Loss and expense – Contractor's estimates	Para 3	Para 5	Para 5	—	—	—

DB 2016 Description	DB 2016	DB 2011	DB 2005	JCT SBC/Q 2016 CDP clause	JCT ICD 2016	MWD 2016
Part 2						
Acceleration Quotation	Para 4	Para 6	—	Para 2	—	—
Collaborative working	Para 5	Para 7	—	Sch 8, para 1	Sch 5, para 1	Sch 3, para 1
Health and Safety	Para 6	Para 8	—	Sch 8, para 2	Sch 5, para 2	Sch 3, para 2
Cost Savings and value improvements	Para 7	Para 9	—	Sch 8, para 3	Sch 5, para 3	Sch 3, para 3
Sustainable development and environmental considerations	Para 8	Para 10	—	Sch 8, para 4	Sch 5, para 4	Sch 3, para 4
Performance indicators and monitoring	Para 9	Para 11	—	Sch 8, para 5	Sch 5, para 5	Sch 3, para 5
Notification and negotiation of disputes	Para 10	Para 12	—	Sch 8, para 6	Sch 5, para 6	Sch 3, para 6
Transparency	Para 11	—	—	Sch 8, para 7	—	Sch 3, para 7
The Public Contracts Regulations 2015	Para 12	—	—	Sch 8, para 8	—	Sch 3, para 8
Schedule 3 – Insurance Options						
Insurance Option A (New buildings – All Risks Insurance of the Works by the Contractor)						
Contractor to effect and maintain a Joint Names Policy	A.1	A.1	A.1	A.1	Sch 1, A.1	5.4A.1
(2011) Insurance documents – failure by Contractor to insure	—	A.2	A.2	—	—	—
Use of Contractor's annual policy – as alternative	A.2	A.3	A.3	A.2	Sch 1, A.2	—
Loss or damage	A.3	A.4	A.4	A.3	Sch 1, A.3	—
(2005) Terrorism cover – premium rate changes	—	—	A.5	—	—	—

DB 2016 Description	DB 2016	DB 2011	DB 2005	JCT SBC/Q 2016 CDP clause	JCT ICD 2016	MWD 2016
Insurance Option B (New Buildings – All Risks Insurance of the Works by the Employer)						
Employer to effect and maintain a Joint Names Policy	B.1	B.1	B.1	B.1	Sch 1, B.1	5.4B.1
(2011) Evidence of Insurance	—	B.2	B.2	—	—	—
Loss or damage	B.2	B.3	B.3	B.2	Sch 1, B.2	—
Insurance Option C (Joint Names Insurance by the Employer of Existing Structures and Works in or Extensions to them)						
Existing Structures and contents – Joint Names Policy for Specified Perils	C.1	C.1	C.1	C.1	Sch 1, C.1	5.4C.1
The Works – Joint Names Policy for All Risks	C.2	C.2	C.2	C.2	Sch 1, C.2	—
(2011) Evidence of Insurance	—	C.3	C.3	—	—	—
Loss or damage	C.3	C.4	C.4	C.3	Sch 1, C.3	—
Schedule 4 – Code of Practice	Sch 4	Sch 4	Sch 4	Sch 4	—	—
Schedule 5 – Third Party Rights	Sch 5	Sch 5	Sch 5	Sch 5	—	—
Schedule 6 – Forms of Bonds	Sch 6	Sch 6	Sch 6	Sch 6	Sch 3	—
Schedule 7 – JCT Fluctuations Option A	Sch 7	Sch 7	Sch 7	Sch 7	Sch 4	Sch 2
Schedule 7 – JCT Fluctuations Option B	Sch 7	Sch 7	Sch 7	—		—
Schedule 7 – JCT Fluctuations Option C	Sch 7	Sch 7	Sch 7	—		—

Appendix B

FIDIC Yellow Book Reconciliation

FIDIC Conditions of Contract for Plant Design & Build descriptions (Yellow Book) – 2nd Ed [2017]	2017 Clause	1999 Clause
1 – General Provisions		
Definitions	1.1	1.1
Interpretation	1.2	1.2
Notices and other communications	1.3	1.3
Law & Language	1.4	1.4
Priority of documents	1.5	1.5
Contract Agreement	1.6	1.6
Assignment	1.7	1.7
Care and supply of documents	1.8	1.8
Errors in the Employer's Requirements	1.9	1.9
Employer's use of the Contractor's Documents	1.10	1.10
Contractor's use of the Employer's Documents	1.11	1.11
Confidentiality	1.12	1.12
Compliance with Laws	1.13	1.13
Joint and Several Liability	1.14	1.14
Limitation of Liability	1.15	17.6
Contract Termination	1.16	—
2 – The Employer		
Right of Access to the Site	2.1	2.1
Assistance	2.2	
Employer's Personnel and Other Contractors	2.3	2.3
Employer's Financial Arrangements	2.4	2.4
Site Data and Items of Reference	2.5	
Employer-supplied Materials and Employer's Equipment	2.6	4.20

Design and Build Contracts, First Edition. Guy Higginbottom.
© 2024 John Wiley & Sons Ltd. Published 2024 by John Wiley & Sons Ltd.

FIDIC Conditions of Contract for Plant Design and Build descriptions (Yellow Book) – 2nd ed. (2017)	2017 Clause	1999 Clause
3 – The Engineer		
The Engineer	3.1	
Engineer's Duties and Authority	3.2	3.1
The Engineer's Representative	3.3	
Delegation by the Engineer	3.4	3.2
Engineer's Instructions	3.5	3.3
Replacement of the Engineer	3.6	3.4
Agreement or Determination	3.7	3.5
Meetings	3.8	
4 – The Contractor		
Contractor's General Obligations	4.1	4.1
Performance Security	4.2	4.2
Contractor's Representative	4.3	4.3
Subcontractors	4.4	4.4
Nominated Subcontractors	4.5	4.5
Co-operation	4.6	4.6
Setting Out	4.7	4.7
Health & Safety Obligations	4.8	4.8
Quality, Management, and Compliance Verification Scheme	4.9	4.9
Use of Site Data	4.10	4.10
Sufficiency of the Accepted Contract Amount	4.11	4.11
Unforeseeable Physical Conditions	4.12	4.12
Rights of Way and Facilities	4.13	4.13
Avoidance of Interference	4.14	4.14
Access Route	4.15	4.15
Transport of Goods	4.16	4.16
Contractor's Equipment	4.17	4.17
Protection of the Environment	4.18	4.18
Temporary Utilities	4.19	4.19
Progress Reports	4.20	4.21
Security of the Site	4.21	4.22
Contractor's Operations on Site	4.22	4.23
Archaeological and Geological Findings	4.23	4.24
5 – Design		
General Design Obligations	5.1	5.1

FIDIC Conditions of Contract for Plant Design and Build descriptions (Yellow Book) – 2nd ed. (2017)	2017 Clause	1999 Clause
Contractor's Documents	5.2	5.2
Contractor's Undertaking	5.3	5.3
Technical Standards and Regulations	5.4	5.4
Training	5.5	5.5
As-Built Documents	5.6	5.6
Operation and Maintenance Manuals	5.7	5.7
Design Error	5.8	5.8
6 – Staff and Labour		
Engagement of Staff and Labour	6.1	6.1
Rates of Wages and Conditions of Labour	6.2	6.2
Recruitment of Persons	6.3	
Labour Laws	6.4	6.4
Working Hours	6.5	6.5
Facilities for Staff and Labour	6.6	6.6
Health & Safety of Personnel	6.7	6.7
Contractor's Superintendence	6.8	6.8
Contractor's Personnel	6.9	6.9
Contractor's Records	6.10	6.10
Disorderly Conduct	6.11	6.11
Key Personnel	6.12	—
7 – Plant, Materials and Workmanship		
Manner of Execution	7.1	7.1
Samples	7.2	7.2
Inspection	7.3	7.3
Testing by the Contractor	7.4	7.4
Defects and Rejection	7.5	7.5
Remedial Work	7.6	7.6
Ownership of Plant and Materials	7.7	7.7
Royalties	7.8	7.8
8 – Commencement, Delays and Suspension		
Commencement of Works	8.1	8.1
Time for Completion	8.2	8.2
Programme	8.3	8.3
Advance Warning	8.4	—
Extension of Time for Completion	8.5	8.4

FIDIC Conditions of Contract for Plant Design and Build descriptions (Yellow Book) – 2nd ed. (2017)	2017 Clause	1999 Clause
Delays caused by Authorities	8.6	8.5
Rate of Progress	8.7	8.6
Delay Damages	8.8	8.7
Employer's Suspension	8.9	8.8
Consequences of Employer's Suspension	8.10	8.9
Payment for Plant and Materials after Employer's Suspension	8.11	8.10
Prolonged Suspension	8.12	8.11
Resumption of Work	8.13	8.12
9 – Tests on Completion		
Contractor's Obligations	9.1	9.1
Delayed Tests	9.2	9.2
Retesting	9.3	9.3
Failure to Pass Tests on Completion	9.4	9.4
10 – Employer's Taking Over		
Taking Over the Works and Sections	10.1	10.1
Taking Over Parts	10.2	10.2
Interference with Tests on Completion	10.3	10.3
Surfaces Requiring Reinstatement	10.4	10.4
11 – Defects After Taking Over		
Completion of Outstanding Work and Remedying Defects	11.1	11.1
Cost of Remedying Defects	11.2	11.2
Extension of Defects Notification Period	11.3	11.3
Failure to Remedy Defects	11.4	11.4
Remedying Defective Work off site	11.5	11.5
Further Tests after Remedying Defects	11.6	11.6
Right of Access after Taking Over	11.7	11.7
Contractor to Search	11.8	11.8
Performance Certificate	11.9	11.9
Unfulfilled Obligations	11.10	11.10
Clearance of Site	11.11	11.11
12 – Tests After Completion		
Procedure for Tests after Completion	12.1	12.1
Delayed Tests	12.2	12.2

FIDIC Conditions of Contract for Plant Design and Build descriptions (Yellow Book) – 2nd ed. (2017)	2017 Clause	1999 Clause
Retesting	12.3	12.3
Failure to Pass Tests after Completion	12.4	12.4
13 – Variations and Adjustments		
Right to Vary	13.1	13.1
Value Engineering	13.2	13.2
Variation Procedure	13.3	13.3
Provisional Sums	13.4	13.5
Daywork	13.5	13.6
Adjustments for Changes in Laws	13.6	13.7
Adjustments for Changes in Cost	13.7	13.8
14 – Contract Price and Payment		
The Contract Price	14.1	14.1
Advance Payment	14.2	14.2
Application for Interim Payment	14.3	14.3
Schedule of Payments	14.4	14.4
Plant and Materials intended for the Works	14.5	14.5
Issue of IPC [Interim Payment Certificate]	14.6	14.6
Payment	14.7	14.7
Delayed Payment	14.8	14.8
Release of Retention Money	14.9	14.9
Statement at Completion	14.10	14.10
Final Statement	14.11	—
Discharge	14.12	14.12
Issue of FPC [Final Payment Certificate]	14.13	14.13
Cessation of Employer's Liability	14.14	14.14
Currencies of Payment	14.15	13.4 / 14.5
15 – Termination by Employer		
Notice to Correct	15.1	15.1
Termination for Contractor's Default	15.2	15.2
Valuation after Termination for Contractor's Default	15.3	15.3
Payment after Termination for Contractor's Default	15.4	15.4
Termination for Employer's Convenience	15.5	—
Valuation after Termination for Employer's Convenience	15.6	—
Payment after Termination for Employer's Convenience	15.7	—

FIDIC Conditions of Contract for Plant Design and Build descriptions (Yellow Book) – 2nd ed. (2017)	2017 Clause	1999 Clause
16 – Suspension and Termination by Contractor		
Suspension by Contractor	16.1	16.1
Termination by Contractor	16.2	16.2
Contractor's Obligations after Termination	16.3	16.3
Payment after Termination by Contractor	16.4	16.4
17 – Care of the Works and Indemnities		
Responsibility for Care of the Works	17.1	17.2
Liability for Care of the Works	17.2	17.2
Intellectual and Industrial Property Rights	17.3	17.5
Indemnities by Contractor	17.4	17.1
Indemnities by Employer	17.5	17.1
Shared Indemnities	17.6	17.1
18 – Exceptional Events [Force Majeure]		
Exceptional Events [Definition of Force Majeure]	18.1	19.1
Notice of an Exceptional Event [Notice of Force Majeure]	18.2	19.2
Duty to Minimise Delay	18.3	19.3
Consequences of an Exceptional Event [/ Force Majeure]	18.4	19.4
Optional Termination	18.5	19.6
Release from Performance under the Law	18.6	19.7
19 – Insurance		
General Requirements	19.1	18.1
Insurance to be provided by the Contractor	19.2	18.2–18.4
20 – Claims		
Claims	20.1	20.1
Claims for Payment and / or EOT	20.2	
21 – Disputes and Arbitration		
Constitution of the DAAB	21.1	
Failure to Appoint DAAB Member(s)	21.2	20.3
Avoidance of Disputes	21.3	
Obtaining DAAB's Decision	21.4	20.4
Amicable Settlement	21.5	20.5
Arbitration	21.6	20.6
Failure to Comply with DAAB's Decision	21.7	20.7
No DAAB in Place	21.8	20.3

Appendix C

NEC Engineering and Construction Contract (ECC) Reconciliation

NEC Engineering and Construction Contract clauses:	NEC4	NEC3[a]
1 – General		
Actions	10	10
Identified and defined terms	11.1 11.2(1) to 11.2(20)	11.1 11.2(1) to 11.2(19)
– Option A: Priced Contract with Activity Schedule	11.2(21), 11.2(23), 11.2(28), 11.2(29), 11.2(32)	11.2(20), 11.2(22), 11.2(27), 11.2(30)
– Option B: Priced Contract with Bill of Quantities	11.2(22), 11.2(23), 11.2(28), 11.2(30), 11.2(33)	11.2(21), 11.2(22), 11.2(28), 11.2(31)
– Option C: Target Contract with Activity Schedule	11.2(21), 11.2(24), 11.2(26), 11.2(31), 11.2(32)	11.2(20), 11.2(23), 11.2(25), 11.2(29), 11.2(30)
– Option D: Target Contract with Bill of Quantities	11.2(2), 11.2(24), 11.2(26), 11.2(31), 11.2(33), 11.2(35),	11.2(21), 11.2(23), 11.2(25), 11.2(29), 11.2(31), 11.2(33)
– Option E: Cost Reimbursable Contract	11.2(24), 11.2(26), 11.2(31), 11.2(34)	11.2(23), 11.2(25), 11.2(29), 11.2(32)
Interpretation and the law	12.1 to 12.4	12.1 to 12.4
Communications	13.1 to 13.8	13.1 to 13.8
The Project Manager and the Supervisor	14.1 to 14.4	14.1 to 14.4
Early Warning	15.1 to 15.4	16.1 to 16.4
Contractor's Proposals	16.1 to 16.3	—
Requirements for instructions	17.1, 17.2	17[b], 18[c]
Corrupt Acts	18.1 to 18.3	—
Prevention	19.1	19

Design and Build Contracts, First Edition. Guy Higginbottom.

NEC Engineering and Construction Contract clauses:	NEC4	NEC3[a)]
2 – The Contractor's Main Responsibilities		
Providing the Works	20.1	20.1
– Option C	20.3, 20.4	20.3, 20.4
– Option D	20.3, 20.4	20.3, 20.4
– Option E	20.3, 20.4	20.3, 20.4
The Contractor's design	21.1 to 21.3	21.1 to 21.3
Using the Contractor's design	22.1	22.1
Design of Equipment	23.1	23.1
People	24.1, 24.2	24.1, 24.2
Working with the Client and Others	25.1 to 25.3	25.1 to 25.3
2 – The Contractor's Main Responsibilities cont'd		
Subcontracting	26.1 to 26.3	26.1 to 26.3
– Option C	26.4	26.4
– Option D	26.4	26.4
– Option E	26.4	26.4
Other responsibilities	27.1 to 27.4	27.1 to 27.4
Assignment	28.1	—
Disclosure	29.1 to 29.2	—
3 – Time		
Starting, Completion, and Key Dates	30.1 to 30.3	30.1 to 30.3
The programme	31.1 to 31.3	31.1 to 31.3
– Option A	31.4	31.4[d)]
Revising the programme	32.1, 32.2	32.1, 32.2
Access to and use of the Site	33.1	33.1
Instructions to stop or not to start work	34.1	34.1
Take over	35.1 to 35.3	35.1 to 35.3
Acceleration	36.1 to 36.3	36.1 to 36.3[e)]
4 – Quality Management		
Quality management system	40.1 to 40.3	—
Tests and inspections	41.1 to 41.6	40.1 to 40.6
– Option C	41.7	40.7
– Option D	41.7	40.7
– Option E	41.7	40.7

NEC Engineering and Construction Contract clauses:	NEC4	NEC3[a]
Testing and inspection before delivery	42.1	41.1
Searching for and notifying Defects	43.1, 43.2	42.1, 42.2
Correcting Defects	44.1 to 44.4	43.1 to 43.4
Accepting Defects	45.1, 45.2	44.1, 44.2
Uncorrected Defects	46.1, 46.2	45.1, 45.2
5 – Payment		
Assessing the amount due	50.1 to 50.6	50.1 to 50.5[f]
– Option C	50.7, 50.9	50.6
– Option D	50.7, 50.9	50.6
– Option E	50.8, 50.9	50.7
Payment	51.1 to 51.5	51.1 to 51.4, 50.2[g]
Defined Cost	52.1	52.1
– Option C	52.2, 52.4	52.2, 52.3
5 – Payment cont'd		
– Option D	52.2, 52.4	52.2, 52.3
– Option E	52.2, 52.4	52.2, 52.3
Final Assessment	53.1 to 53.4	—
– Option C: The Contractor's share	54.1 to 54.4	53.1 to 53.4
– Option D: The Contractor's share	54.5 to 54.8	53.5 to 53.8
– Option A: The Activity Schedule	55.1, 55.3, 55.4	54.1 to 54.3
– Option C: The Activity Schedule	55.2	54.1 to 54.3
– Option B: The Bill of Quantities	56.1	55.1
– Option D: The Bill of Quantities	56.1	55.1
6 – Compensation Events		
Compensation events	60.1(1) to 60.1(21), 60.2, 60.3	60.1(1) to 60.1(19), 60.2, 60.3
– Option B	60.4 to 60.7	60.4 to 60.7
– Option D	60.4 to 60.7	60.4 to 60.7
Notifying compensation events	61.1 to 61.7	61.1 to 61.7
Quotations for compensation events	62.1 to 62.6	62.1 to 62.6
Assessing compensation events	63.1 to 63.11	63.1 to 63.9[h]
– Option A	63.12, 63.14, 63.16	63.10, 63.12, 63.14
– Option B	63.12, 63.15, 63.16	63.10, 63.13

NEC Engineering and Construction Contract clauses:	NEC4	NEC3[a)]
– Option C	63.13, 63.14	63.11, 63.12, 63.14, 63.15
– Option D	63.13, 63.15,	63.11, 63.13, 63.15
The Project Manager's assessments	64.1 to 64.4	64.1 to 64.4
Proposed instructions	65.1 to 65.3	—
Implementing compensation events	66.1 to 66.3	65.1, 65.2, 65.4[i)]
7 – Title		
The Client's title to Plant and Materials	70.1, 70.2	70.2, 70.2
Marking Equipment, Plant and Materials outside the Working Areas	71.1	71.1
Removing Equipment	72.1	72.1
Objects and materials within the Site	73.1, 73.2	73.1, 73.2
The Contractor's use of material	74.1	—
8 – Liabilities and Insurance		
Client's liabilities	80.1	80.1
Contractor's liabilities	81.1	81.1
Recovery of costs	82.1 to 82.3	—
Insurance cover	83.1 to 83.3	84.1, 84.2
Insurance policies	84.1 to 84.3	85.1 to 85.4
If the Contractor does not insure	85.1	86.1
Insurance by the Client	86.1 to 86.3	87.1 to 87.3
9 – Termination		
Termination	90.1 to 90.4	90.1 to 90.3, 90.5
Reasons for termination	91.1 (R1 to R22) to 91.8	91.1 (R1 to R21) to 91.7[j)]
Procedures on termination	92.1, 92.2 (P1 to P4)	92.1, 92.2 (P1 to P4)
Payment on termination	93.1, 93.2	93.1, 93.2
– Option A	93.3	93.3
– Option C	93.4, 93.6	93.4, 93.6
– Option D	93.5, 93.6	93.5, 93.6

NEC Engineering and Construction Contract clauses:	NEC4	NEC3[a)]
Resolving and Avoiding Disputes		
Option W1 (HGCRA 1996 does not apply)		
Resolving Disputes	W1.1	W1.1
The Adjudicator	W1.2	W1.2
The adjudication	W1.3	W1.3
The tribunal	W1.4	W1.4
W2 (HGCRA 1996 applies)		
Resolving disputes	W2.1	W2.1
The Adjudicator	W2.2	W2.2
The adjudication	W2.3	W2.3
The tribunal	W2.4	W2.4
W3 (DAB, HGCRA 1996 does not apply)		
The Dispute Avoidance Board	W3.1	—
Resolving potential disputes	W3.2	—
The tribunal	W3.3	—
Option X1: Price Adjustment for Inflation (Options A to D)		
Defined terms	X1.1	X1.1
Price Adjustment Factor	X1.2	X1.2
Price adjustment Options A and B	X1.3	X1.4
Price adjustment Options C and D	X1.4	
Compensation events	X1.5	X1.3
Option X2: Changes in the Law		
Changes in the law	X2.1	X2.1
Option X3: Multiple Currencies (Options A and B)		
Multiple currencies	X3.1, X3.2	X3.1, X3.2
Option X4: Ultimate Holding Company Guarantee		
Ultimate holding company guarantee	X4.1, X4.2	X4.1

NEC Engineering and Construction Contract clauses:	NEC4	NEC3[a)]
Option X5: Sectional Completion		
Sectional completion	X5.1	X5.1
Option X6: Bonus for Early Completion		
Bonus for early Completion	X6.1	X6.1
Option X7: Delay Damages		
Delay damages	X7.1 to X7.3	X7.1 to X7.3
Option X8: Undertakings to the Client or Others		
Undertakings to the Client or Others	X8.1 to X8.5	—
Option X9: Transfer of Rights		
Transfer of rights	X9.1	—
Option X10: Information Modelling		
Defined terms	X10.1	—
Collaboration	X10.2	—
Early warning	X10.3	—
Information Execution Plan	X10.4	—
Compensation Events	X10.5	—
Use of the information model	X10.6	—
Liability	X10.7	
Option X11: Termination by the Client		
Termination by the client	X11.1, X11.2	—
Option X12: Multiparty Collaboration (not used with X20)[k)]		
Identified and defined terms	X12.1(1) to X12.1(7)	X12.1(1) to X12.1(7)
Actions	X12.2(1) to X12.2(6)	X12.2(1) to X12.2(6)
Collaboration	X12.3(1) to X12.3(9)	X12.3(1) to X12.3(9)
Incentives	X12.4(1), X12.4(2)	X12.4(1), X12.4(2)

NEC Engineering and Construction Contract clauses:	NEC4	NEC3[a)]
Option X13: Performance Bond		
Performance Bond	X13.1	X13.1
Option X14: Advanced Payment to the Contractor		
Advanced payment	X14.1 to X14.3	X14.1 to X14.3
Option X15: The Contractor's Design		
The contractor's design	X15.1 to X15.6	X15.1, X15.2[l)]
Option X16: Retention (not used with Option F)		
Retention	X16.1 to X16.3	X16.1, X16.2[m)]
Option X17: Low Performance Damages		
Low performance damages	X17.1	X17.1
Option X18: Limitation of Liability		
Limitation of liability	X18.1 to X18.6	X18.1 to X18.5
Option X20: Key Performance Indicators (not used X12)		
Incentives	X20.1 to X20.5	X20.1 to X20.5
Option X21: Whole Life Cost		
Whole life cost	X21.1 to X21.3	—
Option X22: Early Contractor Involvement (Options C and E)		
Identified and defined terms	X22.1	—
Forecasts	X22.2	—
Proposals for Stage Two	X22.3	—
Key persons	X22.4	—
Notice to proceed to Stage Two	X22.5	—
Changes to the budget	X22.6	—
Option X22: Early Contractor Involvement (Options C and E) cont'd		
Incentive payment	X22.7	—

NEC Engineering and Construction Contract clauses:	NEC4	NEC3[a]
Option Y(UK)1: Project Bank Account		
Definitions	Y1.1	Y1.1
Project bank account	Y1.2 to Y1.4	Y1.2 to Y1.4
Named suppliers	Y1.5, Y1.6	Y1.5, Y1.6
Payments	Y1.7 to Y1.11	Y1.7 to Y1.11
Effect of payment	Y1.12	Y1.12
Trust deed	Y1.13	Y1.13
Termination	Y1.14	Y1.14
Option Y(UK)2: Housing Grants Construction and Regeneration Act 1996		
Definition	Y2.1	Y2.1
Dates for payment	Y2.2	Y2.2
Notice of intention to pay less	Y2.3, Y2.4	Y2.3
Suspension of performance	Y2.5	Y2.4
Option Y(UK)3: The Contracts (Rights of Third Parties) Act 1999		
Third party rights	Y3.1 to Y3.3	Y3.1
Option Z: Additional Conditions of Contract		
Additional conditions of contract	Z1.1	Z1.1

a) NEC3 Engineering and Construction Contract (April 2013).
b) NEC3, clause 17 (Ambiguities and inconsistencies).
c) NEC3, clause 18 (Illegal and impossible requirements).
d) NEC3, Options A and C include clause 31.4.
e) NEC3, clause 36.3 for Options A, B, C, and D, clause 36.4 for Option E.
f) NEC3, clause 50 (Assessing the amount due) does not provide for the contractor submitting an application for payment (as per NEC4 – clause 50.2).
g) NEC3 clause 50.2 contains a similar clause to NEC4's 51.5.
h) Save for NEC3 Option E (Cost reimbursable contract) clause 63.14, NEC 3 does not contain the equivalent to NEC4's clause 53.2 (contractor and project manager may agree on lump sums to assess the change to the prices).
i) NEC3 Options A, B, C, and D – clause 65.4, Option E – clause 65.3.
j) NEC3 clause 91 does not contain an equivalent to NEC4's clause 91.8 (allowing the employer (client) to terminate if the contractor commits a corrupt act).
k) NEC3 Option X12 is referred to as 'Partnering'.
l) NEC3 Option X15 (Limitation of the contractor's liability for his design to reasonable skill and care) does not contain equivalent provisions to NEC4 Option X15, clauses X15.3 to X15.6.
m) NEC Option X16 does not contain provision for a retention bond (as NEC4 Option X16, clause X16.3).

Index

a

Acceleration   22, 152, 153, 154, 155, 157, 167, 187, 215, 219, 300, 301, 388
  quotation for   153, 155, 167, 219, 300, 301, 378
Acceptance   29, 38, 40, 70, 85, 86, 87, 93, 155, 158, 160, 161, 169, 170, 176, 194, 204, 238
Accepted Contract Amount   42, 127, 129, 137, 140, 382
Accepted Programme   39, 85, 108, 188, 189, 190, 191, 194, 204, 211, 212, 247
Access   22, 33, 36, 37, 40, 69, 76, 89, 136, 138, 145, 159, 180, 182, 184, 189, 198, 199, 209, 216, 226, 291, 294, 296, 297, 311, 323, 357, 371, 384
Access date   40, 138, 142, 189
Access route   184, 226, 349, 382
Access to and use of the site   182, 216, 226, 235, 291, 294, 296, 297, 311, 323, 381, 388
Actions   387
Activity schedule   9, 12, 42, 43, 57, 84, 85, 101, 108, 133, 138, 176, 197, 208, 209, 210, 261, 262, 276, 308, 309, 310, 333, 347, 387, 389
Adjudication   9, 174, 301, 304, 312, 313, 315, 316, 320, 321, 324, 325, 326, 327, 329, 331, 350, 367, 377, 391
  appointment of adjudicator   326
Advance Payment   71, 119, 125, 126, 128, 129, 134, 168, 170, 361, 372, 385
Advance payment bond   119, 125, 134

Advance Payment Certificate   128, 134
Advance Payment Guarantee   129, 134, 361
Advance Payment Security   128
Advance warning   141, 150, 283, 284
Agency   25, 66, 68, 69
Agreement or Determination   38, 171, 173, 174, 184, 185, 186, 222, 223, 224, 246, 290, 306, 316, 317, 318, 319, 323, 329, 382
All Risks Insurance   136, 270, 378, 379
Ambiguities and inconsistences   13, 28, 33, 70, 82, 188, 190, 192, 195, 280, 351
Amendments   4, 5, 6, 11, 31, 36, 56, 87, 90, 105, 108, 123, 174, 337, 344, 347, 348, 349, 350, 351, 352, 353, 354, 355, 356, 357, 360
Amount due   41, 112, 138, 164, 175, 176, 210, 246, 261, 262, 263, 264, 276, 285, 298, 306, 307, 308, 310, 311, 312, 321, 333, 334, 389, 394
Antiquities   219, 354, 371
Applicable laws   39, 238, 327, 368
Application for loss and/or expense   45, 167, 183, 191, 228, 229, 251, 252, 257, 298, 353
Application for payment   102, 129, 163, 164, 165, 166, 169, 170, 171, 175, 176, 258, 349, 354, 355, 385, 394
Appointments   56, 57, 58, 59, 60, 61, 62, 65, 66, 69, 94, 125, 316, 340, 341, 342, 344, 345, 361, 371
  of DAAB   322, 323
  of DAB   319, 321

Design and Build Contracts, First Edition. Guy Higginbottom.
© 2024 John Wiley & Sons Ltd. Published 2024 by John Wiley & Sons Ltd.

Approval   15, 32, 37, 39, 65, 70, 87, 104, 135,
    138, 169, 203, 247, 264, 330, 349, 350,
    352, 357, 363, 365
Arbitration   9, 38, 301, 305, 312, 315, 316,
    319, 321, 322, 323, 324, 325, 326, 327,
    328, 329, 331, 349, 367, 377, 386
Arbitration Act 1996   325
Arbitrator, appointment   325
Archaeological findings   189, 190, 226, 382
Architect   2, 23, 55, 58, 59, 60, 61, 62, 64, 65,
    69, 110, 112, 231, 232, 270, 271, 288,
    341, 343, 345
Architects Act 1997   58, 377
Articles of Agreement   10, 11, 94, 133, 232,
    367
As-built information, records   38, 64, 251,
    265, 266, 270, 273, 289, 295, 370,
    383
Ascertainment   200, 228, 229, 252, 299, 373
Assessing the amount due-amount due   41,
    138, 262, 298, 389, 394
Assessment date   129
Assignment   246, 353, 375, 381, 388
Assistance
    by consultants   65
    by employer   381
Assistant of Engineer   25, 151, 170, 255, 382.
    see Engineer's Representative
Attestation   8, 10, 133
Authority, of Employer's representatives   25,
    68, 69, 70, 382

b

Bankruptcy   248, 263
Base Date   34, 185, 219, 351
Best endeavours   146, 147, 153, 219
Bill of Quantities   12, 30, 41, 42, 43, 44, 45,
    101, 133, 138, 176, 188, 190, 197, 200,
    206, 207, 208, 209, 210, 247, 262, 276,
    308, 309, 333, 362, 363, 387, 389
Bond   7, 9, 15, 37, 46, 49, 57, 78, 103, 117,
    119, 125, 126, 127, 128, 129, 130, 131,
    132, 134, 136, 167, 177, 207, 264, 272,
    348, 349, 355, 361, 373, 375, 379, 393,
    394

Bonus for early completion   57, 277, 392
Boundaries (site)   33, 34, 351, 352
Breach of contract   24, 27, 33, 99, 102, 106,
    108, 111, 120, 121, 125, 126, 127, 137,
    145, 146, 164, 190, 195, 226, 227, 243,
    244, 245, 246, 247, 248, 250, 253, 264,
    285, 288, 331, 334, 342, 344, 348, 353,
    356
Brief   2, 3, 30, 55, 61, 66, 67, 80, 83, 341
Building Act 1984   23
Building Information Modelling (BIM)   9,
    11, 36, 37, 39, 66, 76, 79, 146
Building Regulations 2010   23, 145
Business day   131, 165, 248

c

CDM personnel   10, 37, 367
CDM Regulations 2015   9, 22, 35, 37, 59, 182,
    246, 247, 265, 289, 371
Certification   32, 68, 76, 349, 357, 363, 364
CESMM4   30, 31, 32, 33, 41, 44, 45, 77, 78,
    362, 363
Change in the law or legislation   57, 171,
    185, 203, 226, 385, 391
Changes   22, 27, 28, 33, 34, 38, 50, 51, 64, 70,
    103, 111, 167, 181, 182, 183, 184, 188,
    189, 190, 197, 198, 199, 200, 201, 216,
    219, 233, 253, 270, 300, 344, 360, 371,
    373, 377. *see also* Variations
Changes in Cost (FIDIC)   171, 203, 210,
    385
Change to Completion or Key Date   155,
    189, 192, 193, 194, 196, 203, 205, 207,
    211, 225, 285, 333
Change to the Prices   41, 85, 155, 192, 193,
    194, 196, 203, 204, 205, 207, 208, 210,
    225, 229, 285, 307, 333
Charges   46, 47, 174, 175, 277, 292, 369
CIC Collateral Warranty   125
CIC Novation Agreement   341, 345
CIMAR   325
City of London Law Society Novation
    Agreement   341, 345
Civil commotion   219, 247, 256, 354

Claims   45, 84, 126, 127, 129, 130, 131, 145,
        152, 175, 181, 184, 186, 190, 195, 213,
        221, 222, 223, 224, 225, 227, 228, 229,
        235, 236, 258, 268, 275, 279, 280, 283,
        284, 291, 292, 294, 304, 306, 317, 318,
        319, 328, 331, 332, 353, 374, 386
    of continuing effect   317, 318
    definition of   221
    Fully Detailed   317
    Notice of   222, 223, 283, 318, 353
    for payment   175, 184, 186, 221, 225, 235,
        258, 275, 283, 304, 306, 332, 386
    under Performance Security   317, 361
Clearance, of Site   44, 89, 295, 384
Clerk of works   25, 232
Climatic conditions   149
Code of Practice   13, 233, 379
Collaboration   57, 84, 88, 392
Collateral warranties   7, 9, 37, 78, 97, 103,
        106, 117, 119, 120, 121, 122, 123, 124,
        348, 349, 355, 365, 375, 376
Commencement Date   9, 34, 122, 127, 135,
        136, 137, 141, 159, 169, 246
Commissioning tests   159, 230, 243, 265, 266,
        267, 330, 363, 364
Common law damages, principle of   226, 227
Communications   32, 158, 188, 193, 220, 222,
        254, 255, 256, 259, 301, 311, 357, 368,
        381, 387
    period for reply   29, 38, 66, 93, 108, 158,
        160, 189, 204
Compensation   225, 226, 232, 278, 292
Compensation events   28, 29, 32, 41, 43, 50,
        51, 57, 70, 82, 84, 86, 101, 103, 108,
        145, 151, 152, 158, 160, 177, 186, 187,
        188, 189, 190, 191, 192, 193, 194, 195,
        196, 203, 204, 205, 206, 207, 208, 209,
        210, 211, 212, 213, 214, 221, 225, 227,
        229, 238, 239, 240, 307, 309, 328, 333,
        334, 353, 357, 361, 389, 390, 391, 392
    assessment
        changes to completion date and key
        dates   193, 196, 205, 221, 222, 225
        changes to the prices   41, 207, 210, 394

        changes to the prices, time and risk
            allowances   40, 51, 205, 208
        failure to assess   204
        failure to give early warning   192
        notification by the Project Manager   194
        programme   211, 212
        by the Project Manager   152, 194, 196,
            203, 204, 205, 207, 208, 210, 212
        use of lump sums   346
        use of schedule of cost components and
            shorter schedule of cost components
            206, 209, 362
    consequences for failing to notify   191, 192
    implementing   196
    notifications
        by the Contractor   191
        by the Project Manager   191
    quotations
        acceptance by the PM   194, 196
        extending time for submission   194
        failure to reply by Project Manager   194,
            195
        reply by Project Manager   194
        revised or alternative quotations   194
        submission   194
        submission procedure   194, 196
Completion   1, 9, 22, 33, 39, 46, 51, 57, 64,
        69, 71, 87, 93, 96, 97, 102, 126, 129,
        130, 133, 137, 138, 139, 140, 141, 142,
        143, 144, 145, 146, 148, 149, 150, 151,
        152, 153, 154, 155, 157, 159, 160, 163,
        177, 178, 182, 187, 189, 192, 193, 194,
        195, 196, 203, 204, 205, 210, 211, 212,
        213, 214, 215, 217, 218, 219, 220, 221,
        224, 225, 227, 229, 230, 231, 232, 234,
        237, 239, 240, 243, 245, 249, 250, 251,
        256, 257, 265, 266, 267, 268, 269, 270,
        271, 272, 273, 275, 276, 277, 278, 279,
        280, 281, 282, 283, 284, 285, 286, 287,
        289, 293, 296, 297, 298, 299, 300, 304,
        317, 333, 334, 349, 350, 356, 357, 363,
        364, 367, 370, 383, 384, 388, 392
Date of Completion   141, 276, 283
    planned   108, 160, 190, 247

Completion (*contd.*)
  Tests on   160, 187, 226, 234, 266, 267, 268, 269, 273, 274, 275, 277, 290, 293, 364, 384
  Time for   9, 39, 129, 137, 140, 141, 148, 150, 154, 159, 205, 224, 280, 283, 293, 317, 383
Completion Date   9, 39, 87, 96, 102, 145, 146, 149, 150, 152, 153, 154, 155, 157, 158, 160, 189, 192, 193, 194, 195, 196, 203, 204, 205, 211, 212, 213, 214, 215, 217, 218, 219, 220, 221, 225, 229, 240, 276, 278, 279, 280, 281, 282, 284, 285, 288, 297, 364, 370
Conclusive evidence   205, 299, 311, 330, 331, 332, 351, 356
Concurrent delays   215, 216, 353, 356
Conditions of Contract
  additional   57, 105, 394
  FIDIC Yellow Book 2017   4
  in general   4, 5, 7, 8, 10, 11, 12, 25, 57, 99, 105, 107, 148, 339, 381, 382, 383, 384, 385, 386, 394
Confidentiality   88, 316, 349, 381
Confirmed Acceptance (Acceleration)   153, 155, 300
Consent   70, 86, 104, 105, 112, 118, 154, 169, 186, 187, 271, 288, 290, 370, 371
Construction Industry Council (CIC)   123, 341, 345
Construction Industry Scheme (CIS)   372
Construction materials, selection of   351
Consultant, appointment   56, 57, 58, 122
Consultants   2, 52, 53, 55, 56, 57, 58, 59, 60, 61, 62, 63, 64, 65, 66, 68, 81, 87, 91, 94, 122, 123, 126, 178, 208, 339, 340, 341, 342, 343, 344, 345, 346, 352, 361, 365
  architect   2, 55, 58, 59, 60, 61, 64, 65
  collateral warranties   122, 123, 352
  conditions of Engagement   56, 58, 59, 64, 66, 123, 162, 177, 178, 340, 341, 342, 343, 346, 365
  engineer   2, 25, 55, 58, 59, 60, 61, 64, 65
  services to be performed   58, 59, 60, 61, 62, 63, 64, 65, 66, 67, 68

Consultant switch (novation)   3, 6, 17, 339, 340, 341, 342, 343, 344, 345, 361
Contract
  Priority of documents   13, 26, 381
  repudiation   111, 244
  signing   8, 10, 131, 133, 134, 135, 137, 246
Contract (Rights of Third Parties) Act 1999   57, 118, 123, 124, 368, 394
Contract Agreement   8, 10, 11, 12, 15, 29, 31, 35, 42, 94, 131, 133, 134, 135, 136, 137, 169, 246, 258, 347, 365, 381
Contract Data   8, 9, 10, 11, 12, 25, 27, 28, 29, 31, 32, 34, 35, 38, 40, 41, 42, 43, 58, 62, 68, 78, 81, 85, 90, 93, 94, 104, 122, 123, 124, 127, 129, 130, 133, 134, 136, 137, 138, 140, 141, 142, 143, 144, 149, 151, 158, 160, 169, 170, 171, 172, 173, 174, 175, 176, 177, 190, 195, 202, 204, 206, 210, 211, 212, 238, 239, 240, 254, 276, 278, 280, 283, 284, 285, 296, 298, 304, 307, 308, 309, 310, 311, 312, 313, 319, 320, 322, 323, 325, 329, 347, 360
Contract Date   258, 262
Contract Documents   4, 9, 10, 11, 12, 13, 22, 29, 32, 35, 42, 78, 87, 110, 126, 140, 146, 265, 347, 369
  relationship between   12, 13
Contract-Governing law   38, 39, 121, 124, 145, 162, 171, 238, 284, 295, 311, 326, 327, 328, 329, 334, 349, 350, 368, 381, 387
Contractor's design   3, 4, 5, 12, 13, 22, 23, 24, 27, 28, 29, 30, 31, 34, 35, 36, 37, 38, 39, 40, 63, 64, 69, 76, 86, 87, 92, 94, 108, 122, 123, 124, 135, 138, 159, 181, 182, 185, 238, 251, 265, 278, 289, 338, 339, 350, 354, 359, 360, 361, 370, 373, 388, 393
  acceptance of   35, 36, 37
  client's use   31, 41
  documents   13, 22, 35, 37, 69, 92, 251, 265, 370
  equipment   35, 388

Contractor's Design Portion (CDP)   5, 6, 108, 360, 368, 369, 370, 371, 372, 373, 374, 375, 376, 377, 378, 379

Contractor's documents
engineer's review of   35, 37, 38, 39
operation and maintenance ('O&M')
Manuals   33, 37, 38, 64, 265, 266, 273, 289, 291, 293, 297, 383
progress reports   24, 39, 66, 67, 149, 150, 160, 161, 162, 278, 320, 382

Contractor's Equipment   24, 33, 35, 36, 40, 52, 83, 92, 141, 145, 151, 160, 184, 185, 189, 227, 235, 239, 256, 257, 258, 261, 262, 273, 295, 338, 350, 353, 382, 388

Contractor's Personnel   150, 151, 383

Contractor's Proposals   3, 5, 9, 11, 12, 13, 22, 23, 26, 27, 32, 33, 34, 35, 36, 37, 39, 40, 41, 47, 49, 52, 60, 78, 81, 82, 87, 92, 133, 135, 146, 149, 179, 182, 232, 233, 339, 347, 350

Contractor's Representative   47, 371, 377

Contractor's share
calculation   308
final assessment   262, 308, 350, 389
payment   298

Contract Particulars   10, 11, 29, 31, 35, 42, 43, 90, 94, 120, 124, 128, 129, 130, 131, 133, 134, 135, 136, 140, 143, 165, 166, 167, 183, 201, 219, 232, 247, 270, 272, 280, 281, 282, 287, 289, 315, 325, 329

Contract price
adjustments of   47, 48, 50, 167, 177, 184, 198, 201, 203, 209, 225, 226, 229, 300, 301, 372, 385
generally   9, 11, 41, 42, 43, 44, 45, 53, 76, 126, 141, 144, 150, 154, 178, 179, 184, 185, 186, 197, 201, 205, 221, 223, 241, 267, 283, 288, 292, 297, 298, 299, 330, 344, 385

Contract Sum   3, 5, 9, 10, 11, 12, 13, 22, 32, 41, 42, 44, 45, 48, 50, 52, 78, 111, 112, 133, 153, 165, 179, 183, 185, 197, 198, 202, 232, 233, 272, 282, 288, 299, 300, 301, 330, 333, 344, 347, 362, 363, 367, 372, 373

Contract Sum Analysis   5, 9, 11, 12, 13, 21, 32, 42, 44, 49, 78, 133, 146, 197, 232, 233, 347, 362, 363, 367

Contra proferentum   280

Co-operation   16, 105, 258, 382

Correction of defects   9, 147, 195, 237, 239, 240, 278, 296, 297, 333, 357

Cost Plus Profit   43, 202, 275, 291

Costs, generally   3, 13, 15, 33, 46, 47, 50, 52, 63, 64, 76, 85, 88, 89, 90, 112, 114, 120, 121, 142, 144, 148, 152, 153, 154, 155, 164, 167, 171, 176, 183, 185, 187, 188, 190, 191, 196, 198, 199, 200, 201, 202, 206, 207, 209, 212, 213, 221, 225, 227, 229, 233, 235, 245, 250, 251, 252, 257, 268, 278, 279, 290, 291, 294, 298, 301, 314, 344, 353, 354, 355, 356, 362, 390

Critical Path   36, 39, 40, 52, 148, 156, 157, 159, 160, 162, 212, 214, 215, 216, 217, 353

Currency of contract   9

d

Damages
common law   195, 226, 227, 228, 250
generally   9, 15, 57, 64, 110, 114, 130, 137, 139, 140, 141, 142, 143, 144, 145, 146, 152, 195, 213, 222, 226, 227, 228, 229, 247, 250, 257, 258, 269, 272, 275, 277, 278, 279, 280, 281, 282, 283, 284, 285, 286, 292, 294, 297, 298, 314, 317, 352, 355, 357, 364, 370, 374, 384, 392, 393
for late completion   64, 278, 279, 280, 281, 282, 283, 284, 285, 286
liquidated   9, 64, 114, 130, 139, 140, 143, 250, 272, 278, 279, 280, 281, 282, 283, 284, 286, 314, 357, 364, 370
low performance   286, 292, 298, 355
unliquidated   279, 285, 286

Date for Completion   159, 160, 281, 283, 284

Date of Completion   141, 276, 283
and outstanding work   172, 272, 273, 274, 275, 277

Daywork   44, 49, 198, 200, 201, 203, 317, 373, 385

Defective Premises Act 1972   23, 270
Defects   7, 8, 9, 12, 22, 25, 39, 40, 64, 67, 68,
       71, 97, 112, 120, 126, 127, 130, 141,
       142, 160, 166, 167, 168, 172, 175, 176,
       177, 181, 183, 184, 229, 230, 231, 232,
       233, 234, 236, 237, 239, 240, 241, 243,
       245, 250, 251, 267, 268, 269, 270, 271,
       272, 273, 274, 275, 276, 277, 287, 288,
       289, 290, 291, 292, 293, 294, 295, 296,
       297, 298, 299, 301, 302, 303, 307, 317,
       320, 330, 333, 334, 350, 355, 357, 363,
       370, 383, 384, 389
    correction period   9, 142, 195, 239, 240,
       278, 296, 297, 333, 357
    date   239, 240, 296, 320, 333
    making good of   13, 130, 168, 245, 250,
       251, 270, 287, 288, 297, 302, 357, 370
    outstanding works   286, 297
    rectifying   22, 102, 130, 166, 183, 232, 233,
       239, 240, 241, 244, 245, 269, 272, 287,
       288, 289, 290, 292, 296, 297, 298, 301,
       357
    remedying, costs   289, 290, 317
    remedying of, in general   289, 290, 293,
       295, 302, 303, 384
    search for   189, 195, 226, 232, 240, 241,
       291, 297
    uncorrected   240, 261, 389
Defects Certificate   175, 177, 195, 239, 296,
       297, 298, 307, 333, 334
Defects Notification Period (DNP)   9, 127,
       130, 141, 172, 222, 223, 224, 268, 274,
       289, 291, 293, 294, 295, 357, 384
    definition of   141, 172
    extension of   222, 291, 384
Deferment of possession   136, 140, 141, 145,
       219, 368
Defined Cost   42, 105, 176, 177, 190, 193,
       205, 206, 207, 208, 209, 210, 261, 262,
       309, 389
    actual   206
    forecasts   206, 207, 209
Delay damages   9, 15, 57, 64, 114, 137, 139,
       140, 141, 142, 144, 145, 146, 152, 213,

       222, 258, 275, 277, 278, 279, 283, 284,
       285, 286, 298, 317, 324, 384, 392
Delegation, by Engineer   70, 382
Design and construct   1, 4, 107
Design-rejection   28, 38, 85
Determination, by Engineer   38, 171, 173,
       174, 184, 185, 186, 222, 223, 224, 246,
       290, 306, 316, 317, 318, 319, 323, 329,
       382
Development control requirements   247
Disallowed Cost   176, 206
Discharge   222, 303, 305, 306, 333, 385
Discrepancies   5, 13, 28, 36, 94, 102, 157, 219,
       338, 339, 351, 369
Dispute Adjudication Board (DAB)   312,
       315, 316, 319, 320, 321, 322, 391
Dispute Avoidance/Adjudication Board
       (DAAB)   246, 304, 305, 315, 316, 319,
       322, 323, 324, 386
Divergences   27, 33, 34, 351, 369
Dividing date   206, 207, 212

e

Early warning   138, 150, 151, 192, 392
Early warning notices (EWN)   151, 152, 278
Electronic, communications and documents
       91, 95, 170, 218, 255, 304, 368
Employer
    generally   1, 2, 3, 5, 6, 7, 9, 10, 11, 12, 14,
       15, 16, 21, 22, 23, 24, 25, 26, 27, 28, 29,
       31, 32, 33, 34, 35, 37, 40, 41, 43, 45, 47,
       48, 50, 51, 52, 53, 55, 59, 60, 61, 62, 64,
       66, 67, 68, 69, 73, 75, 76, 77, 78, 79, 80,
       81, 82, 83, 84, 86, 87, 88, 89, 90, 91, 92,
       93, 94, 95, 96, 97, 98, 99, 100, 103, 104,
       105, 106, 107, 108, 109, 110, 111, 112,
       113, 114, 117, 118, 119, 120, 121, 122,
       123, 124, 125, 126, 127, 128, 129, 130,
       131, 132, 133, 134, 135, 136, 137, 138,
       139, 140, 145, 146, 147, 148, 149, 150,
       151, 152, 153, 154, 155, 159, 162, 163,
       164, 165, 166, 167, 168, 169, 171, 172,
       173, 178, 179, 180, 181, 182, 183, 184,
       188, 191, 196, 197, 198, 201, 205, 211,
       213, 214, 215, 216, 218, 219, 220, 221,

222, 223, 224, 225, 226, 227, 228, 229,
230, 231, 232, 233, 234, 235, 236, 237,
241, 243, 244, 245, 246, 247, 248, 249,
250, 251, 252, 253, 254, 255, 256, 257,
258, 259, 260, 265, 266, 267, 268, 270,
271, 272, 273, 274, 275, 277, 278, 279,
280, 281, 282, 283, 284, 286, 287, 288,
289, 290, 291, 292, 293, 294, 295, 297,
299, 300, 301, 302, 306, 307, 314, 315,
318, 322, 323, 330, 331, 332, 333, 337,
338, 339, 340, 341, 342, 343, 344, 345,
346, 347, 348, 349, 350, 351, 352, 353,
354, 355, 356, 357, 359, 360, 361, 362,
363, 364, 367, 368, 370, 374, 376, 379,
381, 385, 386, 394

JCT DB 2016   25, 26, 27, 33, 35, 38, 50, 68,
69, 104, 107, 111, 112, 113, 114, 119,
120, 121, 129, 130, 131, 134, 135, 136,
139, 140, 147, 153, 154, 155, 165, 166,
167, 168, 181, 182, 183, 197, 198, 201,
218, 219, 220, 221, 228, 229, 230, 232,
233, 234, 241, 245, 246, 247, 248, 249,
250, 251, 252, 253, 265, 266, 270, 271,
272, 277, 281, 282, 286, 287, 288, 297,
299, 300, 330, 331, 332, 367, 368, 370,
374, 376, 379

Employer's agent   9, 10, 25, 31, 56, 61, 68, 69,
71, 220, 271, 345, 362, 367, 371

Employer's Representative   21, 22, 24, 25, 26,
34, 40, 55, 56, 60, 61, 68, 69, 98, 101,
103, 104, 106, 149, 232, 247, 271, 285,
297, 315, 345

duties   70, 71

Employer's Agent (JCT DB 2016)   25, 70,
71, 232, 247, 271, 297

Engineer (FIDIC)   9, 25, 31, 34, 38, 39, 47,
50, 55, 61, 65, 68, 69, 70, 71, 107, 109,
111, 115, 122, 127, 128, 134, 136, 141,
144, 148, 149, 150, 151, 153, 154, 160,
161, 169, 170, 171, 172, 173, 174, 181,
184, 185, 186, 187, 190, 196, 200, 201,
202, 203, 222, 223, 224, 226, 227, 235,
241, 254, 255, 257, 258, 266, 267, 268,
273, 274, 275, 277, 283, 284, 289, 290,
291, 293, 294, 295, 297, 303, 304, 305,

306, 309, 316, 318, 323, 324, 333, 345,
382

the Project Manager (NEC4)   25, 70, 71,
285, 297

subcontractors   103, 104, 106

the Supervisor (NEC4)   9, 16, 25, 188, 191,
193, 195, 207, 212, 237, 238, 239, 240,
260, 276, 296, 307, 334, 387

Employer's Requirements   3, 9, 10, 11, 12,
13, 22, 23, 24, 26, 27, 28, 29, 30, 31, 32,
33, 34, 35, 36, 38, 39, 40, 41, 43, 44, 45,
46, 47, 48, 49, 50, 51, 52, 60, 63, 77, 78,
79, 81, 82, 84, 86, 87, 92, 93, 94, 100,
102, 106, 107, 109, 114, 131, 133, 135,
136, 137, 145, 146, 158, 160, 162, 165,
170, 179, 181, 182, 185, 197, 226, 230,
232, 233, 235, 265, 266, 269, 273, 293,
295, 317, 330, 334, 338, 339, 344, 346,
347, 350, 351, 352, 359, 360, 362, 363,
367, 369, 381

errors in   23, 34, 82, 226, 317, 351, 381

Engineer   2, 9, 25, 31, 34, 38, 39, 43, 47, 50,
55, 56, 58, 59, 60, 61, 65, 68, 69, 107,
109, 111, 113, 122, 123, 127, 128, 134,
136, 141, 144, 148, 150, 151, 153, 154,
158, 160, 161, 169, 170, 171, 172, 173,
174, 181, 184, 185, 186, 187, 190, 196,
200, 201, 202, 203, 222, 223, 224, 226,
227, 235, 236, 237, 241, 243, 254, 255,
257, 258, 266, 267, 268, 273, 274, 275,
277, 283, 284, 289, 290, 291, 293, 294,
295, 303, 304, 305, 306, 316, 317, 318,
319, 323, 324, 329, 333, 345, 382

Equipment   24, 33, 35, 36, 39, 40, 52, 83, 92,
141, 145, 151, 160, 184, 185, 189, 227,
235, 239, 256, 257, 258, 261, 262, 273,
287, 295, 311, 338, 350, 353, 381, 382,
388, 390

Errors in pricing   77, 92, 177

Exceptional Events   190, 226, 236, 255, 256,
386

Exceptionally adverse weather   219, 225, 357

Extensions of time   16, 24, 33, 38, 45, 51, 71,
111, 112, 131, 141, 144, 148, 150, 152,
156, 179, 183, 184, 186, 187, 190, 213,

Extensions of time (*contd.*)
   214, 218, 219, 220, 221, 222, 225, 226,
   228, 229, 233, 235, 280, 281, 285, 317,
   351, 353, 354, 355, 356, 357, 360, 383

f

Facilities   35, 37, 99, 147, 189, 195, 200, 237,
   239, 251, 256, 257, 277, 293, 323, 353,
   356, 382, 383
Factory acceptance testing   231
Fees   44, 46, 47, 49, 57, 76, 96, 202, 278, 321,
   343, 344, 369
Final Account   8, 22, 67, 68, 69, 245, 298,
   299, 302, 303, 306, 355
Final amount due   210, 307, 308, 310, 311,
   312, 333, 334
Final payment   69, 131, 165, 183, 253, 282,
   288, 301, 302, 303, 304, 305, 307, 308,
   330, 332, 334, 354, 372, 373, 385
Final Payment Certificate (FPC)   303, 304,
   305, 306, 317, 332, 334, 385
Final Statement   8, 205, 299, 300, 301, 302,
   303, 304, 305, 306, 330, 331, 332, 333,
   368, 373
Fitness for purpose   58
Float   40, 156, 159, 162, 215, 216
Fluctuations   9, 15, 167, 171, 201, 210, 225,
   298, 300, 379
Force majeure   188, 219, 247, 255, 354, 386

g

General Conditions (FIDIC)   4, 10, 11, 13,
   30, 122, 124, 125, 141, 172, 315
General Provisions (FIDIC)   381
Good faith   16, 90, 315
Goods   23, 24, 27, 31, 36, 37, 40, 48, 56, 92,
   102, 109, 119, 125, 128, 131, 134, 151,
   180, 185, 201, 202, 230, 232, 233, 234,
   235, 252, 257, 258, 330, 334, 338, 351,
   353, 356, 368, 369, 373, 382
Ground conditions   15, 36, 45, 48, 49, 357

h

Head office overheads   15, 44, 49, 50, 52, 76,
   77, 81, 91, 200, 201, 202, 206, 227

Heads of claim   227
Health and Safety (H&S)   3, 7, 31, 32, 33, 37,
   76, 78, 79, 93, 102, 146, 150, 159, 238,
   264, 265, 354, 378, 382, 383
Housing Grants Construction and
   Regeneration act 1996 (as amended)
   129, 162, 279, 311, 394

i

Inaccuracy   338, 360
Inadequacy   5, 27, 36, 82, 94, 102, 219, 338,
   339, 360
Inconsistencies   13, 28, 33, 82, 188, 190, 351,
   360
Increased costs   375
Insolvency   108, 112, 128, 245, 247, 248, 251,
   253, 254, 261, 262, 263, 356, 376
   of Contractor   245
   of Employer/Client   245
Inspections   22, 40, 47, 62, 63, 67, 103, 150,
   159, 195, 231, 232, 233, 235, 237, 238,
   239, 240, 269, 290, 355, 388
Instructions   24, 28, 50, 84, 102, 105, 108,
   109, 110, 111, 112, 142, 154, 155, 170,
   181, 182, 183, 184, 185, 186, 188, 189,
   190, 191, 192, 193, 194, 196, 204, 207,
   212, 219, 233, 234, 236, 237, 241, 244,
   245, 246, 260, 261, 262, 264, 266, 267,
   288, 290, 291, 292, 300, 317, 330, 357,
   361
Interest   166, 174, 177, 285, 301
Interim Payment   40, 113, 129, 136, 162, 163,
   165, 166, 168, 169, 170, 173, 174, 175,
   176, 183, 185, 233, 258, 355, 372, 385
   assessment   177
   assessment date   169, 175, 176
Interim Payment Certificate (IPC)   113, 129,
   134, 169, 170, 171, 172, 173, 174, 227,
   277, 305, 317, 385

j

JCT ICD 2016   90, 99, 107, 108, 109, 110, 111,
   112, 113, 114, 367, 368, 369, 370, 371,
   372, 373, 374, 375, 376, 377, 378, 379

JCT MWD 2016   5, 367, 368, 369, 370, 371, 372, 373, 374, 375, 376, 377, 378, 379
JCT SBC/Q 2016 (CDP)   30, 108, 112, 153, 163, 197, 207, 208, 271, 288, 300, 339, 360, 362, 367, 368, 369, 370, 371, 372, 373, 374, 375, 376, 377, 378, 379

k

Key Date   142, 143, 144, 145, 149, 157, 159, 189, 190, 192, 193, 195, 205, 207, 211, 212, 215, 225, 261
Key Personnel   9, 383

l

Late payments   166, 174, 177, 227, 301, 314
Legal proceedings   329, 331, 367
Letter of Acceptance   10, 11, 12, 42, 131, 134, 136, 137, 322, 347
Letter of Tender   10, 12, 42, 132, 294, 347
Limitation Act 1980   295, 334
Liquidated Damages   9, 64, 130, 139, 140, 143, 250, 272, 279, 280, 281, 282, 284, 286, 357, 364, 370
Listed Items   131, 167, 373
Loss and/or Expense   24, 71, 103, 111, 112, 114, 167, 179, 187, 191, 196, 212, 213, 221, 225, 227, 228, 229, 232, 251, 252, 257, 298, 314, 331, 351, 353, 355, 373, 377
  assessment   228, 229
Loss of opportunity   84, 90, 353
Loss of profit   84, 90, 257, 353
Low performance damages   269

m

Main Option Clauses   11, 12, 31, 41, 42, 57, 78, 85, 101, 176, 177, 189, 190, 197, 205, 206, 210
  Option A (Priced Contract with Activity Schedule)   31, 41, 42, 43, 70, 71, 90, 108, 133, 176, 197, 208, 210, 261, 262, 270, 387, 389
  Option B (Priced Contract with Bill of Quantities)   31, 42, 43, 71, 84, 133, 176, 197, 208, 209, 210, 262, 387, 389

  Option C (Target Contract with Activity Schedule)   31, 42, 43, 71, 78, 85, 90, 138, 176, 197, 210, 262, 276, 277, 308, 309, 310, 387, 389
  Option D (Target Contract with Bill of Quantities)   31, 42, 43, 71, 138, 176, 197, 210, 262, 276, 277, 308, 309, 387, 389
Maintenance manuals   33, 37, 64, 266, 273, 289, 293, 383
Making Good   13, 130, 168, 245, 250, 251, 270, 287, 288, 297, 302, 357, 370
Materials   3, 22, 23, 24, 27, 31, 32, 33, 36, 37, 39, 40, 48, 52, 56, 59, 60, 83, 92, 102, 109, 119, 125, 128, 131, 134, 145, 147, 151, 167, 171, 173, 180, 185, 189, 200, 201, 202, 206, 209, 227, 230, 231, 232, 233, 234, 235, 236, 237, 238, 239, 252, 257, 260, 261, 269, 289, 292, 311, 317, 330, 334, 338, 343, 351, 353, 356, 368, 369, 373, 381, 383, 384, 385, 390
  off site   52, 119, 125, 128, 131, 134, 138, 232, 235, 369, 373, 384
Mediation   8, 315, 316, 321, 322, 328, 329, 331, 377
Methods of measurement   30, 43, 51, 77, 188, 190, 210
Method statement   3, 36, 37, 93, 105, 183, 185
Milestone certificate   141, 144
Milestones   9, 141, 142, 143, 144, 190
Minimum requirements   180
Mitigation   152, 215
Model Services Agreement (FIDIC)   56, 58, 122, 123, 178, 361, 365
Mutual trust and co-operation   16, 104, 105, 151, 191

n

Named sub-contractor   13, 99, 100, 106, 107, 108, 110, 111, 112, 113, 114, 115, 377, 394
Negligence   109, 252, 283, 306, 328, 333
Negotiation   73, 81, 82, 84, 88, 89, 90, 279, 315, 329, 378

New Rules of Measurement (NRM2)   30, 31, 32, 35, 41, 43, 44, 45, 46, 47, 48, 49, 50, 51, 52, 60, 77, 197, 362, 363

Nominated Sub-contractors   40, 100, 106, 107, 108, 109, 110, 111, 112, 113, 114, 115, 159, 382

Notice of Claim   222, 223, 224, 283, 318, 353

Notice of Completion of Making Good   130, 168, 270, 288, 297, 302, 370

Notice of Demand (Bond)   131

Notice of Dissatisfaction (NOD)   304, 315, 316, 318, 319, 323, 324

Notices

generally   16, 38, 86, 88, 90, 103, 109, 111, 112, 121, 122, 126, 127, 129, 130, 131, 136, 137, 141, 144, 150, 161, 163, 164, 166, 168, 170, 173, 174, 175, 186, 191, 194, 196, 218, 219, 220, 221, 222, 223, 224, 229, 231, 234, 235, 236, 237, 243, 246, 248, 249, 252, 253, 254, 255, 256, 257, 259, 260, 266, 267, 268, 270, 272, 273, 274, 281, 282, 283, 284, 286, 288, 289, 290, 291, 292, 293, 294, 295, 297, 299, 300, 301, 302, 303, 304, 314, 315, 316, 318, 319, 323, 324, 326, 327, 331, 332, 353, 355, 357, 369, 370, 372, 377, 385, 386, 393, 394

of no-objection   38, 226, 266, 267, 273, 293

of objection   109

Notice to Commence   137

Notice to Correct   237, 246, 254, 385

Novation   3, 6, 339, 340, 341, 342, 343, 344, 345, 361

Nuisance   32, 349

o

Omission of work   179, 182, 184, 194, 198, 200, 204, 219, 221, 292

Opening up and testing   13, 219, 240, 355

Others   31, 57, 65, 85, 99, 100, 102, 122, 123, 125, 126, 159, 188, 212, 237, 276, 292, 310, 354, 365, 388, 392

p

Parent Company Guarantee   78, 119, 125, 134, 361

Partial Possession   133, 139, 140, 141, 145, 168, 271, 272, 274, 276, 280, 282, 286, 357, 370

Particular Conditions   10, 11, 13, 43, 68, 104, 121, 122, 130, 134, 140, 169, 170, 172, 202, 322, 323, 361, 365

Partnering   16, 394

Patent defects   166, 270, 272, 363

Patents   167, 369

Pay less notice   164, 166, 170, 173, 174, 175, 252, 253, 281, 282, 286, 332, 353

Payment

Interim (periodic)   9, 163, 165, 167, 169, 229, 234

Interim (stage)   131, 163, 165, 167, 229, 233

on termination   252, 253, 258, 261, 390

Payment notice   69, 129, 163, 164, 166, 170, 174, 175, 252, 282, 284, 301, 372

People rates   210

Performance bonds   15, 57, 119, 125, 126, 127, 128, 375, 393

Performance Certificate   289, 291, 294, 295, 297, 303, 304, 305, 332, 333, 334, 384

Performance Damages   292, 294, 393

Performance indicators (including Key Performance Indicators)   9, 57, 378, 393

Performance Security   127, 129, 131, 133, 134, 137, 173, 246, 255, 257, 290, 305, 306, 317, 361, 382

Performance specification   31, 63, 76, 78, 92, 346, 360

Period for reply   29, 38, 66, 93, 108, 158, 160, 189, 204

Permanent Works   141, 151, 184, 234, 257, 273, 274, 284

Person in charge (foreman, site manager)   315

Planning permission   352

Plant   22, 24, 26, 36, 40, 46, 47, 83, 92, 131, 141, 145, 147, 151, 171, 173, 184, 185,

201, 202, 206, 209, 216, 227, 231, 234, 235, 236, 238, 239, 257, 260, 261, 265, 266, 269, 289, 290, 292, 295, 311, 317, 334, 338, 350, 353, 360, 383, 384, 385, 390

Pollution   32, 349

Possession   1, 22, 135, 136, 226, 280, 350, 364, 368

Date of   9, 134, 135, 136, 139, 140, 142, 143, 159, 165, 368

Deferment of   136, 140, 141, 142, 143, 219, 228, 368

Postponement   70, 231, 280, 371

Practical completion (JCT)   126, 130, 168, 183, 218, 219, 220, 232, 245, 265, 266, 269, 270, 271, 272, 273, 275, 278, 280, 281, 282, 287, 299, 300, 334, 363, 370

Practice, generally accepted (good)   39, 75, 145, 156, 157, 215, 216, 231, 271, 351

Preliminaries   15, 30, 31, 32, 35, 44, 46, 47, 49, 50, 51, 77, 81, 91

Prevention by Employer   184, 188, 190, 219, 225, 387

Priced contracts with activity schedule (NEC4)   43, 57, 84, 108, 176, 197, 208, 261, 262, 387

Priced contracts with bill of quantities (NEC4)   42, 43, 176, 188, 197, 207, 208, 262, 387

Price for work done to date   176, 177, 262, 309

Principal Contractor   367

Principal Designer   56, 367

Priority of documents   13, 26, 381

Proceedings   79, 323, 326, 328, 329, 331, 332, 334, 367

Procurement   1, 2, 3, 4, 5, 6, 15, 17, 36, 40, 55, 59, 60, 63, 67, 73, 74, 75, 80, 82, 91, 159, 337, 339

Professional Fees   44, 46, 47, 49, 57, 76, 96, 278, 321, 343, 344, 369

Professional Indemnity (PI) insurance   9, 25, 36, 57, 64, 101, 120, 138, 349, 375

Professional Services Agreement (ACE)   56, 58, 123, 178

Professional Services Contract (NEC4)   56, 58, 122, 178, 260, 340, 361, 365

Professional Services Contract (RIBA)   56, 58, 61, 178, 271, 340, 341, 361

Profit   3, 15, 43, 44, 50, 52, 77, 81, 84, 88, 90, 91, 104, 200, 201, 202, 207, 208, 227, 228, 257, 268, 275, 285, 291, 353

Programme   32, 36, 37, 39, 40, 45, 51, 53, 57, 61, 63, 64, 66, 67, 71, 78, 79, 85, 86, 87, 90, 93, 94, 96, 102, 104, 105, 106, 108, 109, 113, 115, 133, 135, 138, 141, 142, 144, 145, 146, 147, 148, 149, 150, 154, 155, 156, 157, 158, 159, 160, 161, 162, 176, 178, 179, 185, 186, 188, 189, 191, 194, 201, 203, 204, 211, 212, 213, 214, 215, 216, 217, 218, 220, 227, 237, 247, 261, 265, 266, 267, 268, 283, 293, 314, 320, 349, 350, 352, 353, 364, 383, 388

detail of   36, 40, 51, 52, 64, 102, 113, 144, 154, 155, 156, 157, 158, 159, 160, 161, 162, 178, 179, 185, 214, 215, 216, 217, 218, 266, 293, 364, 383, 388

and rate of progress   144, 145, 148, 149

rejections of   149, 161

revisions of   148, 149, 155, 158, 161, 162, 211, 267, 320

Progress reports   39, 66, 149, 150, 160, 161, 162

Project bank account   57, 394

Project Manager   9, 13, 16, 24, 25, 27, 28, 29, 31, 35, 38, 39, 40, 47, 55, 56, 61, 66, 68, 69, 84, 85, 86, 93, 101, 105, 108, 126, 128, 129, 135, 142, 149, 151, 152, 154, 155, 160, 161, 175, 176, 177, 188, 190, 191, 192, 193, 194, 195, 196, 203, 204, 205, 207, 208, 209, 210, 211, 212, 238, 239, 240, 259, 260, 261, 262, 276, 278, 285, 288, 296, 298, 307, 308, 310, 311, 313, 320, 327, 333, 345, 361, 387

Provisional Sums   44, 48, 49, 50, 51, 77, 91, 107, 108, 110, 111, 171, 181, 197, 201, 202, 203, 219, 301, 354, 371, 373, 385

Public liability insurance   9, 36

q

Quality   3, 15, 22, 24, 25, 32, 36, 47, 55, 62,
64, 67, 69, 76, 79, 86, 87, 102, 109, 145,
146, 151, 182, 216, 232, 314, 330, 334,
340, 342, 343, 351, 356
Quality management   57, 63, 64, 78, 91, 92,
142, 143, 195, 230, 231, 234, 235, 237,
238, 239, 240, 296, 382, 388
Quantity surveyor   2, 47, 53, 55, 56, 59, 87,
91, 209, 345, 362
Quantum meruit   96
Quotations
    acceleration   152, 153, 155, 167, 219, 300,
    301, 378
    compensation events (NEC4)   188, 190,
    191, 193, 194, 195, 196, 203, 204, 205,
    307, 389
    to reduce the 'Prices' (NEC4)   288, 296
Quotations (FIDIC)   202, 203

r

Rate of progress   71, 144, 145, 147, 148, 149,
150, 152, 384
Reasonable skill and care   56, 58, 59, 102,
346, 351, 356, 394
Recitals   10, 11, 133, 232, 360, 367
Records   46, 78, 150, 177, 213, 216, 217, 223,
266, 273, 284, 289, 291, 295, 297, 383
Rectification Period   9, 22, 139, 140, 142, 143,
232, 245, 270, 272, 287, 288, 302, 355,
357
Regularly and diligently   146, 246
Release from performance   386
Relevant Date (partial possession, JCT DB
2016)   272, 278
Relevant date for Completion (FIDIC)   283,
284
Relevant Event   110, 188, 189, 190, 218, 219,
220, 221, 233, 370
Relevant Matter   95, 198, 219, 227, 228, 331,
373
Relevant Part (defects-FIDIC)   168
Relevant part (partial possession-JCT DB
2016)   168, 272, 278, 282, 370

Remedial work   121, 226, 234, 236, 237, 287,
288, 289, 290, 291
Repeat tests and inspections   290
Repudiation   111, 112, 121, 123, 244
Resolving and Avoiding Disputes
    ('W' clauses-NEC4)   311, 312, 313,
    316, 319, 320, 322, 327, 328, 329, 391
Retention   9, 128, 134, 140, 141, 152, 166,
167, 168, 170, 172, 173, 177, 227, 229,
252, 261, 270, 272, 275, 276, 277, 288,
295, 298, 355, 361, 373, 385, 393, 394
release of   140, 172, 272, 295, 385
Retention Bond   119, 125, 130, 134, 136
Retesting   267, 290, 294, 298, 299, 384, 385
RIBA Code of Professional Conduct (2018)
    342
RIBA Plan of Work 2020   2, 3, 6, 21, 22, 30,
46, 47, 55, 58, 60, 61, 64, 66, 68, 73, 80,
81, 83, 84, 115, 133, 240, 243, 244, 269,
338, 339, 341, 343, 344, 345, 359, 360
    Stage 0 (Strategic Definition)   55, 61, 62,
    66, 67, 339, 341
    Stage 1 (Preparation & Briefing)   2, 3, 55,
    61, 67, 83, 341
    Stage 2 (Concept Design)   2, 3, 30, 55, 61,
    67, 83, 115, 338, 339
    Stage 3 (Spatial Co-ordination)   2, 3, 30,
    47, 55, 60, 61, 67, 73, 80, 83, 84, 338
    Stage 4 (Technical Design)   2, 3, 21, 22, 30,
    47, 55, 60, 61, 67, 68, 73, 80, 81, 83, 84,
    115, 133, 339, 341, 359, 360
    Stage 5 (Manufacturing & Construction)
    2, 3, 21, 22, 55, 61, 64, 67, 68, 69, 133,
    240, 244, 269, 344
    Stage 6 (Handover)   2, 3, 22, 55, 61, 67, 68,
    133, 243, 271
    Stage 7 (Use)   2, 22, 55, 60, 61, 62, 67, 84,
    243
RIBA Professional Services Contract (2020)
    56, 57, 58, 60, 61, 62, 63, 64, 65, 66, 68,
    83, 123, 178, 340, 341, 361
Risk allocation   6, 14, 86, 91, 100, 132
Risk Register   64, 66, 138, 151
Royalties   66, 138

S

Samples   32, 36, 231, 234, 235, 237, 239, 383

Schedule of defects   183, 245, 370

Schedule of Payments (FIDIC)   169, 170, 317, 385

Schedule of Performance Guarantees (FIDIC) 185, 294

Schedule of Rates and Prices (FIDIC)   7, 9, 42, 43, 78, 201, 202, 347, 362

Schedule of Services (consultant appointments)   60, 62, 65, 66, 68, 78, 271, 362

Schedules (FIDIC)   10, 12, 23, 32, 42, 43, 258

Schedules (JCT DB 2016)   10, 11, 12, 13, 35, 43, 92, 107, 110, 111, 112, 113, 124, 125, 126, 128, 130, 131, 133, 134, 135, 136, 153, 154, 155, 158, 219, 233, 270, 315, 361, 362, 365, 377, 378, 379

Schedules of Cost Components (NEC4)   9, 12, 42, 43, 58, 78, 176, 206, 209, 347, 362

Scheme for Construction Contracts (England and Wales) Regulations 1998 (as amended)   163, 165, 252, 253, 327

Scope (NEC4)   9, 12, 13, 24, 27, 28, 29, 30, 31, 34, 35, 38, 39, 40, 50, 68, 77, 78, 79, 81, 82, 84, 85, 86, 87, 92, 94, 99, 108, 113, 114, 126, 128, 133, 138, 160, 161, 162, 179, 183, 185, 188, 189, 190, 192, 195, 207, 210, 237, 238, 239, 240, 265, 269, 275, 276, 288, 296, 297, 338, 347, 360, 362

Secondary Option Clauses   9, 11, 12, 31, 57, 85, 86, 122, 123, 125, 126, 129, 138, 141, 142, 143, 149, 175, 177, 178, 196, 211, 240, 259, 269, 276, 277, 278, 284, 285, 286, 298, 308, 310, 333, 361, 365, 391, 392, 393, 394

Sectional completion   57, 93, 138, 139, 140, 141, 142, 143, 149, 153, 157, 159, 172, 211, 240, 267, 270, 271, 272, 273, 278, 281, 282, 284, 285, 287, 299, 364, 367, 392

Set-off   353

Setting out   32, 226, 369, 382

Site

access route to   37, 180, 184, 349

clearance of   44, 89

operations on   32, 180

security of   89

unforeseen physical conditions   45, 189, 226, 317, 357, 382

Site acceptance testing   231

Site Data   381, 382

Site Information (NEC4)   12, 30, 31, 77, 78, 347

Special Provisions (FIDIC)   10, 11, 43, 121, 122, 125, 130, 134, 136, 140, 141, 143, 172, 315, 316, 322, 326

Specification   3, 9, 13, 24, 27, 30, 31, 34, 36, 47, 48, 59, 63, 78, 94, 98, 106, 107, 114, 145, 184, 198, 199, 208, 216, 225, 234, 287, 288, 346, 360

Specified Perils   190, 219, 354, 379

Stage payments   131, 163, 167, 229, 233

Starting date   9, 137, 138, 151, 175, 320

Statutory authorities, body, undertakers   46, 52, 139, 190, 219, 247, 350, 356, 357

Statutory dispute resolution   321, 326, 329

Statutory obligations   2, 7, 15, 99, 231, 330

approval / compliance   35, 350

divergence   59, 369

powers   219, 354

Statutory Requirements   23, 33, 34, 351, 352

Sub-contracting, Employer's approval   104

Sub-contractors

Contractor's objection to named or nominated subcontractor   108, 109

domestic   103, 104, 105, 106

early involvement   114, 115

named or nominated sub-contractor   106, 107, 108, 109, 110, 111, 112, 113, 114

Sub-contracts

FIDIC   99, 105, 106, 109, 110, 113, 114, 122, 202, 365

generally   40, 52, 63, 97, 98, 99, 100, 101, 102, 103, 104, 105, 106, 107, 108, 109, 110, 115, 117, 118, 124, 202, 261, 349, 356, 361, 365

Sub-contracts (*contd.*)
  JCT (DBSub/C, ICSub/NAM/C) 99, 100,
    105, 106, 107, 109, 110, 111, 112, 113,
    114, 121, 122, 124
  NEC4 99, 101, 105, 106, 108, 122, 124,
    261, 365
Supervisor 9, 16, 25, 188, 190, 191, 193, 195,
    207, 212, 237, 238, 239, 240, 260, 276,
    296, 297, 334, 387
Supplemental Provisions (FIDIC) 143, 144,
    190, 364
Supplemental Provisions (JCT) 11, 39, 43,
    107, 108, 110, 111, 113, 142, 153, 154,
    158, 183, 186, 196, 201, 219, 315, 362,
    363, 367, 377
Supply of Goods and Services Act 1982 58,
    109, 146
Supporting documents 113, 170, 172, 173,
    284, 299, 300, 301, 304, 353
Suspension of the contractor's obligations
    164, 165, 166, 167, 174, 177, 190, 219,
    226, 292, 372, 386, 394
Suspension of the works 70, 136, 141, 226,
    247, 249, 292, 383, 384
Sustainability and Secure Buildings Act 2004
    23

t

Take over 120, 189, 274, 276, 285, 334, 350,
    388
Taking Over 126, 127, 141, 172, 187, 231,
    234, 265, 272, 273, 274, 275, 276, 277,
    283, 285, 286, 302, 334
  access rights to Site after 226, 291, 384
  defects after 289, 290, 293
  Parts 141, 142, 226, 273, 274, 276, 280,
    284, 286, 317, 384
  and surfaces requiring reinstatement 273,
    275, 384
  and Tests on Completion 234
Taking Over Certificate 141, 169, 172, 184,
    187, 236, 241, 267, 268, 272, 273, 274,
    275, 289, 290, 293, 306, 357, 384
  and outstanding work 172, 273, 274, 275,
    277, 286, 297

Temporary Works 46, 89, 184, 200, 227, 256,
    257, 258
Tendering 34, 53, 62, 73, 75, 76, 77, 79,
    80, 81, 82, 91, 95, 114, 115, 132, 205,
    381
Termination 57, 71, 108, 111, 112, 113, 114,
    121, 136, 165, 175, 241, 243, 244, 246,
    247, 248, 249, 250, 251, 253, 255, 256,
    257, 258, 259, 260, 261, 262, 263, 264,
    307, 317, 353, 357, 374, 376, 385, 390,
    392, 394
  consequences 250, 251, 257, 258, 376
  by contractor 136, 246, 251, 253, 356, 376,
    386
  for convenience 247, 253, 385
  by Employer (Client) 165, 241, 246, 251,
    253, 292, 353, 376, 385, 392
  grounds for 246, 247, 248
  no fault (by either party) 247, 253, 256,
    374, 376, 385
  procedure 248, 249, 250, 251, 255, 256,
    259, 260, 261, 262, 263, 264
  sub-contracting 108, 111, 112, 113, 114
Termination certificate (NEC4) 259, 260,
    307
Termination collateral warranties 121
Testing, generally 7, 13, 25, 32, 40, 63, 67,
    71, 83, 84, 141, 145, 150, 160, 172, 189,
    219, 229, 230, 231, 232, 233, 234, 235,
    236, 237, 238, 239, 240, 241, 243, 265,
    266, 267, 268, 269, 289, 290, 293, 294,
    295, 296, 297, 298, 299, 323, 330, 349,
    355, 363, 364, 383, 389
Tests after Completion 226, 234, 289, 290,
    293, 294, 295, 303, 357, 384, 385
  delays in 226
  failure to pass 294, 385
Tests and inspections (NEC4) 40, 195, 237,
    238, 240, 269, 388
Tests on Completion 160, 187, 226, 234, 266,
    267, 269, 273, 275, 277, 290, 293, 364,
    384
  delays in 268
  failure to pass 267

interference with   226, 268
Third party rights   9, 69, 117, 118, 119, 123,
    124, 356, 375, 376, 379, 394
Time for Completion   9, 39, 129, 137, 140,
    141, 148, 150, 154, 159, 205, 224, 280,
    283, 293, 317, 383
Total of the prices   9, 41, 43, 85, 86, 190, 262,
    308, 309, 310

u

Undertakings   122, 171, 371, 392
Unforeseeable   185, 236, 261, 262, 264, 275,
    276, 317, 357, 382

v

Valuation
    after termination   255, 256, 317, 385
    fair   198, 199, 200, 202
    rules   50, 96, 196, 198, 201, 202, 203, 300,
        373
    of work, variations and changes (inc
        'assessment' of compensation events
        NEC4)-general   5, 22, 25, 41, 43, 50,
        96, 111, 152, 166, 167, 173, 177, 178,
        193, 194, 196, 197, 198, 199, 200, 201,

202, 203, 204, 205, 206, 207, 208, 209,
    210, 229, 233, 234, 255, 256, 262, 276,
    277, 300, 306, 307, 308, 310, 311, 312,
    313, 314, 317, 321, 350, 372, 373, 389,
    395
Variations   27, 28, 34, 38, 43, 50, 51, 64, 67,
    84, 87, 94, 108, 111, 147, 151, 153, 154,
    161, 178, 179, 180, 181, 182, 183, 184,
    185, 186, 187, 188, 189, 190, 196, 197,
    198, 199, 200, 201, 202, 203, 205, 207,
    208, 216, 225, 226, 235, 244, 290, 291,
    292, 317, 344, 351, 360, 373, 385. *see
    also* Changes

w

Weather   189, 219, 287, 354, 357
Working areas   238, 261, 348
Workmanlike manner   22, 146, 147
Workmanship   23, 24, 27, 31, 32, 37, 106,
    151, 232, 234, 236, 287, 289, 330, 334,
    338, 345, 356, 368, 371
Works information   237

z

Z clauses   5, 13, 105, 108, 113